Marketing Fundamentals

1 2 3 4 5 6 7 8 9 LWI 27 26 25 24 23

ISBN 978-1-26-660641-0
MHID 1-26-660641-6

Cover Image: PopTika/Shutterstock

Interior Image (abstract background):
All credits appearing on page or at the end of the book are considered to be an extension of the copyright page.

Meet Our Lead Content Architect

Lauren Skinner Beitelspacher (PhD, University of Alabama) is an associate professor in the marketing division at Babson College. Her research interests include buyer–supplier relationships, retail management, and the retail supply chain. Her work has been published in numerous scholarly journals including *Journal of Marketing, Journal of Applied Psychology, Journal of Retailing, Journal of the Academy of Marketing Science, Journal of Business Research, and Industrial Marketing Management.* She also has presented her work at numerous conferences and won several "Best Paper in Track" awards at the American Marketing Association, Society for Marketing Advances, and Academy of Marketing Science. Lauren is the co-chair of the Retail and Pricing Special Interest Group for the American Marketing Association and is on the editorial review boards of *Journal of Business Research and International Journal of Physical Distribution and Logistics Management.* Lauren was named one of the Top 40 under 40 Business Professors by Poets & Quants in 2016. In 2017 Lauren received the Dean's Excellence Award in Undergraduate Teaching at Babson College, and in 2018 she received the Babson Faculty of the Year Award.

Acknowledgments

Marketing Fundamentals would not have been possible without the generosity, innovation, and expert contributions of the following:

Lead Subject-Matter Experts, Adaptive

Ted Tedmon, *North Idaho College*

Casey Wilhelm, *North Idaho College*

Lead Subject-Matter Expert, Assessment and Adaptive

Lauren Michelle Brewer, *University of Texas, Arlington*

Lead Subject-Matter Expert, Interactive Reader

Bryan W. Hochstein, *The University of Alabama*

Lead Expert Reviewers

Beth Deinert, *Southeast Community College*

Luke Hopkins, *Florida State University*

Content Contributors

Shanita Baraka Akintonde, *Columbia College Chicago*

Catharine Curran-Kelly, *University of Massachusetts, Dartmouth*

Sung H. Ham, *The George Washington University*

AnnMarie Hughes, *Brookdale Community College*

Adam Mills, *Loyola University, New Orleans*

Courtney T. Pham, *Missouri State University*

Ric Sweeney, *University of Cincinnati*

Deborah Utter, *Boston University*

Content Reviewers

Ajay T. Abraham, *Seattle University*

Wendi Achey, *Northampton Community College*

Thomas Ainscough, *University of South Florida*

Daniel Allen, *Utah State University*

Grace Ambrose, *University of Wisconsin, Milwaukee*

Brian J. Baldus, *California State University, Sacramento*

Wayne E. Ballentine, *Prairie View A&M University*

Schuyler Banks, *SUNY–Erie*

Frank Barber, *Cuyahoga Community College*

Darrell E. Bartholomew, *Penn State Harrisburg*

Robert Battle, *Nassau Community College*

Leta Beard, *University of Washington*

Jill Bernaciak, *Cuyahoga Community College*

Parimal Bhagat, *Indiana University of Pennsylvania*

Curtis R. Blakely, *Ivy Tech Community College–Richmond*

William F. Bodlak, *Robert Morris University*

Nicholas J. Bosco, *Suffolk County Community College*

Henry C. Bradford III, *Massasoit Community College*

Fabienne Cadet, *St. John's University*

Rae Caloura, *Providence College*

Amy Caponetti, *Pellissippi State Community College*

Mary Carlson, *College of DuPage*

Shari Carpenter, *Eastern Oregon University*

Samit Chakravorti, *Western Illinois University*

Anindya Chatterjee, *Slippery Rock University*

Vincent Cicchirillo, *Saint Xavier University*

Christina Chung, *Ramapo College of New Jersey*

Cathleen Cogdill, *Northern Virginia Community College*

Kimberly Conrad, *West Virginia Wesleyan College*

Gary M. Corona, *Florida State College at Jacksonville*

Oliver Cruz-Milan, *Texas A&M University–Corpus Christi*

Brent Cunningham, *Jacksonville State University*

Amy Danley, *Wilmington University*

Meredith David, *Baylor University*

Chandan Desarkar, *University of Albany*

Alan Dick, *University at Buffalo*

Laura Downey, *Purdue University*

Yancy D. Edwards, *Shippensburg University*

Maurice Elliard, *Albany State University*

Dawn Fairchild, *Delta College*

Phyllis Fein, *Westchester Community College*

Don Fisher, *Dixie State University*

Erica Fleisher, *Harrisburg Area Community College*

Renée J. Fontenot, *Georgia College & State University*

Anthony Fruzzetti, *Johnson & Wales University*

Pat Galitz, *Southeast Community College*

Meredith Gander, *East Carolina University–College of Business*

Greg Gerfen, *Western Michigan University*

Jeanette Gile, *Fox Valley Technical College*

John Gironda, *Nova Southeastern University*

Keith Gosselin, *Mount Saint Mary's University*

Robin Grambling, *University of Texas at El Paso*

Teresa Greenlees, *Western Michigan University*

Melodi Guilbault, *New Jersey Institute of Technology*

Pranjal Gupta, *University of Tampa*

Andrew W. Gump, *Liberty University*

Audrey Guskey, *Duquesne University*

John Hadjimarcou, *The University of Texas at El Paso*

Terry Hall, *Davenport University*

Jamey Halleck, *Marshall University*

Sara Hanson, *University of Richmond*

Frank Harber, *Indian River State College*

Robert Harrison, *Western Michigan University*

Stephen He, *West Virginia University*

Judith Hoffman, *Davenport University*

Dan Horne, *Providence College*

Jianping Huang, *Jacksonville State University*

Roy Iraggi, *New York City College of Technology*

James Jarrard, *University of Wisconsin, Platteville*

Sungwoo Jung, *Columbus State University*

Ahmad Kareh, *Salt Lake Community College*

Ioannis Kareklas, *University at Albany, State University of New York*

Hamid Kazeroony, *Inver Hills Community College*

Kacy Kim, *Elon University*

John Kinnett, *Columbus State University*

Heather Kirkwood, *SUNY–Farmingdale State College*

Stephen Konrad, *Mount Hood Community College*

Karen Koza, *Western Connecticut State University*

Jamie Lambert, *Ohio University*

Karen Lancendorfer, *Western Michigan University*

Jane Lang, *East Carolina University*

Mildred Lanier, *Jefferson State Community College*

Jon Lezon, *University of Arkansas, Fayetteville*

Marilyn Liebrenz-Himes, *The George Washington University*

Xuefeng Liu, *Loyola University Maryland*

Yan Liu, *Southern Connecticut State University*

Temo Luna-Nevarez, *Sacred Heart University*

Alicia Lupinacci, *Tarrant County College*

Trina Lynch-Jackson, *Ivy Tech Community College–Lake County Campus*

Ahmed Maamoun, *University of Minnesota, Duluth*

Anne Magi, *University of Illinois at Chicago*

Derine McCrory, *Mott Community College*

Carter McElveen, *Clemson University*

Enda McGovern, *Sacred Heart University*

Kevin McMahon, *University of Colorado at Boulder*

Robert McMillen, *James Madison University*

Robert Meyer, *Parkland College*

Tim Mittan, *Southeast Community College*

William Montford, *Jacksonville University*

Gerardo J. Moreira, *Sacred Heart University*

Paula Morris, *Salisbury University*

Amit Mukherjee, *Stockton University*

Lakshmi Nagarajan-Iyer, *Middlesex County College*

Aidin Namin, *University of Idaho*

Mary Judene Nance, *Pittsburg State University*

Nik Nikolov, *Appalachian State University*

Lisa Novak, *Mott Community College*

Luke Nowlan, *KU Leuven*

Louis Nzegwu, *University of Wisconsin, Platteville*

Lois Bitner Olson, *San Diego State University*

Beng Soo Ong, *California State University, Fresno*

Nick Ostapenko, *Keller Graduate School of Management*

KE Overton, *Houston Community College*

Lisa Palumbo, *University of New Orleans*

Becky Parker, *McLennan Community College*

Thomas Passero, *Owens Community College*

Michael Pepe, *Siena College*

Joel Petry, *Southern Illinois University Edwardsville*

Wesley Pollitte, *St. Edward's University*

Annie Quaile, *Full Sail University*

Mohammad Rahman, *Shippensburg University*

Greg Rapp, *Portland Community College*

Mohammed Rawwas, *University of Northern Iowa*

Sherilyn Reynolds, *San Jacinto College*

William Rice, *California State University, Fullerton*

Paul Rinaldi, *Brookdale Community College*

Dawn Robinson, *Jefferson Community College*

Doug Ross, *Franklin University*

Carol Rowey, *Community College of Rhode Island*

Don Roy, *Middle Tennessee State University*

Alberto Rubio, *University of the Incarnate Word*

Kimberly Ruggeri, *Cleveland State University*

Amber Ruszkowski, *Ivy Tech Community College*

Elise Sautter, *New Mexico State University*

Leah Schneider, *University of Oregon*

Mary Schramm, *Quinnipiac University*

Eric Schulz, *Utah State University*

Anna Shadrick, *College of Staten Island*

Sarah Shepler, *Ivy Tech Community College–Terra Haute*

Judy A. Siguaw, *East Carolina University*

Birud Sindhav, *University of Nebraska at Omaha*

Allison Smith, *West Kentucky Community & Technical College*

Reo Song, *California State University, Long Beach*

Laurel Steinfield, *Bentley University*

Randy Stuart, *Kennesaw State University*

Ric Sweeney, *University of Cincinnati*

Ronda D. Taylor, *Ivy Tech Community College*

Ramendra Thakur, *University of Louisiana, Lafayette*

Claudia Townsend, *Miami Business School*

Juliett Tracey, *Palm Beach State College*

Ann E. Trampas, *University of Illinois–Chicago*

Rebecca VanMeter, *Ball State University*

Ann Veeck, *Western Michigan University*

Maria Vittoria, *Westchester Community College*

Elisabeth Wicker, *Bossier Parish Community College*

Tiffany Williams, *Davenport University*

Rick Wills, *Illinois State University*

Mark Young, *Winona State University*

Gail M. Zank, *Texas State University*

James Zemanek, *East Carolina University*

Board of Advisors

A special thank you goes out to our Board of Advisors:

Thomas F. Frizzell Sr., *Massasoit Community College*

Mary Jane Gardner, *Western Kentucky University*

Lisa Harris, *Southeast Community College*

Prashanth Pilly, *Indian River State College*

Ann R. Root, *Florida Atlantic University*

Andrew Thoeni, *University of North Florida*

Joe Tungol, *College of DuPage*

Donna Wertalik, *Virginia Tech*

Jefrey R. Woodall, *York College of Pennsylvania*

Table of contents

Chapter 9: Communicating Value

Appendix: The Marketing Plan

1 Marketing Overview

What To Expect

You may think that marketing is the advertising that you see online or while streaming content. Some think that marketing is the salesperson who sells you a car or a cell phone. Marketing is so much more! This lesson will provide a broad perspective of its function and importance to a firm.

Chapter Topics

- **1-1** Overview of Marketing
- **1-2** Value Creation
- **1-3** Marketing Ethics

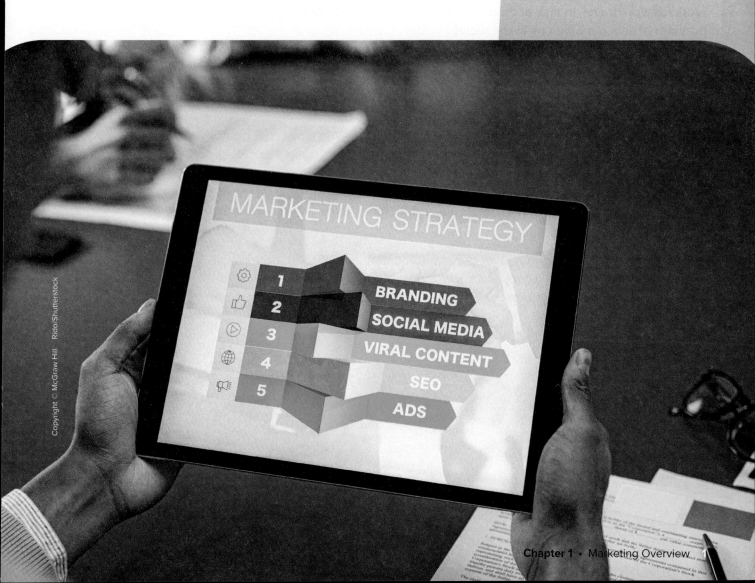

Copyright © McGraw Hill   Rido/Shutterstock

Introduction: Overview of Marketing

Airbnb: Marketing to a Worldwide Audience

You've no doubt heard of Airbnb—the behemoth e-commerce business that enables users to rent their properties to others, with Airbnb earning a tidy profit along the way. You may not, however, know the story of Airbnb, which is one of innovation and tenacity sprinkled with a little bit of good luck.

Airbnb got its start in 2007, when founders Joe Gebbia and Brian Chesky moved from New York City to San Francisco. The friends were struggling to pay rent and hustling to find ways to earn extra money. When an industrial design conference came to town, Gebbia and Chesky saw an opportunity to capitalize on visitors who weren't able to obtain hotel rooms. They offered these hotel-less guests an "Air Bed and Breakfast" for $80 a night, literally using air beds on their living room floors. Thus, Airbnb was born.

According to its website, Airbnb now has over 5.6 million listings worldwide in over 100,000 cities and 220 countries. The company also has about 4 million hosts, earning almost $10,000 a year for their Airbnb offerings. Airbnb now offers specialty categories for travel including Airbnb Luxe for luxury retreats and Airbnb Work for work travel. In fact, according to *Fortune,* Airbnb is now the largest provider of accommodations in the world.

How did Airbnb grow so quickly to become such a global brand? The company has built its brand by carefully developing a community and a sense of belonging. To do so, Airbnb must communicate effectively with two audiences: its hosts and its guests. Airbnb's hosts are a critical part of the company's business model as they are the ones offering the space for travelers. Not only must Airbnb recruit quality hosts, but it also must retain them and ensure consistency and quality across all accommodations. To incentivize hosts, Airbnb offers a user-friendly platform as well as competitive financial compensation. To attract guests, Airbnb works to ensure that their experience will be safe and clean, and that their expectations will match with reality.

Airbnb relies on word of mouth, user-generated content, and strong branding, as well as a carefully curated digital and social media strategy, to attract both hosts and guests to share unique experiences globally.

Airbnb combines an innovative idea with a user-friendly interface and many forms of marketing to attract both hosts and guests alike. Search is prominently featured and allows users to explore by location, dates, and/or number of guests easily.

Overview of Marketing

Welcome to marketing! Wherever your life and career take you, you can be assured that knowing how to implement marketing principles will be an important part of professional and personal success. Some of you reading this lesson will become marketing majors, but most of you will major in another subject, perhaps not even related to business. Either way, your ability to apply marketing principles in your future career or to market yourself will be critical to your success. Because of its importance, you are encouraged to approach this class as more than just a collection of concepts to be forgotten at the end of the semester.

By the end of this lesson you will be able to

- Define marketing.
- Explain why marketing is important, both within and outside the firm.
- Describe the role of marketing in organizations.
- Define marketing strategy.
- Describe the role of a marketing manager.

Defining Marketing

Marketing is a sum of many parts. Airbnb uses marketing elements such as advertising, sales, marketing research, promotion, pricing, service quality, and logistics to better serve you, the customer. Airbnb (like all good marketers) organizes these elements to best create value for the customer.

Defined formally, **marketing** is an organizational function and set of processes for creating, communicating, and delivering value to customers and managing customer relationships in ways that benefit the organization and its employees, customers, investors, and society as a whole. In other words, marketing is all about the creation and promotion of products (goods, services, or ideas) that serve the needs of customers and the sale of those products, which help firms to remain profitable and sustainable.

The Importance of Marketing

Marketing within Firms

Most firms exist because they have the potential to create a profit. However, to earn profits, most firms must successfully make their way through a minefield of competition, societal change, economic uncertainty, technological disruption, evolving customer demand, and regulation (to name a few factors facing business).

These factors are part of the **marketing environment,** which comprises internal and external factors that affect a firm's ability to succeed.

Because marketing drives sales and revenue, many consider it the most important function of a firm. The argument is that without product sales that result from marketing, firms would have no reason to exist. Without marketing, the other functions of business would not be needed. Do you agree?

All internal functions of the firm are important. However, marketing plays a leading role, generating the sales required to make other critical functions, namely accounting, finance, human resources, and operations/technology, necessary.

Accounting

The accounting function of a firm is tasked with accurately and properly keeping track of the firm's incomes and expenditures. Accounting also handles important tasks such as tax planning and preparation.

Marketing's Importance to Accounting: Without the efforts of marketing (the correct products, prices, promotion, and place to sell), there would be no inflows to track, cash flow to pay expenses, or taxes to be calculated.

Finance

The finance function of a firm is responsible for accurately forecasting the future financial inflows and outflows of the firm. Finance makes sure that required capital is available for future expansion and growth and that the firm's assets are returning value to the firm.

Marketing's Importance to Finance: The role of finance is related to expectations of return on the efforts of marketing. Without successful marketing, there is no need to expand a business or invest in capital expenditures if the firm is not selling products and generating revenue.

Human Resources

Human resources (HR) is a vital part of any firm. HR is responsible for attracting, onboarding, and maintaining employees that work for the firm. HR is also charged with making sure that the workplace is safe and conducive to productivity.

Marketing's Importance to HR: The need for employees and safe work environments is driven by the marketing efforts of the firm that generate sales and revenue.

Operations/Technology

The operations/technology functions work to produce or facilitate a firm's products and services. From manufacturing to information technology, a smooth and efficient process management system helps a firm to operate more smoothly.

Marketing's Importance to Operations/Technology: The operations and technology functions work with marketing, but they are also dependent on the efforts of marketing to provide the need for the function—without sales, no products would need to be produced.

Although each function of a firm is important, marketing is the "fuel" that keeps all the functional areas going. If marketing is succeeding, all functional areas are able to operate effectively. If marketing is failing, all other functions must compensate. Thus, we urge you to take this class seriously, regardless of what area of business interests you.

Marketing Outside of Firms

What if no area of business interests you; why should you care about marketing?

Maybe you plan to work for a nonprofit organization. Even nonprofit organizations face environmental factors that impact their success. For example, you may be required to successfully recruit new donors, educate communities, and secure grant funding—in an increasingly competitive nonprofit environment. To do this effectively, you will be wise to focus on marketing principles (value, satisfaction, product, etc.).

Not all careers have a direct tie to marketing, but securing a career almost always does. Everything you do to prepare and apply for a job is related to marketing. In almost all cases, you will first develop the required skills and abilities (creating the right *product*), consider what salary you can expect (setting your *price*), determine where you want to work (establishing *place*), and produce a professional résumé for your application and interview (*promoting* yourself). If all goes well, you will "sell" your new employer on you and get the job.

In either the profit or nonprofit world, you will need to find some way to differentiate either your firm or yourself from others to demonstrate value to customers or employers. In all settings, marketing principles are important because they will help you to accomplish your goals.

Marketing's Role in Organizations

Marketing's specific role within an organization is to understand its customers and markets in an effort to develop strategies to sell products or services that produce profit.

First, marketers focus on the needs and wants of the firm's customers. By developing products that match customers' needs and wants, marketers create value and improve the chances that consumers will purchase their firm's products.

Marketers are also responsible for scanning the environmental factors within and outside the firm to identify strengths, weaknesses, opportunities, and threats. Marketers attempt to position the firm's products to capitalize on changing market strengths and opportunities and to minimize weaknesses and threats.

Finally, marketers develop strategies that address the price, place, and promotional mix of the firm's products to maximize sales revenues. In short, effective marketing drives customer satisfaction and firm profitability.

Think about the basics of Airbnb's marketing and how it drives the firm's success. The marketers at Airbnb start with a focus on the needs of their customers. Upon gaining a good understanding of these needs, marketers develop specific strategies and plans to offer new or improve current rental listings in light of changing customer needs. This is a main function of marketing because when a product or service is correctly matched to what customers need and/or want, it is more likely that customers will purchase it. In addition, properly designed products and services make customers happy, increasing satisfaction, positive reviews, and future purchases.

For the past 15 years, Airbnb has been applying these basic principles to create win–win situations for travelers and hosts in over 220 countries and regions.

Meeting Customer Needs

- Airbnb has two customer markets: guests and hosts. Guests are looking for unique places to stay and hosts aim to monetize their property.
- Airbnb's app was designed with the brand message in mind: Make guests feel welcome.
- More than 200 million guests use the platform.

Scanning the Marketing Environment

- Study rental markets including hotels, hostels, and bed and breakfasts to anticipate needs.
- Offer unique or "experience" stays that cannot be found by the competition.
- Continue building a network of hosts to increase the number of rentals available to guests.
- Scan the environment for new location opportunities.

Implementing Effective Marketing Strategy

- **Product:** Airbnb's products are its website and app that connect customers looking for rental accommodations to individuals wanting to rent their property.
- **Price:** Airbnb charges a service fee from the hosts.
- **Place:** The company began in San Francisco and now operates in more than 220 countries.
- **Promotion:** Airbnb's campaigns emphasize that rentals are homes, not just accommodations, highlighting guests interacting with their surroundings.

Achieving Successful Marketing Outcomes

- In the early days, company founders went door-to-door taking professional pictures for listings, resulting in doubled revenue.

- Homeowners monetize unused space in their homes and renters find price-competitive short-term housing rentals.

- The Made Possible by Hosts campaign educates people on the experiences that guests receive from hosts.

- Advertising focuses on the unique experiences that guests have when staying in an Airbnb rental.

The Marketing Strategy

If your career is within a marketing department, you will have some role in the firm's marketing strategy. A **marketing strategy** is the set of actions taken to accomplish organizational objectives. A successful marketing strategy can lead to higher profits, stronger brands, larger market share, and other desired outcomes for stakeholders of the firm.

Many marketing strategies exist, but all should be implemented within a strategic plan. **Strategic planning** is the process of thoughtfully defining a firm's objectives and developing a method for achieving those objectives. Firms must continually undertake the task of strategic planning. Shifting conditions, including changing customer needs and competitive threats, ensure that what worked in the past will not always work in the future, thus requiring firms to modify their strategy.

Strategic planning helps ensure that marketers will select and execute the right marketing mix strategies to maximize success. The **marketing mix,** which is the foundation of modern marketing, is a combination of activities related to product, price, place, and promotion (often referred to as the four Ps of marketing) that represent areas a firm can adjust to influence demand for its good, service, or idea.

Airbnb uses the four Ps to strategically manage its business. For Airbnb, the products are its website and app that connect guests looking for accommodations with hosts wanting to rent their properties. The company operates in a market that offers hosts the value proposition of "earning money to fund their own passions."

Because most hosts do not have experience setting prices, Airbnb offers an estimate based on similar listing type, guest capacity, and location. Airbnb then charges the hosts a service fee on each rental.

Airbnb began as an "air bed and breakfast" in the living room of a San Francisco home in 2007. Since that time, accommodation types have expanded to include castles, boats, and other unique "experience" properties in more than 220 countries.

Airbnb communicates the value proposition of "living like a local" through advertisements via digital media, outdoor advertising, and television. The company also engages in public relations by donating to various causes such as offering free accommodations to displaced individuals from natural disasters.

Marketing strategy makes sense for Airbnb, but does marketing strategy apply to your life? Imagine starting college and just randomly taking classes because they are interesting or easy, or your friends are taking them. You take a full load each semester, get good grades, and at the end of four years you have accomplished little, except building student loan debt. Hopefully you have a better strategy and plan. Perhaps you have a checklist of courses required to graduate in your selected field. You developed this plan by selecting a major early and determining the required courses you need. Without the specific objectives of your degree program and a strategy for balancing your classes with the personal and professional demands on your time, you likely will not succeed in achieving your desired result. This example is true of a firm's marketing efforts. Without strategy and planning, firms can spend a lot of money with little effect.

The Role of the Marketing Manager

The **marketing manager** is responsible for short- and long-term marketing operations and strategy:

- *Operational activities* include implementing promotional campaigns, graphic design, social media, and marketing research (to name a few).
- *Strategy activities* include developing product value propositions, marketing objectives, forecasts, budgets, and monitoring results.

In firms that market only a few products and/or operate in only a few markets, one marketing manager may oversee all of the marketing implementation duties of the firm. In firms that manage more complex situations, such as a large product portfolio or diverse markets, multiple marketing managers may be employed, each responsible for a different aspect of the firm's overall marketing operations and strategy.

Often, a marketing manager may be considered *brand manager* or *segment manager* depending on what portion of a firm's business they manage. At Airbnb, for example, marketing manager positions exist in areas such as growth, internal communications, platform development, user experience development, and payment processing, among others.

Responsible Marketing

Airbnb: Is It Always Best for the Community?

Airbnb, as discussed earlier in this lesson, has transformed the travel industry. With over 7 million listings in 220 countries around the world, Airbnb is ubiquitous. According to its website, Airbnb hosts can earn significant income on their rentals. Additionally, Airbnb touts that the guests that stay in Airbnb locations spend money in local restaurants, shops, and sights, thus bolstering the local economy.

However, not everyone is cheering on the growth of Airbnb. For example, homeowners in residential neighborhoods may not enjoy the constant stream of travelers visiting the home right next to them. There are countless examples of complaints from Airbnb neighbors ranging from mild noise complaints to theft and property damage. Currently, many jurisdictions do not require homebuyers to disclose that they are going to use a home as an Airbnb rental property when they purchase, much to the chagrin of neighbors.

Additionally, Airbnb does not often pay the same taxes as hotels. In many metropolitan cities, tax revenue generated from hotels is a huge source of local income. However, Airbnb often does not collect lodging taxes, thus allowing them to offer a lower price than hotels and retaining income from local municipalities.

Airbnb has no doubt disrupted the traditional travel model and made travel more accessible, and often more comfortable, for many travelers. But does the good always outweigh the bad? You decide.

Reflection Questions

1. What, if any, responsibility does Airbnb have to local residents? Should homebuyers and homeowners be required to disclose that the property is a short-term rental?

2. Should Airbnb pay the same lodging taxes and fees as hotels? What are the implications for the Airbnb host and Airbnb guest?

Lesson 1-2
Value Creation

This lesson focuses on value creation. Creating, communicating, delivering, and capturing value for customers is the ultimate goal of a firm's marketing efforts. When customers perceive value in a firm's products, they are likely to purchase and repurchase products. Over time, value is what builds a customer's loyalty to products and brands. For firms, customer loyalty is what drives sustainable profits. The concept that value leads to long-term success is clear in the case of Kroger. For Kroger, a continued commitment to creating value is evident through its product selection, online ordering and delivery options, and personalized recommendations and incentives. By creating value for consumers, Kroger continues to have phenomenal and sustained growth.

By the end of this lesson you will be able to

- Define customer value.
- Define the marketing mix, which includes listing the four Ps.
- Distinguish between consumer needs and consumer wants.
- Describe how marketers create value for a product or service.
- Identify strategies for communicating value to customers.
- Describe how marketers deliver value to customers.
- Identify strategies for capturing value from customers.

Kroger: Adding Value through Groceries and More

When someone says the word "value," we often think of something that is valuable or meaningful to us—something we really want or need. Value is, in effect, "What you get for what you give." In other words, we exchange something we value—money, perhaps—to get something of value in return—something that satisfies our wants and needs.

When we consider our basic needs, food and nourishment are among the most important. Most of us in the United States depend on our local grocery stores to provide us with the nutrition we need every day. But as our wants and needs

Kroger adds value by providing more than what goes into your grocery bags.

have evolved over the years, so too have grocery stores and supermarkets. Consider Kroger. Based in Cincinnati, Ohio, it is the nation's largest retail grocery chain, with more than 2,800 stores in 35 states bearing the Kroger, Fred Meyer, Ralphs, and other brand names.

Barney Kroger opened his first store in 1883 "to serve customers through food freshness, low prices, and innovation." He went on to develop a chain of stores that provided one-stop shopping for more than just groceries. Today, Kroger customers can purchase not just groceries but wine, food for in-store dining, and coffee through national brand coffee shops, among other things. They can also take care of their banking and financial needs in Kroger stores. Add to that its 1,500+ fuel centers and 2,200+ pharmacies, and Kroger adds convenience to its offerings.

Kroger also has its own line of competitively priced, quality private-label brands that are distinct to Kroger and offered nowhere else, helping to meet consumers' needs. Kroger creates, captures, and delivers value through its wide selection of well-priced products and services in locations convenient to where we work and live.

Kroger doesn't add value just by creating stores for consumers to visit. It also offers the option of bringing the grocery store to you through its delivery and curbside pickup services. Whereas Barney Kroger used to saddle up his horse to deliver groceries locally, today Kroger partners with a variety of technology and delivery-based companies to let you order your food online and get it where you want it, when you want it.

Although consumer tastes may change over time, people often have the same or similar needs when it comes to food and associated items, and many customers like getting recommendations based on their purchasing habits. Kroger thus adds additional value to customer relationships by utilizing the services of its own retail data science, insights, and media company, 84.51°. By mining consumer data, the company provides consumers with highly personalized, specific purchase recommendations based on their recent behavior, offering coupons for future purchases and building on customer relationships that create loyalty in the long run.

Defining Customer Value

Customer value refers to the perceived benefits, both monetary and nonmonetary, that customers receive from a product (a good, a service, or an idea) compared to the costs (money, effort, and time) associated with obtaining that product. Simply put, value is what customers receive from a product less what they give in order to obtain it.

The harder part is *understanding* value. The term *value* has many potential meanings. Some think of value in reference to a "value menu" of low-price fast-food offerings. Others may think of value in regard to a higher-priced item of very high quality. In this case, value is not associated with "rock bottom" pricing, but rather the balance of benefits received from a product versus what is expended to get it. A good example is the Apple iPhone 11, which had an introductory price of $999

and up. This price doesn't seem like a value price (many competitors have lower prices), but Apple customers pay the price because they see value in it. Apple's customers' willingness to consistently purchase new iPhone products, despite a relatively high price, indicates that they perceive value in using the product that exceeds the cost of obtaining the product.

Value is an equation of benefit minus cost. To calculate value, a customer should subtract the cost of a product from its benefit. But calculating this equation is often anything but easy. First, quantifying the benefits of a product can be difficult. Think about the iPhone. How do you calculate its benefits? In most cases, assigning a solid number to the benefit of a product is nearly impossible.

Similarly, the actual cost of a product is often more than just the initial price. Many products have up-front costs and require other spending. A good example is a vacation. The benefits of a vacation may be high—perhaps you need a rest and relaxation break—but what are the costs? You can easily add up the costs of plane tickets, hotel stays, Uber rides, dining, museum fees, and the many expenses associated with travel. But many costs are overlooked. For example, missing work, school, or family all have a cost. Eating a lot of food on a cruise ship may lead to weight gain; what will it cost to lose that weight later? Think of other costs associated with a vacation. Do you still see value in the trip? Sometimes the true costs, when thoroughly examined, outweigh the benefits we receive. However, if we believe that the benefits outweigh the costs, then value creation can occur.

Value creation occurs when consumers use products or services that satisfy their needs and wants. For value creation to occur, exchange must also occur. **Exchange** is when a buyer and seller trade things of value so that each is better off as a result.

Firms such as Kroger initiate their part of the exchange of value by creating and offering products—in-store and online—that provide potential value to consumers. Consumers then get to complete the exchange of value by purchasing products and services.

But how can marketers orchestrate value for customers? It is clear that *products* are a part of how value is created because products can be designed and modified to meet customer needs and create value. In addition, marketers have control over the *price* (how much?), *place* (where sold?), and *promotion* (how will consumers know?) of their products. These four elements of marketing, known as the *four Ps,* are described in the next sections.

The Four Ps of the Marketing Mix

The primary way that marketers generate value is through decisions on product, price, place, and promotion. These elements are known as the four Ps of the **marketing mix.** The marketing mix represents everything that a firm can adjust to affect value and influence demand for its products.

Successful marketing managers make strategic decisions focusing on how a specific element of the marketing mix, such as discounting prices or changing the product's packaging, will affect value. With more value, firms can gain advantages over competitors and achieve long-term success.

Product

The discussion of marketing mix typically begins with the product because, without it, a firm has few, if any, decisions to make when it comes to price, place, or promotion.

Product is a specific combination of goods, services, or ideas that a firm offers to consumers. Kroger is a company that provides both products and services, both in-store and online. Its wide range of physical goods allows consumers to select the groceries and other items they want or need on a daily basis. Kroger also offers services, including online ordering, delivery, and a variety of in-store services, all adding value to complement the traditional "supermarket" experience.

Some other examples of products include:

- The Rubik's Cube: The Rubik's Cube is considered one of the top-selling toys of all time, providing children and adults with hours of entertainment.

- Drones: Sales of drones are increasing exponentially as more and more consumers are buying them for a hobby.

- Blow-Dry Bars: Beauty and spa treatments are examples of intangible products (*services*). Blow-dry services appeal to women and men around the world who desire to get their hair styled quickly and conveniently.

A product can also be an idea. For example, for Google, an idea such as a wireless and hands-free home can be considered a product.

Price

Price is the amount of something—money, time, or effort—that a buyer exchanges with a seller to obtain a product.

Pricing is typically the easiest and most common marketing mix element to change, making it a powerful tool for firms looking to quickly adjust their market share or revenue. Setting a price is therefore one of the most important strategic decisions a firm faces because it relates to the customer's assessments of value and quality associated with a product.

For example, Kroger offers many national and international brands consumers know and trust. But Kroger also offers its own line of private-label store brands that aren't offered anywhere else. Looking at price alone, you might think that the national brands are a higher-quality product. But since many Kroger brands are similar to or the same as national manufacturer brands, you might reconsider which brand provides more value to you, the consumer—is it the lower price for a similar item, or is it the comfort in trusting a well-known national brand? Firms must carefully consider price and walk the fine line between value and quality assessments by consumers.

Consider, also, if you are choosing a location for a haircut. The price of the haircut includes the actual cost of the haircut, but also the time it takes you for the appointment. More importantly, the cost of the haircut includes how it makes you feel. Think about the feeling of getting a bad haircut. That is a high cost that many of us are not willing to pay!

Place

Place includes the activities a firm undertakes to make its product available to potential consumers. Companies must be able to distribute products to customers where they can buy and consume them without difficulty. Place includes location, distribution, inventory management, and even where to put an item in a store.

Place is important because, even with the right product at the right price, if customers can't easily purchase the product, they'll likely find a substitute. Think about it: If you order a Coke but a restaurant has only Pepsi, will you accept Pepsi as a substitute? (Most people would say yes.)

Consider this: You can travel to some of the most remote towns in the world and find a McDonald's. This is possible because McDonald's focuses heavily on the place element of its marketing mix.

With grocery stores like Kroger, we often first think of physical stores in neighborhood shopping centers. But as consumer lifestyles have changed, so too has the definition of "place." Consumers now have the option of shopping virtually—ordering their groceries online, paying online, and arranging a delivery date and time that meets their busy needs. This puts Kroger in competition with companies like Amazon, which offers online ordering through Amazon Fresh or Whole Foods Market.

Promotion

The promotion element of the marketing mix is what most people think of when asked to describe marketing. **Promotion** refers to all the activities that communicate the value of a product and persuade customers to buy it.

The different promotional resources that marketers use make up the promotional mix. Four promotional elements are used by marketers: advertising, sales promotion, personal selling, and public relations. Each of these elements benefits from social media and mobile technology that enable faster and more effective delivery of promotional messages. If you've ever seen television commercials, been prompted by a persistent online ad, talked with a salesperson, been pinged by a mobile app while in proximity to a store, or used a coupon (a form of sales promotion), you've been on the receiving end of a promotional activity.

As is the case with every element of the marketing mix, successful promotion involves the firm's ability to integrate its promotional activities in a way that maximizes the value of each. An integrated approach to promotion, where all promotion elements "fit" well together, is important to the success of a promotional strategy. Kroger's consistent use of color, naming, and logo schemes for each of its brands successfully creates an integrated approach to promoting its products and services.

In recent years, the way a firm executes its promotional activities has evolved. Today, firms of all sizes and from all industries communicate quickly and directly with their customers using a variety of online and digital tools and social media platforms.

But how do marketers know the correct mix of the four Ps that will create value? The answer lies in understanding the difference between needs and wants and adjusting the four Ps accordingly based on specific customer variations for different products. Thus, we next explore the difference between consumer needs and wants.

Product

A specific combination of goods, services, or ideas that a firm offers to consumers.

Price

The amount of something—money, time, or effort—that a buyer exchanges with a seller to obtain a product.

Place

The activities a firm undertakes to make its product available to potential consumers. Place includes location, distribution, inventory management, and where to put an item in a store.

Promotion

All the activities that communicate the value of a product and persuade customers to buy it.

Needs versus Wants

Organizations desire to satisfy the *needs* and *wants* of customers. Many of us use these terms interchangeably in casual conversation. Have you ever said "I really *need* a new phone" when your current phone is actually functioning well? What you really meant to say was "I really *want* a new phone." For marketers, understanding customers' needs *and* wants—and the distinction between the two—is important.

Needs are states of felt deprivation. Consumers feel deprived when they lack food, clothing, shelter, transportation, and safety. Marketers do not create needs; they are a basic part of our human makeup. Regardless of whether you ever see an ad, talk to a salesperson, or receive an e-mail from an online retailer, you will still need food, water, shelter, and transportation.

Whereas a need is something that you *have to have* for survival, a **want** is something that you *would like to have.* Most of us make decisions beyond fulfilling our basic needs and more in response to fulfilling our wants. Rarely do we see a marketing campaign appealing to our basic needs. We don't see food manufacturers suggesting, "Well, you have to eat to survive; might as well eat this!"

Because most customers behave based on their wants, most marketers create products and services to appeal to those wants. These wants are typically shaped by a customer's personality, family, job, background, and previous experiences. For example, if you need a certain book but you want it easily, Google Books will make the purchase of that book easy for you—essentially matching your need with your want. In another example, clearly people need shelter from the weather; a marketer will work to turn that need into a want for a specific home—a tiny house, condo, or mansion.

The distinction between needs and wants is important. For example, everyone *needs* food and water to survive. Many consumers, however, *want* specialty foods and beverages to satisfy a variety of tastes.

The better a firm understands the difference between customers' needs and wants, the more effectively it can target its message and convince customers to buy its products. This approach, when done well, works for the simple reason that the firm's offering will meet its customers' needs and wants better than the offerings of its competitors. Likewise, effective marketing happens when a firm is able to align the four Ps with its customers' needs and wants.

Let's remember that *marketing* is an organizational function and a set of processes for *creating, communicating, delivering,* and *capturing* value for customers—all of which are critical to a firm's success. We explore how marketers accomplish each of these goals in the sections that follow, starting with creating value.

Creating Value

Marketers are constantly looking for new ways that products can create value for consumers. Recall that a *product* is the specific combination of goods, services, or ideas that a firm offers to its target market. This product enriches the lives of consumers with designs, features, and functions that they need *and* want.

But how do marketers get consumers to use their goods and services? When marketers successfully match products to consumers' needs and wants, consumers are more likely to purchase and use those products—and that is where value is created.

The simple fact that the phrase "Google it" is so common is testament to the fact that Google has been successful at creating value. "Googling" can fulfill both needs (e.g., finding a doctor when sick) and wants (e.g., researching a tropical vacation).

Matching products with wants and needs may seem simple, but it isn't, as demonstrated by the fact that more than 80 percent of all new products fail. If customers perceive that the product is not meeting their needs and wants, they will not buy the product and thus it will not succeed. Consider Orbitz soda, introduced in 1997. Although Orbitz had an appearance that appealed to young kids (it looked like a lava lamp), customers could not get past its taste, which many said was reminiscent of "Pine-Sol or cough syrup." Within a year, Orbitz was removed from retailers' shelves. Orbitz proved unsuccessful because it seems the company did not conduct taste tests to confirm that the flavors were ones consumers actually wanted and liked, before introducing it to the market.

Even Google has failed at creating value. Have you heard of Google Buzz, for instance? Probably not, because the social networking site did not meet consumer needs and wants. Consumers didn't see value in using the product, so it failed within a year of launch.

Orbitz soda was considered the "lava lamp of soft drinks." Though its look was appealing, its cough-syrup taste was not and the product failed.

Communicating Value

Once a company has created a valuable product (designed with customer needs and wants in mind), it must *communicate* the product's potential value to customers.

Have you ever been to a new restaurant and found that its food, service, and prices were excellent? Is that restaurant still in business? If not, the restaurant likely failed to communicate the value it offered. Business history is littered with failed companies that designed a valuable product or service but lacked the capability to let customers in the market know about it.

Conversely, consider Google, which in recent years launched an exclusively Google-based phone called the Pixel. The Pixel is designed to provide value to customers by incorporating an advanced camera, premium quality construction, and an affordable price. Google is communicating the value the phone provides to users by spending hundreds of millions of dollars on promotion. This communication is designed to let people know why (according to Google) the Pixel offers superior value compared to Samsung and Apple products. Given the level of communication, Google hopes to inform consumers of the value created by the Pixel, and in turn drive sales of the product.

Marketers use various promotional media to communicate value, including online, mobile, streaming, print, and outdoor advertising (such as billboards). However, regardless of the communication channel, most communication strategies focus on one or more of the four Ps.

Delivering Value

The value process doesn't end at communicating value; marketers must also *deliver* the value that they communicate. In order to ensure the value is deliverable to customers, the product or service must be physically available or easily accessible to customers. Think about Diet Coke. Isn't it remarkable to think that you can buy Diet Coke at a grocery store in Chicago, a mall in San Francisco, a restaurant in Miami, a gas station in rural Arkansas, and practically everywhere in between? Millions of people throughout the world buy and enjoy Diet Coke every day. This phenomenon is made possible by Coca-Cola's ability to physically deliver its product to countless places. In this example, Coca-Cola's use of a sophisticated delivery network, which is an aspect of the place "P," is a requirement of delivering value.

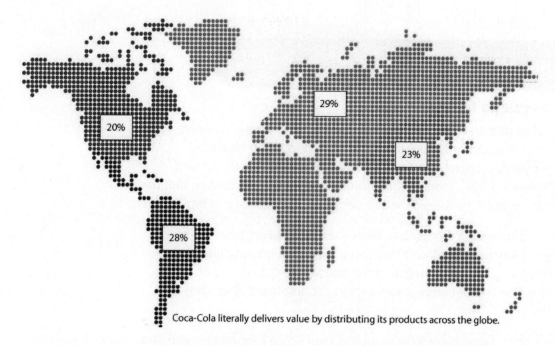
Coca-Cola literally delivers value by distributing its products across the globe.

Tech firms like Google have the same delivery complexities as other firms but often do not deliver to a physical store. Instead, they use digital channels. Think about Google's products and how their value is delivered to you. Google's search engine, Google Drive, Gmail, Google Play apps, and various Google Docs are never physically delivered to your home or business. For example, does a UPS driver deliver your Gmail? Obviously not! Advances in electronic delivery have revolutionized how value is delivered. Now value can be created and realized instantly, as opposed to in the past where you had to wait until a product was delivered (perhaps by UPS), or until you could pick it up.

Capturing Value

In addition to creating, communicating, and delivering value, firms must also **capture value.** The purpose of the first three value components is to generate exchange with customers. For firms to capture value they must identify ways in which they can gain in some way through the exchange.

Price is the primary way that firms can capture value from customers. In order for marketing to occur, there must be an exchange. An exchange occurs when the buyer trades something of value, usually money, to obtain a product or service. Price is the sacrifice the buyer makes to obtain a product. Usually price is monetary, but it can also include time, information, or effort.

Marketers have to be extremely thoughtful about price. Price is often a signal of quality. If an item is priced too low, it might be perceived as low quality. Conversely, if an item is priced too high, consumers might remove it from their consideration set. Marketers must determine the appropriate price that customers are willing to pay in order to best capture value for that product or service.

Responsible Marketing

Misleading Language to Inflate Value

Brands are supposed to market their products in a responsible way. In many countries, there are rules and regulations that firms must follow in how they communicate their value proposition to customers. For example, in the United States, the FTC (Federal Trade Commission) monitors companies to ensure that their communications are truthful, have evidence to back up their claims, and aren't unfair.

Yet, brands have to distinguish themselves from the competition, often using inflated language, claiming things like "this product will change your life" with the expectation that the consumer realizes that their whole life will not change. However, there is a fine line between distinguishing yourself and engaging in misleading language.

While brands want their consumers to understand their value proposition and see their products as superior than their competition, they have to be careful of using misleading, or less-than-truthful, language.

Examples of misleading language include:

- Brands like Reebok and Sketchers, in the early 2010s, launched toning sneakers that claimed to tone your legs better than regular sneakers. These claims went unsupported, and the brands were forced to issue millions of dollars of refunds.

- Naked Juice was forced to retract their claims of being "100% fruit" and "all natural" after synthetic sources of fiber and ascorbic acid were found in their products. Similarly, Kashi brand cereal (owned by Kellogg) had to stop using the term "all natural" as well.

- Weight loss product, Sensa, claimed that it enhanced the smell and taste of food, making users feel fuller and more satisfied faster. The FTC said that these claims were unfounded and the company had to pay $26 million in a settlement.

Reflection Questions

1. The above examples provide some very clear-cut cases of misleading language. However, consider the class action lawsuit against Red Bull for the drink's claim that Red Bull "gives you wings." While customers didn't actually believe that the drink could give them wings, they did think it implied more caffeine or energy than what the drink provided. At what point do you think language extends beyond metaphoric to misleading?

2. What are other ways that brands can distinguish themselves from their competitors while informing consumers about the value proposition?

3. How much of the burden of deciphering misleading language should be on the consumer?

Lesson 1-3
Marketing Ethics

You've just read about a few of the ethical issues marketers face. In this lesson, we look at a number of issues related to marketing ethics. This lesson also offers a process for making decisions on ethical issues. In addition, we explore marketing's role in corporate social responsibility as well as sustainable marketing.

By the end of this lesson you will be able to

- Describe the ethical issues in marketing.
- Identify the ethical values in marketing.
- Explain the process for making ethical marketing decisions.
- Define the concept of corporate social responsibility.
- Define sustainable marketing.
- Describe sustainable marketing strategies.
- Identify the major social criticisms of marketing.

Marketing with a Conscience

When you want to buy something, do you consider how sustainable it is before you make a purchase? According to the National Retail Federation (NRF), almost 50 percent of Gen Zers (those born in the mid-1990s and on) choose brands that they believe to be eco-friendly and socially responsible. In addition to being socially responsible, modern consumers want firms to behave in an ethical manner and to adhere to basic ethical standards. Ethical standards are therefore a critical component of a successful marketing strategy. Making ethical choices not only makes good business sense, it can also generate profits, enhance customer loyalty, and foster better relationships with stakeholders. The following examples show a range of ethical issues a firm might encounter.

Scenario 1:

The COVID-19 pandemic changed many industries. However, few were affected as much as the restaurant industry. With delayed government assistance, many restaurants struggled to stay afloat as in-person dining shut down and supply chain constraints caused delays in food shipments. Consumers were encouraged to order takeout from their favorite restaurants to boost sales and help them stay afloat. Unfortunately, many consumers used third-party delivery apps like Uber Eats and Grubhub, which charge restaurants transaction fees, commission fees, delivery fees, and subscription fees. In fact, in many cases, an order through a delivery app can cost a restaurant up to 30 percent of the price of the order. And to appear as if they have a wider variety of options, some delivery apps list restaurants on their apps without the restaurants' permission. Are these apps behaving in a predatory way? Should consumers use them?

Scenario 2:

Critics of fast-food restaurants often lament the inclusion of toys with the purchase of kids' meals. In fact, in 2018, parents in Canada sued McDonald's over the inclusion of toys in its Happy Meals, saying that the service violated provincial laws designed to protect children from advertising. The case cited that not only were the toys linked to popular films, they were also displayed at eye level for children, making them more irresistible. Other critics also suggest that the inclusion of toys or other incentives in kids' meals rewards bad eating behavior and contributes to childhood obesity. What do you think is the fast-food industry's responsibility?

Scenario 3:

A brand is interested in building its client portfolio. In order to reach a new set of customers, the company buys a list of e-mail addresses from a service company. While these people did not give their consent to be contacted, the Federal Trade Commission (FTC) says that the company is legally allowed to e-mail people without their consent. This could potentially hurt the brand's image, but it could also help the company reach new customers. If you worked for the brand, what would you do?

Scenario 4:

In 2002, Sony Ericsson launched a campaign in seven U.S. cities to promote one of its cell phones. The firm hired 120 actors to pretend to be tourists. These "tourists" then asked passers-by to take their picture with the camera included in the cell phone (a new technology at the time). The actors would then talk up the features of the phone and camera. Passers-by were basically given a sales demonstration without knowing it. Many critics of this guerrilla marketing tactic suggested that there were more transparent ways customers could try a product in a natural setting without the underlying deceit.

Scenario 5:

In 2011, Walmart introduced a line of cosmetics called GeoGirl. The target market for this product was girls as young as 9 years old. Walmart tried to reach this audience, as well as their parents, by promoting the environmentally friendly production of the cosmetics. This product and campaign were criticized by audiences for (1) emphasizing the appearance of young girls and (2) greenwashing the marketing efforts (claiming to be green when they actually were not). The products have since been discontinued. What do you think of Walmart's tactics?

Scenario 6:

In 2011, Patagonia launched the "Don't Buy This Jacket" campaign to highlight the amount of pollution caused by apparel and fashion production. The ad suggested to customers that they shouldn't buy clothes if they don't need them. However, sales of the jacket soared. Some critics of Patagonia claim that it was a marketing plan designed to drive sales. What do you think?

DON'T BUY THIS JACKET

Patagonia's "Don't Buy This Jacket" campaign was designed to reduce consumption and help the environment.

Ethical Issues in Marketing

Ethical decision making should be a key component of all successful businesses. **Ethics** are moral standards expected by society. Marketers should clearly understand the norms and values expected of them and act in a way that reflects their company, their profession, and themselves in a positive, ethical manner.

The consequences of not adhering to an ethical code can be serious. Ignoring ethical considerations has destroyed some of the largest companies over the past two decades. Job seekers, especially college graduates, sometimes face a challenging job market due to the unethical behavior of many firms. For example, Arthur Andersen was a leading recruiter of college graduates throughout the 1990s, until a series of unethical decisions by a limited number of employees led to its demise, causing many people, including new graduates, to lose their jobs. Not only do unethical marketing practices hurt customers, but they also hurt employees and society as a whole.

Marketers may confront decisions that will boost short-term sales at the expense of the long-term reputation of the company. Employees may have to choose between the short-term benefit of a sale and the potential long-term damage to their personal brand if they don't adhere to ethical standards.

Product-Related Ethical Questions

- What safety risks, especially for children and older adults, might a product pose?
- Should environmentally friendly ingredients and packaging be used even if it costs more?
- Are we maintaining the highest privacy standards in our website and with our data collection?

Place-Related Ethical Questions

- Should jobs be outsourced to other members of the supply chain?
- Are we working with other firms that have the same ethical standards we do?
- What opportunities for personal gain might tempt a firm's suppliers?

Price-Related Ethical Questions

- Should the firm charge customers different prices based on their ability to pay?
- Should the firm increase prices due to a lack of local competition?
- Should the firm lower prices on unhealthy products to make them more accessible and attract a bigger audience?

Promotion-Related Ethical Questions

- Does the advertising message represent the product's benefits honestly?
- Does the promotional strategy incorporate violence, sex, or profanity that may be inappropriate for some audiences?
- Does the advertising message attack competing products rather than highlight the benefits of the firm's product?

Ethical Values in Marketing

The American Marketing Association (AMA) is the leading organization for promoting and advancing marketing thought. The AMA has a Code of Ethics to promote the highest standard of ethics for its community members and the marketing discipline as a whole. Within this statement of ethics are six core ethical values that the AMA believes are the most important for championing ethics in business and marketing. The six core ethical values are:

- **Honesty:** Marketers are expected "to be forthright in dealings with customers and

stakeholders." This means that marketers are always truthful, stand behind the claims of their products and services, and honor their commitments and promises.

- **Responsibility:** Ethical marketers "accept the consequences of our marketing decisions and strategies." Responsible marketers serve the needs of their customers and communities. Responsible marketers acknowledge the vulnerable market segments such as children, older adults, and others who may be disadvantaged, and recognize that they have special commitments to these audiences. Responsible marketers always consider the ethical, environmental, and social consequences of their actions.

- **Fairness:** Marketers are expected "to balance justly the needs of the buyer with the interests of the seller." Fairness involves using clear and unambiguous communication with customers in all promotions. Fairness means resisting using manipulative sales tactics to entice customers. Fairness also means honoring customers' privacy and avoiding potential conflicts of interest.

- **Respect:** Ethical marketers "acknowledge the basic human dignity of all stakeholders." Marketers know that respect means listening to the needs and voices of all customers regardless of differences in race, gender, sexual orientation, or socioeconomic status. Respect means treating everyone as we wish to be treated.

- **Transparency:** Ethical marketers "create a spirit of openness in marketing operations." Transparency involves clear communication with all constituents of an organization. Transparency includes disclosing information and ensuring that there is never a conflict of interest in any of a firm's business dealings.

- **Citizenship:** Marketers are expected "to fulfill the economic, legal, philanthropic and societal responsibilities that serve stakeholders." Marketers are expected to be good citizens, contribute to their communities, protect the environment, engage in charitable activities, and seek supply chain partners that have similar values.

Ethical Decision-Making Framework

Despite the positive impact ethical decision making can have on a firm, the ethical choice is not always clear. This section outlines an ethical decision-making framework—shown below—that can help you navigate difficult marketing challenges. You can apply this systematic framework to ethical challenges you may face in your career.

Let's take a look at each step in the framework.

Step 1: Determine the facts in an unbiased manner.

First, determine the factual elements of a specific problem without letting any potential bias influence the decision. Fundamental factors such as an individual's history, background, and experience can influence how we review and interpret the facts at hand. We must make a conscious effort to determine the relevant information in an unbiased way.

Step 2: Identify the ethical issue at hand.

It's possible to avoid ethical problems if the ethical issue is clearly defined. The rest of the framework is relevant only if the issue itself is clearly defined.

Step 3: Identify the stakeholders impacted by the decision.

Stakeholders can be both external and internal and include the firm's employees, customers, suppliers, shareholders, and the community in which the firm operates. Identify and consider each group as part of the ethical decision-making framework.

Step 4: Consider all available alternatives.

After the relevant stakeholders have been identified, all parties should brainstorm alternatives. Different groups often view issues through different perspectives, and brainstorming can lead to creative, ethical, and useful solutions.

Step 5: Consider how the decision will affect the stakeholders.

Managers sometimes refer to this step as "seeing through a problem to the other side." This means that we should consider ahead of time how the decision will affect all stakeholders. For example, makers of Nutella marketed the product as part of a balanced breakfast for children, even though it was loaded with sugar. Ultimately, Nutella was sued for this and made to give $20 per jar to anyone who purchased Nutella believing it to be a "healthy breakfast."

Step 6: Discuss the pending decision with the stakeholders.

Seek feedback on potential decisions from stakeholders. It is often impossible to fully appreciate all of the dynamics of an ethical decision without getting input from those who will be affected. Many business problems can be avoided if a thoughtful discussion occurs when the decision is still pending.

Step 7: Make the decision.

Once the issue has been discussed with the relevant parties, make a final decision based on the stated criteria. Making decisions that impact others can be a stressful and challenging task, but using this decision-making framework can ensure thoroughness and thoughtfulness in arriving at the decision.

Step 8: Monitor and assess the quality of the decision.

The economy, regulatory environment, and consumer opinions are always changing and developing. It was not that long ago that smoking cigarettes on planes and in offices was considered perfectly ethical. Today, because we have more information about the dangers of secondhand smoke, laws prevent people from smoking in public places. Firms will face many ethical challenges in the years ahead that we cannot even begin to predict now. It will be incumbent upon all business professionals, especially marketers, to monitor and assess whether the

STEP 1
Determine the facts in an unbiased manner

STEP 2
Identify the ethical issue at hand

STEP 3
Identify the stakeholders impacted by the decision

STEP 4
Consider all available alternatives

STEP 5
Consider how the decision will affect the stakeholders

STEP 6
Discuss the pending decision with the stakeholders

STEP 7
Make the decision

STEP 8
Monitor and assess the quality of the decision

Source: Laura Hartman and Joseph Desjardins, *Business Ethics: Decision-Making for Personal Integrity and Social Responsibility* (New York: McGraw-Hill, 2011), pp. 47–57.

decisions they've made still represent the right and ethical choice for the firms, their consumers, and society as a whole.

Minicase

Tanisha works for a sporting goods store called East/West Adventures. Tanisha recently signed up for the newsletter for her competitor, another sporting goods store called North/South Adventures, because she wanted to keep up with what her competition was doing from a promotional perspective. One day, Tanisha receives an e-mail newsletter from North/South Adventures. All of North/South Adventures' contacts are accidentally included in the e-mail rather than having been blind copied. What should Tanisha do?

Step 1: Determine the facts in an unbiased manner.

What are the facts?

- Tanisha was trying to get information about her competition by signing up for its newsletter.
- Tanisha inadvertently received the list of customers for her competition.
- North/South Adventures mistakenly sent out all the e-mail addresses of its customers.

Step 2: Identify the ethical issue at hand.

What are the ethical issues at hand?

There are two ethical decisions at hand:

- Should Tanisha use the e-mail contacts of her competition?
- Should Tanisha alert North/South Adventures of its mistake?

Step 3: Identify the stakeholders impacted by the decision.

Who are the stakeholders?

The stakeholders in this are:

- North/South Adventures employees
- North/South Adventures customers
- East/West Adventures employees (especially Tanisha)
- East/West Adventures customers

Step 4: Consider all available alternatives.

What are the available alternatives?

There are two potential alternatives:

- Alternative A: Tanisha could use the list to send all of North/South Adventures' customers a 25 percent off coupon to try East/Coast Adventures.
- Alternative B: Tanisha could tell North/South Adventures of its mistake, offering it an opportunity to correct it.

Step 5: Consider how the decision will affect the stakeholders.

How will the decision affect the stakeholders?

Each decision has multiple implications for stakeholders.

If Tanisha chooses Alternative A, she might improve sales for East/West Adventures. However, she runs the risk of violating the privacy of North/South Adventures' customers. She also runs the risk of creating a harmful relationship with her competitors. Alternatively, if she chooses Alternative B, she risks losing market share to North/South Adventures.

Step 6: Discuss the pending decision with the stakeholders.

With whom should the ultimate decision be discussed?

Tanisha discusses the decisions with some of her colleagues at East/West Adventures.

Step 7: Make the decision.

What should the decision be?

Tanisha decides on Alternative B. She alerts North/South Adventures of its mistake and decides not to reach out to its customers.

Step 8: Monitor and assess the quality of the decision.

How will the quality of the decision be monitored and assessed?

Tanisha notices that East/West Adventures does not enjoy an increase in sales. However, its relationship with North/South Adventures is strengthened and the two stores combine efforts to engage in outreach events in their community.

Corporate Social Responsibility

Consumers are increasingly making purchase decisions in part because of a firm's reputation for corporate social responsibility. **Corporate social responsibility (CSR)** refers to an organization's obligation to maximize its positive impact and minimize its negative impact on society.

Organizations today are forced to confront a new economic reality: it is no longer acceptable to experience economic prosperity in isolation from those stakeholders (customers, communities, employees, etc.) who are impacted by the organization's decisions. Today's firms must accept responsibility for balancing profitability with social well-being when determining their success. Success begins with the quality of the relationships that a company develops with its customers and other stakeholders. These relationships are at the heart of CSR.

The following are some examples of ways that CSR is reshaping the way firms do business:

- **LEGO:** LEGO has launched the Sustainable Materials Center. The purpose of this center is to develop and implement sustainable alternatives to existing materials by 2030.

- **LinkedIn:** LinkedIn created a nonprofit called LinkedIn for Good. This group works to connect underserved audiences to economic opportunity.

- **Ben & Jerry's:** Ben & Jerry's began the Ben & Jerry's Foundation in 1985 to allocate pretax profits to philanthropy. Now, the foundation awards almost $2 million annually to fund community action, social change, and sustainability initiatives.

CSR has been shown to benefit companies in many ways, including employee retention, brand image, and overall success. Within many organizations, the marketing department is often responsible for the ideas and strategies comprising a firm's CSR program. However, in order to succeed, CSR must be adopted and enacted by all of the functional areas within a firm.

As with ethical decision making, CSR should consider multiple stakeholders. For firms to reach all of the people affected by their business practices, they should consider four dimensions of CSR: economic, legal, ethical, and philanthropic.

- **Economic dimension:** For-profit firms have an obligation to their stakeholders to be profitable. Without profits, a business cannot survive. A failed business hurts employees, investors, and communities.

- **Legal dimension:** Firms have a responsibility to understand and obey the laws and regulations of the communities in which they do business. They must follow local, state, and federal laws. Beyond this, U.S. companies are also subject to the laws and regulations of the foreign countries in which they do business.

- **Ethical dimension:** The ethical challenges facing marketers come from many different places. Inevitably, they involve more gray areas than the legal dimension. Marketers are responsible for a number of choices with an ethical dimension and will be held accountable for making the right decision. Marketers who take the time to identify the ethical issue at hand and consider how their decision will impact each of the firm's stakeholders are far more likely to make the right decision and successfully resolve the problem.

- **Philanthropic dimension:** Marketers understand that giving back to the community is not only the right thing to do but also a great way to get the firm's name, product, or promotion out to consumers at a reasonable cost. Corporate philanthropy is the act of organizations voluntarily donating some of their profits or resources to charitable causes.

Economic Dimension

"The LEGO Group is a privately held company based in Billund, Denmark. The company is still owned by the Kirk Kristiansen family who founded it in 1932. It is the LEGO® philosophy that 'good quality play' enriches a child's life—and lays the foundation for later adult life. We believe that play is a key element in children's growth and development and stimulates the imagination and the emergence of ideas and creative expression. All LEGO products are based on this underlying philosophy of learning and development through play."

Source: Courtesy of the LEGO Group

Legal Dimension

"Our corporate policy framework introduces high standards that reflect our core values. The LEGO Group is committed to doing business with integrity—the LEGO Way—and in adherence to anti-bribery and anti-corruption laws. As a company, and as individuals, we must never compromise our compliance to high ethical standards."

Source: Courtesy of the LEGO Group

Ethical Dimension

"In our daily work we strive to uphold the highest ethical business standards and business practices, with respect for human and labour rights, while doing everything we can to promote and protect our employees' well-being and safety. We will never sacrifice our values or purpose for the sake of short-term profit."

Source: Courtesy of the LEGO Group

Philanthropic Dimension

"Together with UNICEF, we want to strengthen child protection governance in the LEGO Group. We are the first in our industry to establish a global partnership with UNICEF, which is a clear statement of our commitment to promote and implement the Children's Rights and Business Principles in our business and to drive awareness of how corporations can generate positive change for children."

Source: Courtesy of the LEGO Group

Understanding Sustainable Marketing

Today more than ever, marketers recognize that, beyond the moral and ethical implications, adopting sustainable strategies has become an essential element of a firm's CSR efforts and one that contributes to long-term competitive advantage. **Sustainable marketing** is the process of creating, communicating, and delivering value to customers through the preservation and protection of the natural systems that provide the natural resources upon which our society and economy depend.

One of the easiest ways for marketers to engage in sustainable marketing is to seek ways to cut costs using sustainable practices as a guideline. Firms can choose from a wide range of strategies and ideas, from developing different packaging to using less energy.

Levi's, for example, focuses heavily on sustainability. Marketers at Levi's were among the first in the apparel industry to conduct a life-cycle assessment of the firm's major products. Between when the cotton was grown in the fields and the end of the product's life, Levi's found that manufacturing had the least impact on water and energy use. Meanwhile, growing the cotton used a great deal of water. Armed with this information, Levi's decided to partner with the Better Cotton Initiative to educate farmers on how to grow cotton with less water. Levi's now uses 12 percent of low-water cotton it its jeans.

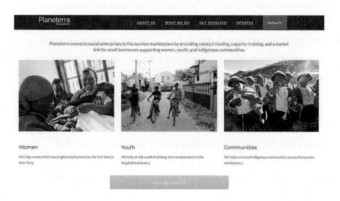

Planeterra Foundation combines CSR and sustainability into its business model. The foundation uses money funded by travel and tourism to support communities and build social enterprises. Planeterra's purpose is "to improve people's lives by creating and supporting social enterprises that bring underserved communities into the tourism value chain."

Source: Planeterra

In addition to having sustainable initiatives, many companies have sustainability engrained in their ethos. Often sustainability and CSR are aligned because most firms believe that protecting the environment is a big component of CSR. This is particularly true of the Planeterra Foundation. Since it was established in 2003, this nonprofit organization has contributed millions of dollars toward programs that focus on empowering women through education and job training; conserving cultures through such social enterprises as arranging homestays for travelers interested in learning more about indigenous cultures; helping at-risk youth develop the skills they need to create positive life paths; and contributing to organizations that help protect the health of our oceans.

Consumers are increasingly demanding that firms become more transparent and adhere to higher sustainability and CSR standards. Firms often assume that only the "younger" generations care about such things, but in reality there is an increasing demand across all the generations. Consumers are choosing to buy resource-conserving products from the shelves and boycott the products of companies that damage the environment. Environmentalism is a movement of citizens, government agencies, and the business community that advocates the preservation, restoration, and improvement of the environment.

Sustainable Marketing Strategies

Sustainable marketing (also called *green marketing*) activities can be divided into three levels: tactical greening, quasi-strategic greening, and strategic greening. These three categories of marketing activities represent the various degrees to which a firm can adopt an environmental focus.

Sustainable Marketing Activities

1. **Tactical greening:** Tactical greening involves implementing limited change within a single area of the organization, such as purchasing or advertising. Tactical activities represent relatively small actions aimed at instituting environmentally friendly practices within an organization. For example, Whole Foods might decide to stop doing business with suppliers that do not meet the company's environmental or recycling requirements.

2. **Quasi-strategic greening:** Quasi-strategic greening usually involves more substantive changes in marketing actions, such as redesigning the firm's logo or overhauling a product's packaging. Telecommunications provider Sprint engaged in quasi-strategic greening when it began using 100 percent recycled materials in all its branded packaging. In addition, it now uses soy inks and environmentally friendly adhesives and coatings.

3. **Strategic greening:** Strategic greening requires a holistic approach that integrates and coordinates all the firm's activities on environmental issues across every functional area. It represents a fundamental shift in the way the firm markets its products. For example, Nestlé has reformulated certain

products to decrease their environmental impact without affecting their taste, nutrition, or consumer appeal. Beyond this, the company has begun training farmers on good environmental stewardship and funding scientific research into producing sustainable cocoa and coffee.

These three categories are governed by the overall environmental strategy a company chooses to implement.

Environmental Marketing Strategies

There are five types of environmental marketing strategies marketers can choose to implement, depending on their competitive advantage and the overall marketing strategy they have adopted.

Strategy 1: Eco-Efficiency

Marketers seeking to reduce costs and the environmental impact of their activities typically pursue a strategy of **eco-efficiency.** This strategy involves identifying environmentally friendly practices that also have the effect of creating cost savings and driving efficiencies throughout the organization.

Strategy 2: Beyond Compliance Leadership

Most marketers who adopt a **beyond compliance leadership** strategy focus on communicating to stakeholders the company's attempts to adopt environmentally friendly practices. Marketers who select this strategy want to show customers that the company does more than the competition to implement an environmental strategy. Unlike with an eco-efficiency strategy, companies that employ a beyond compliance leadership strategy typically care more about differentiating themselves from competitors than about keeping costs low.

Strategy 3: Eco-Branding

An **eco-branding** strategy focuses on creating a credible green brand. For this strategy to be effective, consumers must recognize a noticeable benefit from their purchase. The eco-branding strategy tends to succeed in industries in which significant barriers to imitation exist. To achieve differentiation as part of an eco-branding strategy, the environmental improvement, such as the technology involved in developing a desirable electric car, should be difficult to imitate.

Strategy 4: Environmental Cost Leadership

Firms seeking a price premium for their environmentally friendly products often adopt an **environmental cost leadership** strategy. Green products sometimes cost more to produce than traditional products. Thus a leadership strategy that also seeks to lower costs may be the only way for a company to pay for its ecological investments and generate a profit for its other stakeholders.

Strategy 5: Sustainable Value Innovation

A final strategy firms can pursue is **sustainable value innovation**. This strategy entails reshaping the industry through the creation of differential value for consumers and through making contributions to society in the form of both reduced costs and reduced environmental impact. Firms that engage in this strategy do not aim to outperform the competition in an existing industry but to create a new market space. In doing so, they hope to make the competition irrelevant by giving the consumer more value per product at a lower price.

The benefits of an environmental marketing strategy extend to virtually all of a firm's stakeholders when the strategy is effectively integrated with the firm's general marketing plan.

With many consumers committed to "going green," environmentally focused organizations often benefit from favorable public opinion and loyal customers

Social Criticisms of Marketing

Consumers may have concerns about how firms serve their interests. The following are some of the major criticisms of marketing that are voiced:

- **High prices:** Critics suggest that marketers price items higher than they need to be. Often higher prices are blamed on high distribution costs, advertising costs, and excessive markups.

- **Deceptive prices:** Marketers are often accused of deceptive pricing that encourages customers to believe they are getting a better value than they actually are. Deceptive pricing practices involve falsely advertising sale prices or using other deceptive techniques.

- **Deceptive packaging:** Some marketers might use packaging to exaggerate the contents of the package, making consumers think they are getting more than they actually are. Some packages might have labels that mislead the customer about the contents. For example, firms giving nutrition information might use a smaller serving size to underrepresent the fat or calorie content.

- **Deceptive promotion:** Perhaps one of the biggest criticisms of marketing is deceptive promotion. With this, firms try to amplify a product's features and performance. For example, weight loss pills often show dramatic weight loss results to signal to consumers that they could expect similar results.

- **High-pressure selling:** Sometimes a salesperson's compensation might be tied directly to how much he or she sells. This practice causes some salespeople to

Lilly Pulitzer and Target collaborated together to offer an affordable line of cheerful women's wear, children's wear, and home goods. Within minutes (in some locations), all of the inventory was sold out, leaving shoppers feeling frustrated, confused, and furious.

engage in high-pressure selling that persuades consumers to buy more than they intended.

- **Shoddy or unsafe products:** For many consumers, quality control is a major concern. Firms that aim to cut costs by reducing quality can damage their brand irreparably. In 2016, the FDA cracked down on Castle Cheese, which makes Parmesan cheese for many national and retailer brands. Castle Cheese had been using an unsafe amount of cellulose (or wood pulp) in its Parmesan cheese.

- **Planned obsolescence:** In order to create a feeling of excitement and exclusivity, marketers might purposely cause their products to run out before demand is satisfied, referred to as *planned obsolescence.* This frequently happens with fashion items and newly launched technology products. Target had partnered with Lilly Pulitzer in an exclusive collaboration that instantly sold out. Some loyal Target customers were furious with the retailer over the limited availability of Lilly Pulitzer items. It can also involve limiting the lifespan of a product rather than its availability.

- **Marketing to children:** Marketers, especially food marketers, often work in an ethical gray area. Young people are a growing audience for marketers as they are increasing in their buying power. Plus, firms are trying to build brand loyalty at a young age. However, children are a vulnerable audience and susceptible to manipulation.

- **Materialism:** Marketers are frequently blamed for an increased focus on materialism. Materialism encourages consumers to focus on the acquisition and possession of objects and things that they may not need. Materialism also causes many consumers to believe that their worth, and the worth of others, is based only on possessions.

Unfortunately, there are some marketers who push the ethical boundaries and operate in a gray area. However, for the most part, many firms and marketers adhere to a strict ethical code and try to minimize the above examples from happening. Consumers also have quite a bit of protection, knowledge, and resources available to them should they run into any marketers they believe to be behaving unethically.

Responsible Marketing

Walk It and Talk It: DEI Ethics in Marketing

You may have heard the acronym DEI. It is short for diversity, equity, and inclusion, and it has become a cornerstone to many recent societal pushes toward change. Consumers are more socially and politically conscious than ever before, and brands understand the need to stand in solidarity with customers to fight against racial and social injustice. Many companies are addressing DEI issues both internally within their own ranks and in terms of their messaging to external customers. In fact, in a recent survey, companies reported spending 8.9 percent more on DEI initiatives over the last year, with larger enterprises spending more.

Yet, movements like #BlackLivesMatter and #StopAsianHate may not be areas of expertise for a company's marketing team. Thus, identifying and connecting with key stakeholders who can offer a different perspective is essential. Failure to do so can have consequences.

Take, for example, Starbucks, which recently learned a DEI ethics lesson. It was reported in June 2020 that Starbucks employees were banned from wearing Black Lives Matter paraphernalia such as t-shirts, pins, or accessories. The company cited the move as an attempt to thwart violence. The Starbucks brand subsequently faced an eruption of a different kind—angry consumers. The group created a #BoycottStarbucks hashtag to communicate their concern. As a result, Starbucks quickly reversed its decision and established its own Black Lives Matter apparel line.

Ethical best practices in DEI encourage marketers to remain flexible during uncertain times. They are asked to "walk the walk and not just talk the talk." Many experts agree that the avoidance of groupthink and engagement with external groups that represent diverse stakeholders is key to creating a more just and inclusive environment.

Reflection Questions

1. In the wake of violence against people of color, many brands are conflicted over the appropriate ways to show support without alienating their customer base. What would be the best approach to creating a win–win scenario?

2. How should marketers address mistakes made when trying to incorporate ethical DEI best practices?

Marketing Overview: Test

1. Internal and external factors that affect a firm's ability to succeed are collectively known as the firm's

    A. marketing environment.

    B. marketing mix.

    C. marketing strategy.

    D. marketing plan.

    E. marketing function.

2. As it relates to customers, the aim of marketing is to

    A. provide customers with value.

    B. increase profits by ensuring customers pay the highest price possible for a product.

    C. accurately and truthfully advertise the product.

    D. generate repeat sales from each and every customer.

    E. reduce the number of customer complaints.

3. Which statement best differentiates how marketing benefits accounting versus finance?

    A. Marketing efforts give accounting inflows to track, cash flow to pay expenses, and taxes to be calculated, whereas these same efforts give finance a reason to expand business or invest in capital expenditures.

    B. Marketing efforts give finance inflows to track, cash flow to pay expenses, and taxes to be calculated, whereas these same efforts give accounting a reason to expand business or invest in capital expenditures.

    C. Marketing efforts give accounting a way to track a safe work environment for employees, whereas it gives finance a way to fund it.

    D. Marketing efforts give finance a way to track a safe work environment for employees, whereas it gives accounting a way to fund it.

    E. Marketing efforts give accounting a means to produce or facilitate a firm's products and services, whereas they give finance a means to run smoothly and efficiently.

4. The combination of activities that represent everything a firm can do to influence demand for its good, service or idea is referred to as the marketing

    A. plan.

    B. concept.

    C. mix.

    D. matrix.

    E. model.

5. Of the *four* Ps of the marketing mix, the P that focuses on making sure that the organization is using the most effective means of letting consumers know about its offerings is known as

    A. product.

    B. price.

    C. packaging.

    D. place.

    E. promotion.

6. The organizational function and set of processes for creating, communicating, and delivering value to customers and managing customer relationships in ways that benefit the organization and its employees, customers, investors, and society as a whole is referred to as

    A. marketing.

    B. promotion.

    C. integrated marketing communications.

    D. public relations.

    E. publicity.

7. The job of marketers is to focus on providing products that fulfill customers' _____, which in turn will satisfy their underlying _____.

   A. wants; needs

   B. needs; wants

   C. aspirations; goals

   D. goals; aspirations

   E. demand; value

8. A successful marketing _____ can lead to higher profits, stronger brands, and larger market share.

   A. strategy

   B. objective

   C. plan

   D. proposal

   E. environment

9. A key aspect of strategic planning is the thoughtful defining of a firm's _____ and how to achieve them.

   A. objectives

   B. public relations

   C. market segments

   D. target markets

   E. prices

10. Select the statement that accurately describes the relationship between strategic planning and the marketing mix.

    A. Strategic planning helps to ensure that marketers will select and execute the right marketing mix strategies to maximize success.

    B. Strategic planning allows marketers to identify ways for products to sell themselves.

    C. Strategic planning is flexible whereas marketing mix strategies are inflexible.

    D. Strategic planning determines the life cycle of a firm's marketing mix strategy.

    E. Strategic planning establishes which aspects of the marketing mix a firm must constantly monitor.

11. That a brand like Nike is known for its high-performing shoes and other sportswear that improve according to changing customer demands shows that its success starts with which marketing mix element?

    A. product

    B. price

    C. promotion

    D. place

    E. production

12. The combination of activities that represent everything a firm can do to influence demand for its good, service, or idea is called the

    A. marketing mix.

    B. promotional mix.

    C. publicity.

    D. social media marketing.

    E. digital marketing.

13. In today's tough job environment, it is important to know how to _____ yourself effectively in order to reach the professional goals you have set.

    A. market

    B. sell

    C. segment

    D. evaluate

    E. compare

14. A marketing manager is responsible for implementing promotional campaigns, graphic design, social media, and marketing research. These activities are known as _____ activities.

    A. operational

    B. strategic

    C. public relational

    D. promotional

    E. advertising

15. Which of the following is not true regarding corporate social responsibility (CSR) programs?

A. CSR has been shown to improve employee retention.

B. CSR is the sole responsibility of the marketing department.

C. CSR includes not only economic and legal issues but also a focus on ethics and accountability to stakeholders.

D. CSR has been shown to benefit companies by improving their brand image.

E. CSR is an organization's obligation to maximize its positive impact and minimize its negative impact on society.

16. Which of the following is not one of the four dimensions of corporate social responsibility?

A. social

B. philanthropic

C. ethical

D. legal

E. economic

17. Every year the high school hosts a faculty-staff basketball game to raise money for the American Heart Association. Along with a cash donation, the local Subway restaurant donates sandwiches that the high school can sell at the game to earn extra money for the charity. By doing so, Subway is engaged in

A. charitable advertising.

B. corporate volunteerism.

C. philanthropic marketing.

D. stakeholder responsibility.

E. corporate philanthropy.

18. Patagonia, a California-based clothing company, is deeply committed to the environment. The company makes fleece jackets out of recycled bottles, uses solar panels to generate most of the electricity at its headquarters, and even tells its customers in its advertisements to think twice before buying anything, all in an effort to reduce its carbon footprint. This is an example of how Patagonia is engaged in

A. tactical greening.

B. value-driven marketing.

C. sustainable marketing.

D. corporate philanthropy.

E. eco-marketing.

19. Once marketers at Levi's discovered that growing the cotton needed to make its jeans used a great deal of water, it joined the Better Cotton Initiative to teach farmers how to grow cotton with less water. This is an example of Levi's commitment to

A. corporate philanthropy.

B. environmental marketing.

C. value-driven marketing.

D. sustainable marketing.

E. eco-marketing.

20. A clothing manufacturer has decided to buy its cotton only from those growers that grow it organically. This is an example of

A. tactical greening.

B. quasi-strategic greening.

C. global greening.

D. strategic greening.

E. product greening.

2 Marketing Strategy

What To Expect

You know that success rarely happens by accident. This is especially true in the corporate world where everyone is competing for your customers and their own profit. In order to have an advantage, you need a comprehensive plan. Let's take a look at what you can do to gain the edge on the competition.

Chapter Topics

- **2-1** Strategic Planning

Lesson 2-1

Strategic Planning

This lesson focuses on strategic planning. As the upcoming story about Mattel shows, firms work to perfect their strategic planning process so that they can form the basis of a sustainable competitive advantage. Firms that are successful typically get there by bringing together strategy and planning in their marketing efforts.

By the end of this lesson you will be able to

- Define strategic planning and its importance in marketing.
- Differentiate between direct and indirect competition.
- Describe the three basic characteristics of effective marketing strategy objectives.
- Define what it means for a firm to develop a sustainable competitive advantage.
- Identify the four basic categories of marketing growth strategies.
- Explain the importance of using current, relevant, and accurate internal and external data for strategic planning.

Marketing Analytics Implications

- The quality of the strategic plan is dependent on the availability and quality of the data used to develop the plan.
- The firm must have an ongoing process to collect appropriate data from sources inside the firm and external to the firm. It is important for the firm to implement a systematic means of data collection that allows for easy analysis and retrieval of those data.
- In order to identify the firms in the market that represent direct and indirect competitors, firms must engage in environmental scanning to evaluate current and changing market conditions that lead to changes in the competitive environment.
- In order to set strategies, firms must use data to determine what objectives are possible. Often SWOT analysis is used as the basis for this process.

Mattel: It's Still a Barbie World

From the start, Barbie has been intended as an empowerment brand meant to inspire and encourage young girls. Ruth Handler, Barbie's creator and one of Mattel's founders, wanted girls to have choices and to see that they could be anything they wanted to be. Since the doll's debut in 1959, society and culture have changed significantly, and so has Barbie. The doll that was originally sold wearing a black and white knit swimsuit with white sunglasses, hoop earrings, and black open-toed shoes has evolved to be more diverse and inclusive and is now sold in over 200 career options.

2017
The New Crew

2017
Builder Barbie

2016
Barbie Louvre Exhibit

Using strategic planning, Barbie successfully launched new dolls in line with its purpose: girls imagining their future selves through Barbie.

However, the road to transformation was not always paved with sales. Following a 20 percent drop in sales from 2012 to 2014, Lisa McKnight, global head of Barbie & Dolls Portfolio at Mattel, realized the brand had lost its footing. In McKnight's opinion, it was too focused on marketing products and needed a strategic overhaul to refocus on its purpose: girls imagining their future selves through Barbie. The brand's turnaround was driven by three pillars. The first was to focus on girl empowerment, the second was to produce a product that reinforced an inspiring message to girls, and the third was to focus on diversity and inclusivity.

As part of this new strategy, the Barbie brand worked to showcase a multidimensional view of beauty by introducing more skin tones, body types, hairstyles, hair fibers, and eye colors so that the dolls offer a better reflection of the world around them. Barbie also started the Dream Gap Project, a global initiative that aims to tackle self-limiting beliefs in girls—such as not being as smart or capable as boys—that can start as early as age five. The brand's "Inspiring Women" line sought to honor women who have broken boundaries in their respective areas, such as Frida Kahlo, Rosa Parks, and Katherine Johnson, inspiring the next generation of girls. Strategic planning, including learning and adapting to internal and external environments, has been an integral factor for Barbie throughout this process.

Barbie's story demonstrates that planning and strategy is a key element to a company's success. In 2019 diverse dolls made up over 50 percent of all doll sales. Even as the COVID-19 pandemic soared, the company reported sales from the last half of 2020 as its highest in 20 years. Dolls became friends and playmates and were a much-needed break to the rise in screen time.

The Importance of Strategic Planning

A **strategy** is the set of actions taken to accomplish organizational **objectives,** which are specific results marketers aim to achieve from marketing activities. **Strategic planning** is the process of thoughtfully defining a firm's objectives and developing an approach for achieving those objectives. For example, a marketer may create a plan to run a million online advertisements that return sales of 50,000 products within the next month.

Strategic planning is one of the most important tasks that marketers undertake. The strategic role of marketers is to gather information, feedback, and insights from customers, the market, and competitors. Using this information, firms must continually adjust their strategic plans. Shifting conditions, including changing customer needs and competitive threats, mean that what worked in the past will not always work in the future, thus requiring firms to modify their strategy.

For companies like Mattel, strategic planning can greatly increase the likelihood of success. The toy market is somewhat stable and predictable, so strategic planning can be carried out with a longer-term focus. However, the toy market can change greatly in the short term, so marketers must be able to adapt their long-term strategic plans to react to unforeseen market changes (e.g., an unexpected recession).

Uncertainty surrounding the COVID-19 pandemic forced firms to reevaluate their strategic plans beginning with short-term objectives, core strategies, and resources. Many short-term objective changes created a ripple of shifting long-term plans, affecting cash flow over time. The ever-changing circumstances and fluid situation of the pandemic required firms to focus on what they could accomplish immediately, and to postpone other activities until the crisis stabilized.

In both situations, making strategic decisions is critical to firm success. Proper strategic planning improves the likelihood that marketers will select, adapt, and execute the right marketing actions to maximize success in both the short and long term.

Nike

- **NYSE:** NKE
- **Headquarters:** Beaverton, OR
- **Founded:** 1964
- **Founders:** Bill Bowerman, Phil Knight
- **Revenue:** $36.4 billion USD (2018)
- **Subsidiaries:** Converse, Hurley International, Umbro, and more

Nike has a powerful competitive advantage because of its innovative and compelling products and its association with the world's greatest athletes and teams. Nike has built one of the most well-known consumer brands, which it continually works to maintain and grow. Nike's focus is to build, fuel, and accelerate product growth and brand associations based on their competitive advantage.

Under Armour

- **NYSE:** UAA
- **Headquarters:** Baltimore, MD
- **Founded:** 1996
- **Founders:** Kevin Plank
- **Revenue:** $5.2 billion USD (2018)
- **Subsidiaries:** MyFitnessPal, MapMyFitness

Under Armour realizes that it is much smaller than rival Nike, so Under Armour takes an "underdog" approach to marketing strategy. The company differentiates itself from Nike by featuring athletes that have to work extra hard to succeed, often using lesser-known athletes as spokespeople. Under Armour's plan is to continue growing in customer segments where athletes must strive and work hard to reach seemingly unreachable goals and performance levels.

adidas

- **ETR:** ADS
- **Headquarters:** Herzogenaurach, Germany
- **Founded:** 1949
- **Founders:** Adolf Dassler
- **Revenue:** 21.9 billion EUR (2018)
- **Subsidiaries:** Reebok, Runtastic, Five Ten Footwear, Ashworth, and more

adidas takes a traditional approach to strategy by using a three-point plan for sales growth: (1) adidas plans to become the first true "fast" sports company: fast in satisfying consumer needs, fast in internal decision making—gaining market advantage. (2) adidas plans to grow share of mind, share of market, and share of trend in specific cities (growth through geographic targeting). (3) adidas plans to promote consumer engagement with its brand and products (growth through customer focus).

lululemon/athletica

- **NASDAQ:** LULU
- **Headquarters:** Vancouver, Canada
- **Founded:** 1998
- **Founders:** Chip Wilson
- **Revenue:** 2.7 billion USD (2018)

Yoga-wear retailer lululemon has seen steady, positive growth since its inception. The company is relying on marketing research to expand its business through product and market advances. Specifically, lululemon is pursuing a high-growth strategy through a combination of product innovation, store expansion, e-commerce, and growth in the men's business.

Converse

- Subsidiary of Nike since 2003
- **Headquarters:** Boston, MA
- **Founded:** 1908
- **Founders:** Marquis Mills Converse
- **Revenue:** $1.9 billion USD (2018)

Converse's iconic "Chuck Taylor" sneakers have remained essentially unchanged for the past 100 years. They are also one of the most popular sneakers in the world. Converse has fueled this success for so long because its plan is to focus on customers and consistently produce a great product. This simple strategy seems to work for Converse—100 years of growth is impressive!

The Competitive Environment

Strategic planning starts with marketers systematically assessing the competitive environment. To complete this assessment, firms look at both direct and indirect competition.

Direct competitors are firms that compete with products designed around the same or similar characteristics. For example, Mattel's direct competitors include Hasbro, LEGO, VTech, Jakks Pacific, Spin Master, and other similar toy manufacturers. These companies compete directly because they all sell toys that are similar in design and function to those that Mattel sells. One way to think of direct competitors is to list all of the products that you might buy when making a purchase decision. For example, when ordering a pizza, you may check the current specials of Pizza Hut, Domino's, and Papa Johns—all three are in direct competition with one another.

Whereas most direct competition is identifiable, indirect competition is more difficult to recognize and may be overlooked entirely. **Indirect competitors** are firms that compete with products that have different characteristics but serve a similar function. Indirect competitors can take market share away from a firm as trends or consumer preferences change. For Mattel, Nickelodeon, the television network, could be considered an indirect competitor. Television networks do not sell toys, but they do sell entertainment in the form of shows. Thus, Nickelodeon competes for a consumer's entertainment time and budget with Mattel. For another example, recall the previous pizza decision. If you also considered other products, such as fast-food or sandwich delivery, instead of pizza, you were considering indirect competitors to pizza.

MacBook Computer

Direct Competitors: Lenovo, Dell, HP, Asus, Acer

Indirect Competitors: iPhone, iPad, Samsung (tablets and phones), any technology that replaces the need for a person to actually buy a laptop computer (many people don't have computers anymore; they rely on a phone or tablet)

Chipotle

Direct Competitors: Qdoba, Moe's Southwest Grill, Baja Fresh Mexican Grill, Taco Bell

Indirect Competitors: Five Guys Burgers and Fries, BurgerFi, Pita Pit, Zoes Kitchen, Subway, Chick-fil-A, or any business that sells food that people can eat on the go

Target

Direct Competitors: Walmart, Costco, Sam's Club, Dollar General

Indirect Competitors: General (Amazon, Kohl's, Macy's, etc.), grocery (e.g., Kroger, Albertsons, Publix, etc.), clothing (American Eagle, Old Navy, Ann Taylor, etc.), electronics (Best Buy, New Egg, etc.), pharmacy (Walgreens, CVS, etc.), and many, many more

Establishing Marketing Strategy Objectives

A firm's **marketing strategy** is the set of marketing actions (e.g., product design, advertising, promotions, etc.) taken to accomplish marketing objectives (e.g., sales, market share, awareness goals, etc.). Thus, a successful marketing strategy can lead to higher profits, stronger brands, larger market share, and a number of other desired outcomes for the organization.

The effectiveness of the marketing strategy depends in part on the clarity of the marketing objectives the firm has defined. Quality marketing strategy objectives have three basic characteristics:

- **They are specific.** Objectives need to be specific if they are to be of any value. Vague marketing objectives lead to a lack of focus and accountability, whereas specific objectives provide direction and guidance.

  ▶ **Example:** Mattel can set an objective to increase sales. Is this a useful objective? What does increasing sales mean? Is one more sale enough, or is 1 million more sales closer to the right number? A more specific objective would be that Mattel will engage in marketing activities that increase sales of Barbie dolls by 10 percent and American Girl dolls by 15 percent by the end of the following year.

- **They are measurable.** Objectives must be measurable so that marketers know if their strategies are working. A common phrase said in marketing offices is, "If it can't be measured, it can't be managed." Firms want to see a specific return on their marketing investment. Using measurable objectives makes it clear if marketing objectives have been met (or missed), and by how much.

- ▶ **Example:** Mattel could measure the previous specific objectives. A 10 percent increase in Barbie doll sales is an actual number that can be measured.
- **They are realistic.** Objectives need to be realistic so that marketers do not demotivate their organizations with unattainable goals.
  - ▶ **Example:** We don't know if a 10 percent increase in sales is realistic for Mattel, but we might guess that a 1,000 percent increase in Barbie doll sales is unrealistic. It may sound good to investors, but in reality, an established company can't deliver such aggressive growth. Setting unrealistic objectives affects all employees and departments. Imagine being responsible for the 1,000 percent increase—would you be motivated to achieve it? Would this objective lead to overproduction of Barbie dolls? Setting challenging but realistic marketing objectives is a better strategy.

Sustainable Competitive Advantage

Firms strive to develop a sustainable competitive advantage over their competition. A **competitive advantage** is the superior position a product enjoys over competing products if consumers believe it has more value than other products in its category. Firms develop competitive advantage by using strategies that create more value for customers than those of other firms. A product that exhibits characteristics that are rare, valuable, and hard to imitate is likely to have a **sustainable competitive advantage.**

Think of Mattel's Barbie. Today the Barbie doll has many competitors, but none have taken a large portion of Barbie's market share. Barbie has endured because its design and strong brand identity are not easily copied by competitors.

Developing a sustainable competitive advantage requires four elements of excellence: customer, operational, product, and locational excellence.

Customer Excellence

One way that firms can develop a sustained competitive advantage is to focus on customer excellence. **Customer excellence** is a strategy designed to put the customer at the center of all marketing activities.

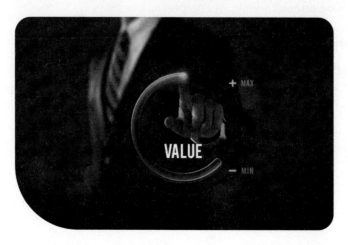

The most fundamental way that marketers create customer excellence is by producing products that provide value, satisfaction, and quality to customers. At a slightly higher level, the goal of marketers is consistent delivery of value, satisfaction, and quality, which result in customer loyalty.

Ultimately, marketers desire to provide long-term customer excellence that transforms customer loyalty into customer relationships. Customer relationships provide reciprocal benefits for both the firm and customer, as each becomes more trusting that the relationship will continue to be beneficial in the long term.

Mattel, striving for customer excellence, overhauled Barbie to include a more diverse and aspirational range of dolls that could serve as role models for girls. Some of the many career options customers can choose from include an astronaut, a doctor, a chef, and a firefighter. These changes seemed to resonate with parents as Mattel reported a 10 percent increase in sales in the quarter following their launch.

Operational Excellence

Firms often rely on their internal operations—how they produce their products—to attain operational excellence. **Operational excellence** is a strategy of focusing operational principles, systems, and tools on improving customer satisfaction with the firm's products and services. In other words, firms can improve how they manage their employees, supply chain, and/or manufacturing processes to maximize customer value and sustain competitive advantage.

Achieving operational excellence requires creating the environment for performing at a maximum level and maintaining operational excellence over time. The table below lists and explains the key factors in creating that type of environment.

Key Factors for Creating an Environment for Maximum Performance	
Factor	Explanation
Leadership	Should be committed to sustained organizational excellence.
Adequate Resources	Resources such as staffing, tools, technology, and so on help achieve operational excellence.
Clear Company Objectives	A clear understanding of company objectives by all functional aspects of the firm contribute to maximum performance.
Non-Conflicting Relationships	Conflict-free relationships between those working in the firm help achieve objectives.
Specific, Measurable, and Realistic Objectives	Specific, measurable, and realistic objectives for each functional unit within the firm contribute to maximum performance.

In 2020 Mattel revamped its supply chain as part of its goal to achieve operational excellence. The company closed factories in China, Indonesia, and Canada to optimize its manufacturing footprint, increase the productivity of its manufacturing infrastructure, and achieve efficiencies across the global supply chain.

Product Excellence

A much more visible strategy is a product excellence strategy. **Product excellence** is a strategy that focuses on the importance of high-quality and value-adding products. This strategy is visible because most of our interaction with a firm is through its products.

Marketers systematically design products to meet customer needs. To do this, product development and redevelopment include a rigorous review of product ideas, customer input, testing, and continual feedback. Ultimately, product excellence is determined by perceptions of the customers and the market. If a product is produced at a high quality, with a focus on adding value, and customers recognize the quality and value, then marketers are successful in following a product excellence strategy.

Mattel's Barbie has been around since 1959 and has more tha99 percent brand awareness worldwide. Over the last 60+ years, Barbie has been on-trend but not trendy, meaning the company has consciously created products, promotions, and messaging that reflect the happenings of the world to maintain its presence in the market and enjoy growth.

Locational Excellence

You may have heard the age-old marketing adage that "location, location, location" is the key to business success. This adage encompasses the final sustainable competitive advantage strategy: locational excellence. **Locational excellence** is a focus on having a strong physical location and/or Internet presence. This strategy focuses on how being in the right location can make all the difference.

Many firms realize that locational excellence is what drives business. To illustrate, consider these real-world examples of companies that value location:

- **Walgreens:** The company locates its pharmacies mainly on corners, in high-traffic areas. In total, the company's stores are located within five minutes of 76 percent of the U.S. population.
- **Starbucks:** The coffee shop places locations within blocks of each other. Starbucks believes that coffee needs to be convenient and easily accessible in urban settings.
- **McDonald's:** Franchisees are required to locate restaurants on corners or in high-traffic areas that allow for convenient parking and highly viable signage.

Even though Toys "R" Us has closed its doors for all brick-and-mortar retail locations throughout the United States, Mattel has still maintained locational excellence by selling Barbie dolls through a variety of other outlets that are accessible and convenient for customers. For example, Barbie can be purchased both in-store and online through big-box retailers such as Walmart and Target and online through Amazon and eBay. Thus, by pairing both physical and online shopping options, Mattel is able to create a competitive advantage based on prominent locations.

Customer Excellence Examples

- **Trader Joe's:** Trader Joe's is a specialty grocery store that competes based on customer service excellence. Customers who visit the store are quickly greeted and store clerks are stationed throughout the store to offer product advice and customer service.

- **Ritz-Carlton:** Ritz-Carlton Hotels specialize in high-end customer service. Guest needs and wants are noted from previous stays, allowing employees to delight customers by providing for needs before the guest even asks. Of course, the hotel charges for this service, but guests who appreciate exceptional service are happy to pay a higher room rate.

Operational Excellence Examples

- **Amazon:** Amazon is known for its convenience, prices, and quick delivery. The basis for success in all these areas is the company's excellence in operations. Amazon has invested in robotics, technology, and research to build a logistics system that is superior to its competitors.

- **Tesla:** This automotive and energy company is known for its environmental efforts as well as its operations. Investments in technology and facilities have helped Tesla build a competitive advantage based on operational excellence.

Product Excellence Examples

- **Bose:** Bose headphones and speakers are designed to be of the highest quality so they can offer the best possible sound. Bose invests heavily in research and development to maintain its product excellence advantage over its competitors.

- **Fossil:** Fossil markets products such as watches, leather goods, jewelry, and accessories that are high in quality and moderately priced. Fossil's focus on high quality differentiates its products from its competitors and drives the success of the brand.

Locational Excellence Examples

- **CVS:** CVS is the largest pharmacy retailer in the United Sates, with more than 9,600 stores. CVS has grown its business based on locational excellence—being conveniently located in high-traffic, growth, and neighborhood areas across the United States.

- **Starbucks:** Starbucks has grown its business by placing locations almost everywhere in the United States. In fact, the company has more than 22,000 locations across the world, which means that if you're thirsty, Starbucks is right around the corner.

Marketing Growth Strategies

A company's marketing strategy can follow various paths based on the product and industry, but most seek to move the product in one of four directions: market penetration, product development, market development, and diversification.

Market Penetration

Coca-Cola: Offering a "4 for $10" special on Coca-Cola and Diet Coke to increase sales in a grocery store that has sold Coca-Cola for decades.

AMC Theaters: Offering a "$5 Tuesday" special on all movies in an attempt to prompt current customers to visit the movie theater.

Product Development

Coca-Cola: Introducing a new line of natural juices, new to the Coca-Cola line, at current grocery store partners.

AMC Theaters: Offering a new product, "AMC IDOL" live talent shows, in current theater locations.

Market Development

Coca-Cola: Offering Coca-Cola products for sale in Cuba, a market that it previously did not serve.

AMC Theaters: Purchasing Carmike Cinemas in an effort to gain theater locations in cities not currently served by AMC Theaters.

Diversification

Coca-Cola: Offering a new line of "Tropical Coolers" in Cuba, to serve local needs in a new market.

AMC Theaters: Marketing a local news and interest program at its theaters in a new, international market that values local programming.

Each of these categories represents the intersection of a strategy related to products and another related to markets. These strategies are discussed in more depth in the sections that follow.

Market Penetration

Market penetration strategies emphasize increased sales of existing goods and services to existing customers. This type of growth strategy often involves encouraging current customers to buy more each time they patronize a store or to buy from the store on a more frequent basis.

For example, marketers at Pizza Hut try to get existing consumers to buy one more pizza each month or add an order of breadsticks to their normal pizza order. They have found success offering the "Big Dinner Box," which includes two medium pizzas and two side items. The product introduced consumers to side dishes, such as wings, pasta, or breadsticks, which they might not have thought to buy from Pizza Hut.

Mattel can use a similar approach to increasing sales of Barbie dolls to current customers through the Barbie Signature membership program, which offers customers savings on purchases and shipping, as well as access to exclusive products and content.

For such a market penetration strategy to succeed, firms often must increase advertising expenses, develop new distribution frameworks, or enhance their social media offerings.

Product Development

Product development strategies involve creating new goods and services for existing markets. A new product can also be an improved product or one with a new feature or innovation.

Dr Pepper used a product development strategy when it introduced Dr Pepper Ten, a 10-calorie soft drink, using a male-targeted marketing campaign with the slogan "It's not for women." Although Diet Dr Pepper's marketing was female-friendly, the Dr Pepper Ten campaign focused on male consumers who enjoyed Dr Pepper but were interested in drinking a beverage with fewer calories. Dr Pepper believed that men were generally unhappy with the taste and image of diet drinks and marketed Dr Pepper Ten as a better-tasting, "manlier" product in an effort to reach its target market in a new way.

Mattel used a product development strategy when it introduced playsets for the Barbie doll, such as dollhouses, furniture, and vehicles. The development of these additional products increased sales for Mattel from current customers that wanted a total play package.

Dr Pepper Ten was marketed as the "manlier" product for those who were interested in drinking a beverage with fewer calories but who disliked the taste and image of diet drinks.

Market Development

Market development strategies focus on selling existing goods and services to new customers. The targeted new customers could be of a different gender, age group, or country.

Globalization is an increasingly critical strategy for virtually any company or industry. The vast majority of the 100 largest American-based companies are rapidly increasing their international presence and aggressively implementing market development strategies throughout the world.

For example, Arkansas-based retail giant Walmart has recently seen its international division Walmart International grow to operate in 27 countries, including China, India, Brazil, and Mexico. A company seeking to expand into foreign markets must have a clear strategy for implementation that maximizes its chances for success.

Walmart's Global Reach

Mattel has followed a market development strategy and has a global presence. For Mattel, market development is used to grow sales by entering new markets or selling to new market segments. The company currently sells Barbie dolls in 150 countries worldwide, so it has largely maximized its growth potential through its market development strategy.

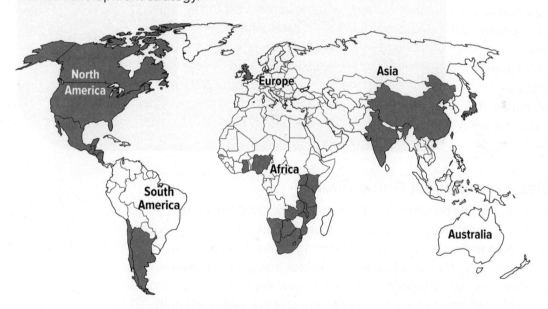

Walmart's globalization strategy has extended its international reach to include operations in 27 countries, including China, India, and Mexico.

Source: Walmart, "Location Facts," http://corporate.walmart.com/our-story/our-locations.

Diversification

Diversification strategies seek to attract new customers by offering new products that are unrelated to the existing products produced by the organization.

Disney has diversified significantly over the past few decades, moving from a company that produced animated movies and ran theme parks to an international family entertainment and media enterprise that owns television channels such as ABC and ESPN as well as independent production companies such as Lucasfilm and sells vacation properties, books, apparel, and international consumer products.

Having developed a positive reputation over many years, the company is poised to further diversify its operations and products to hedge against decreasing sales in some products due to economic conditions.

Mattel followed a diversification strategy when it introduced Barbie-related products for consumers to wear and use, such as bracelets, necklaces, T-shirts, and tote bags. These products are not meant to be played with like consumers play with the Barbie doll, and they can be used by both children and adults.

Marketing Analytics: Using Data Analysis for Strategic Planning

Establishing a Systematic Data Collection Process

A firm must have systems in place to capture all relevant marketing data for use in strategic planning. Often a firm will use marketing information systems (MIS) tied to its customer relationship management (CRM) and other data tracking initiatives. The MIS is used specifically to focus on marketing-related data such as customer sales data, customer feedback data, competitor data, marketing campaign data, and marketing costs. These data are crucial to the strategic planning process and are the basis for the decisions made within the planning process.

Identifying Direct and Indirect Competitors

Market scanning is the process by which a firm constantly monitors the market and collects information not only about its direct competitors, but also about its indirect competitors—that is, competitors in adjacent or emerging markets. A firm often engages outside providers to monitor social media (Twitter, Instagram) as well as traditional media (television, radio) to better understand how the market sentiments toward a firm's products and services are changing. This process allows a firm to understand two key things about its consumers:

1. How they view its products vis-à-vis those of its competitors

2. What other products or services they might want

Because the number of new entrants to a market varies, some firms constantly have to evaluate new entrants. Even those that compete in more stable markets must remain vigilant of such marketplace changes.

Setting Strategic Objectives Based on SWOT Analysis

During the strategic planning process, a firm must determine what its strengths, weaknesses, opportunities, and threats are via a SWOT analysis. This SWOT analysis must be conducted both inside and outside the firm. A SWOT analysis guides the firm to a deeper understanding of what strategies are truly possible because not all firms can be all things to all consumers. A firm must be critical of its ability to deliver value to customers.

A SWOT analysis might reveal that some markets may simply be too expensive for a firm to pursue, or it may reveal that the firm is not capitalizing on its strengths. SWOT analysis must look at the unique value created by the firm or by each division within the firm. Specifically, if a firm has two divisions and one produces a luxury high-end good (for example, Volkswagen's Audi line) and the other a mass-produced, lower-cost good (for example, Volkswagen's Jetta line), then what is considered a *strength* for one division may be a *weakness* for another. The luxury market may want unique products but the mass-produced market may need more practical products.

Responsible Marketing

Strategic Planning and Planned Obsolescence

Patricia works as a planner for a company that manufactures printers. The managers at the company are working on the firm's strategic plan for the following year. One of the company's main goals for the year is to increase the unit and dollar sales of its XYZ4321 printer. Sales have declined in the past few years, but margins on this printer are high.

In strategic planning meetings, one of Patricia's colleagues suggests that they begin implementing connection updates for the printer every 18 months. This would mean that if customers wanted the printer to continue to connect with their network, they would have to purchase a new printer with the updated connection software every 18 months.

Patricia knows from her marketing classes that this strategy is called "planned obsolescence." With planned obsolescence, a product is intentionally created with a finite useful life, making the consumer have to buy that product more frequently. Patricia wants the company to succeed—and she wants to work collaboratively to create the strategic plan—but she is uncomfortable using planned obsolescence as a growth strategy.

Reflection Questions

1. We know that strategic planning involves defining a firm's objectives and a plan for reaching those objectives. What are your thoughts on planned obsolescence as a tool for achieving a growth objective?

2. If you were Patricia, how would you respond in this situation?

3. What are some possible alternatives to planned obsolescence that are more consumer friendly?

Marketing Strategy: Test

1. _____ is the process of thoughtfully defining a firm's objectives and developing a method for achieving those objectives.

   A. Situation analysis

   B. Strategic planning

   C. Marketing strategy

   D. Diversification

   E. Stakeholder management

2. A market penetration strategy

   A. revolves around discontinuing older products in favor of selling new products in new markets.

   B. seeks to attract new customers by offering new products that are unrelated to the existing products produced by the organization.

   C. focuses on selling existing goods and services to new customers.

   D. emphasizes selling more of existing goods and services to existing customers.

   E. involves creating new goods and services for existing markets.

3. Zbar, a mobile phone-manufacturing company, decides to pursue a market development strategy to further its business prospects. In this situation, which of the following actions falls in line with Zbar's latest strategic direction?

   A. Zbar persuades its customers to buy more accessories for their phones.

   B. Zbar enters a price-sensitive market with its low-end phones.

   C. Zbar announces the creation of a new phone.

   D. Zbar tries to get existing customers to buy its new easy-to-use phones for their grandparents.

   E. Zbar tries to get existing customers to buy a new version of their mobile phones for their spouses.

4. A product possesses a competitive advantage when it enjoys a superior position over competing products because

   A. the competing firms are pursuing different marketing strategies.

   B. the product is in the "question mark" product category of BCG matrix.

   C. the manufacturing firm is pursuing a product development strategy.

   D. competing products are manufactured in small numbers.

   E. consumers believe it has more value than other products in its category.

5. Why is effective strategic planning important to a firm?

   A. It improves the likelihood that marketers will select, adapt, and execute the right marketing actions to maximize both short- and long-term success for a firm.

   B. It allows marketers to invest in important marketing activities in order to compete effectively.

   C. It allows marketers to gather information, feedback, and insights from customers, the market, and competitors.

   D. It allows a firm to scan the direct and indirect competition in the competitive environment.

   E. It allows a firm to communicate with its customers.

6. The electronics market is somewhat stable and predictable. For companies like Samsung, strategic planning will improve the likelihood of success.

   A. TRUE

   B. FALSE

7. Elena is leaving work and wants to pick up dinner on her way home. She is debating between picking up stopping at Two Boots Pizza to pick up a slice or stopping off at Subways to pick up a sandwich. In this example, Two Boots Pizza and Subway are considered indirect competitors.

    A. TRUE

    B. FALSE

8. Bouquets and Things has developed a new objective that states, "To increase sales by 5% in the next three months." What is wrong with this objective?

    A. It lacks specificity.

    B. It lacks quality.

    C. It lacks measurability.

    D. It lacks objectivity.

    E. It lacks realism.

9. Which of the following goals meets all the criteria of an effective market strategy objective?

    A. "To increase sales from $1,000 to $2,000 on birthday flower arrangements in 3 months."

    B. "To increase sales of birthday arrangements in several months, making sure the increase equals a quarter of total sales."

    C. "To increase sales by 10% of birthday flower arrangements over time."

    D. "To increase total product sales from $1,000 to $2,000 in a few months."

    E. "To increase sales of birthday arrangements by a quarter of total sales in 3 months."

10. To create sustainable competitive advantage, a firm must demonstrate operational excellence, customer excellence, product excellence, and

    A. competitive excellence.

    B. resource excellence.

    C. price excellence.

    D. locational excellence.

    E. workplace diversity excellence.

11. Companies like Walmart and Target often locate near each other and close to other shopping options. Both companies know that by following this strategy, they are working toward achieving

    A. customer excellence.

    B. operational excellence.

    C. product excellence.

    D. locational excellence.

    E. convenience excellence.

12. The Winter Spree is preparing for the upcoming holiday season by focusing on reviewing product ideas and seeking customer input. This strategy exemplifies which element of sustaining competitive advantage?

    A. product excellence

    B. locational excellence

    C. customer excellence

    D. operational excellence

    E. research excellence

13. Marketing growth strategies include all of the following *except*

    A. market positioning.

    B. market development.

    C. market penetration.

    D. diversification

    E. product development.

14. This type of growth strategy often involves encouraging current customers to buy more each time they patronize a store or to buy from the store on a more frequent basis.

A. market penetration

B. diversification

C. product development

D. market development

E. market positioning

15. This strategy focuses on selling existing goods and services to new customers.

A. market penetration

B. diversification

C. product development

D. market development

E. market positioning

3 Today's Marketplace

What To Expect

For marketers, understanding the forces that exist in markets is often the difference between success and failure. To understand markets, marketers must scan both internal and external environments to detect what is currently happening and what is expected to happen in the future.

Chapter Topics

- **3.1** The Marketing Environment
- **3.2** Global Marketing

The Marketing Environment

This lesson is about marketing environments. As illustrated in the Lucy & Yak example you will soon read, changes in the environment could greatly affect the future success of the company. You will see that if the marketers at Lucy & Yak are able to adapt to marketing environment changes, they will remain successful in the future.

By the end of this lesson you will be able to

- Explain why marketing environments are important.
- Outline how a company's immediate environment affects marketing strategy.
- Summarize the external environmental forces that influence the marketing environment.

Lucy & Yak: Reimagining Sustainability and Fashion

Lucy & Yak is a sustainable clothing brand driven by a purpose: to sell colorful and comfortable, environmentally friendly clothing that is also fashionable. By building a brand community and a social media following, Lucy & Yak has been able to reimagine sustainability and fashion.

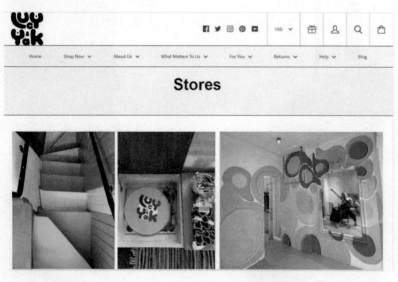

Lucy & Yak is focused on selling sustainable fashion and treating its employees fairly.

How did Lucy & Yak get started? The company was founded by Chris Renwick and Lucy Greenwood. "Yak" was the name they gave to their VW van that they would drive around the UK selling vintage clothing. After realizing they had an eye for fashion, the pair decided to develop their own products, and began with dungarees (aka, pants). They soon realized that fashion is one of the top contributors to environmental degradation.

Lucy & Yak is committed to creating a brand that pays workers a living wage and treats all employees with respect and dignity. The company recently built a new factory that is solar powered and provides proper air-conditioning and ventilation for employees. Most of its apparel is made with organic and recycled materials. It also uses sea shipments to transport merchandise because that form of transportation has the lowest carbon impact. Lucy & Yak also engages in carbon offsetting by planting trees and using renewable materials. Additionally, it learned that many workers who manufacture fashion are not paid a living wage. The company therefore decided to focus on creating fashion while also doing good.

Lucy & Yak recognized an opportunity in the marketplace for environmentally sustainable and ethically sourced fashion that people would support—and be excited to wear.

The Importance of Understanding Marketing Environments

One undeniable truth of marketing is that nothing ever stays the same. The **marketing environment** comprises internal and external factors that affect firms and may help or hurt a product in the marketplace.

Marketing environments are not the same as the physical environment in which we live, but they are similar. To illustrate, think about planning a summer cookout. You invite guests, buy food, and plan activities. However, you have no control over the weather—it may rain. Your only course of action is to plan for rain by having tents or an alternative indoor space.

Marketers usually aren't concerned with rain, but they do face similar, uncontrollable environmental forces in the marketplace. For example, marketers must prepare for economic, political/legal, or social changes that may "rain" on the continued success of their products. Drawing again on the analogy of the summer cookout: If you do not consider poor weather conditions and prepare for them, the cookout will likely fail. Likewise, the marketers at Lucy & Yak must adjust their marketing strategy to anticipate and/or respond to environmental changes—such as changing demographics. This is the difference between sustainable profits and product extinction. Thus the importance of understanding marketing environments for marketers cannot be overstated.

How do marketers forecast what the future holds for their products? Just as we regularly check the weather forecast to plan for a cookout, marketers continually engage in environmental scanning to forecast the future needs of their market. **Environmental scanning** is the act of monitoring developments outside the firm's control with the goal of detecting and responding to threats and opportunities. In short, successful marketers do more than just react; rather, they proactively assess changes and adapt to them before they occur.

The Impact of a Firm's Immediate Environment on Marketing Strategy

Marketing does not occur in a vacuum. Changes occur in marketing environments both inside and outside the firm. Some of these changes occur close to the firm's daily operations (*immediate environment*) and some occur outside the firm (*external environment*). Changes in the external environment are beyond a firm's immediate environment and control. This section focuses on how the immediate environment relates to marketing strategy.

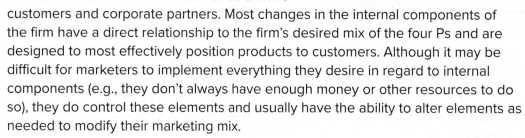

The firm's **immediate environment** is made up of both internal and external factors. *Internal factors* are elements that a company has direct control over, such as employees, products, and manufacturing processes.

A firm's immediate environment comprises the internal structure of the firm as well as a firm's customers and corporate partners. Most changes in the internal components of the firm have a direct relationship to the firm's desired mix of the four Ps and are designed to most effectively position products to customers. Although it may be difficult for marketers to implement everything they desire in regard to internal components (e.g., they don't always have enough money or other resources to do so), they do control these elements and usually have the ability to alter elements as needed to modify their marketing mix.

Customers are also a factor of a firm's immediate environment. To compensate for the external nature of customers, marketers must develop a deep understanding of customers' needs and wants. Gaining this understanding is accomplished through strategic use of marketing research, which is one of the most important tasks that marketers perform. Many firms design their marketing strategy and competitive advantage around their ability to understand customer needs better than their competitors. Armed with customer research information, marketers can focus their strategic efforts to align internal components in a way that satisfies customer needs and wants.

The Impact of the External Environment on Marketing Decisions

Many marketing environment factors occur well outside the direct control of marketers—what is known as the **external environment.** *External factors* include demographic, economic, sociocultural, political/legal, competitive, and technological changes. These factors revolve around a firm, and changes in any of them can have both positive and negative impacts on the firm's marketing activities.

Demographic

Lucy & Yak sells different products for men and women and also offers a unisex line. Lucy & Yak also ensures that its employees and partners receive fair pay and works hard to reduce gender inequity.

Economic

With the economic environment constantly fluctuating, consumers may choose more fast fashion apparel that is less expensive.

Sociocultural

Changing cultural desires for more sustainable and ethical products has led Lucy & Yak to create a shared, collaborative working space among freelancers and entrepreneurs and has led Lucy & Yak to focus its marketing on these attributes.

Political/Legal

As firms are required to adhere to stricter regulations on ethical sourcing and carbon emissions, Lucy & Yak will have to continue to adapt, innovate, and lead the way.

Competitive

Numerous apparel brands are attempting to offer more sustainable apparel, from both start-ups to established brands. Lucy & Yak will have to find ways to differentiate itself.

Technological

Changing technologies allow Lucy & Yak to source materials and manufacture apparel in more sustainable and ethical ways.

The Impact of Demographic Changes on Marketing Decisions

Marketers realize the importance of identifying changes in consumer demographics. *Demographics* are the characteristics of human populations that can be used to identify consumer markets. They include things such as age, gender, education level, and ethnicity, all of which influence the products consumers buy. All of these specific areas relate to individuals (i.e., every person fits into some demographic), but marketers consider them at the population level in the form of large demographic groups (known as *segments*), such as Baby Boomers, Gen Xers, Gen Yers or Millennials, and Gen Zers.

Marketers often use demographic information to help in decision making because demographic details are often easy to identify. The following section explains how changes in demographic characteristics affect marketing decisions.

Age

As mentioned earlier, changes in demographics can affect a firm's marketing strategy. Age-related population trends are important to marketers because they indicate changes in how and what consumers will buy in the future. Each year, the average age of the U.S. population rises, with older adults being the fastest-growing demographic group. The figure below, based on information from the U.S. Census Bureau's Population Clock, illustrates the U.S. population by age for 2005, 2010, and 2015. According to this figure, the percentage of Americans age 65 and older is expected to almost double over the next 50 years.

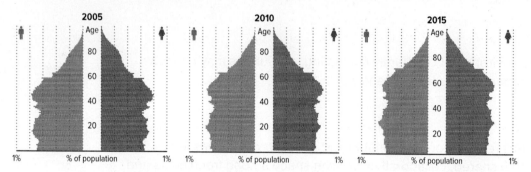

The U.S. Census Bureau's Population Clock. Can you see how the U.S. population is aging? To see the change in real time, visit https://www.census.gov/popclock.

Source: U.S. Census Bureau. "U.S. and World Population Clock." Last modified August 26, 2019. https://www.census.gov/popclock/.

The growing Baby Boomer population could present opportunities for a variety of firms, such as cosmetics manufacturers. Marketers of these firms can consider ways to redefine or rebrand their products as "age-defying" and thus appeal to segments in the population seeking to "look and feel younger."

Gender

One of the most important changes in the United States in recent decades has been the roles, attitudes, and buying habits of men and women in the marketplace. Female consumers now account for 85 percent of all consumers, purchasing everything from cars to groceries to health care. Firms that traditionally targeted men, such as home improvement and hardware stores, must begin to find ways to appeal to women in order to remain competitive. For example, marketers for the home improvement store Home Depot recognized that women were becoming an increasingly large part of their customer base but were largely being ignored by their promotional strategies. In an effort to appeal to female consumers, Home Depot introduced a new line of Martha Stewart products.

Education

During the height of the recession that began in December 2007, the unemployment rate for college-educated workers was approximately half that of the nation as a whole. The United States is pushing to lead the world in college graduation rates by 2020. As a result, the number of professional workers with college degrees is expected to increase significantly in the coming years. Historically, highly educated consumers are more likely to be employed. Educated consumers are also likely to earn significantly more money throughout the course of their lifetimes and comprehend an advertiser's message more readily, making them prime targets for marketing strategies.

Ethnicity

The ethnic composition of the U.S. population is changing rapidly. Projections indicate that by 2050, the Hispanic population in the United States will almost double to more than 127 million, representing 29 percent of the entire U.S. population. The African American population in the United States grew over 12 percent between 2000 and 2010 and now makes up over 13 percent of the total U.S. population. Asian Americans represent approximately 5 percent of the U.S. population but have the highest average family income of all ethnic groups, thereby increasing their purchasing power and importance to marketers. This exciting growth can be an opportunity for firms to adapt their marketing strategies to appeal to a variety of unique tastes and cultures.

Age

Age is a demographic variable that marketers use to design new products and product modifications. For example, there are 76 million Baby Boomers—the generation of children born between 1946 and 1964—retiring at a rate of 10,000 per day. Members of this generation typically possess two things that marketing professionals seek: (1) disposable income and (2) the free time to spend it. Retiring Baby Boomers make up only a quarter of the U.S. population but account for 50 percent of domestic consumer spending.

To reach this demographic, firms are making changes to encourage older Americans to shop at their stores. For example, paint retailer Sherwin-Williams has redesigned its 3,400 stores to make them more comfortable for older shoppers by adding more lighting and seating. Pharmacy CVS Caremark has retrofitted its stores to appeal to older shoppers by lowering shelves and adding carpeting to reduce slipping.*

Gender

Female control over the majority of consumer spending makes women a target for marketers across products. For example, women are responsible for more than half of the new car purchases in the United States, so marketers at Toyota have targeted promotional activities for the Sienna minivan toward female buyers with a marketing campaign titled "Swagger Wagon." The campaign appealed to female consumers who did not want parenthood to take away from their ability to drive a cool car. Delivered via television and YouTube videos, the ads feature a woman and her family describing how the Sienna fits their lifestyle rather than a laundry list of features. "Swagger Wagon" generated over 12 million YouTube hits, and the effects of the ads have endured, as evidenced by the many that still refer to the Sienna as the "Swagger Wagon."*

Education

The increase in the average education level of the United States will give marketers new opportunities. Some are already reaching out to this growing demographic. Knowing that highly educated consumers tend to be environmentally conscious and value technological features, Ford introduced the C-Max hybrid, emphasizing its fuel efficiency and advanced features.*

Ethnicity

Marketers have already taken basic steps, such as advertising in multiple languages, to reach out to different ethnic groups. For example, food company General Mills identified that Hispanic consumers prefer to buy the brands of goods and services they see advertised on television. The company significantly increased the number of ads it ran on Spanish-language media and saw sales of popular General Mills products such as Progresso soup and Honey Nut Cheerios increase with this population segment.* As the ethnic makeup of the United States continues to change, and in the effort to most effectively communicate with each group, marketing professionals will continue studying how each ethnic group responds differently to various types of marketing.

The Impact of Economic Changes on Marketing Decisions

Economic factors influence almost every marketing decision a firm makes. Economic conditions impact consumers' willingness and ability to buy products. Consequently, firms must create, communicate, and deliver value in a way that is appropriate for the current economic climate. Four main economic elements influence marketers: gross domestic product (GDP), income distribution, inflation, and consumer confidence.

Gross Domestic Product

Gross Domestic Product (GDP) refers to the market value of all officially recognized final goods and services produced within a country in a given period. Although not the only economic measure firms should pay attention to, GDP paints a simple picture of the economic health of a nation. GDP per capita is often considered an indicator of a country's standard of living. For example, the GDP per capita in the United States was $62,641 in 2018 compared with $9,698 in Mexico and $9,770 in China. Meanwhile, overall GDP is the most common gauge of the expansion or contraction of an economy. A **recession** occurs when overall GDP declines for two or more consecutive quarters. The U.S. recession that began in December 2007 was characterized as such because GDP declined in both 2008 and 2009.

Income Distribution

How income is distributed across the U.S. population has shifted over the past several decades. This shift has forced marketers to develop new strategies to satisfy consumers at different ends of the spectrum. In 2016, the top 1 percent of the U.S. population earned almost 20 percent of the country's adjusted gross income (and paid 37 percent of total taxes). Meanwhile, the bottom 50 percent of earners earned approximately 12 percent of the country's income (and paid about 3 percent of total taxes). The gap between the top and bottom of the spectrum has grown as income distributions shift.

Inflation

Inflation is an increase in the general level of prices of products in an economy over a period of time. When the general price level rises, each unit of currency (e.g., each U.S. dollar) buys fewer goods and services. Consequently, inflation also reflects erosion in the purchasing power of money. **Purchasing power** is the amount of goods and services that can be purchased for a specific amount of money. For example, if the price of gasoline goes up 10 percent, the amount of gasoline you can purchase for $20 decreases by that same 10 percent. Inflation can impact marketing significantly if prices rise faster than consumer incomes.

Consumer Confidence

Consumer confidence measures how optimistic consumers are about the overall state of the economy and their own personal finances. Consumers purchase more when consumer confidence is high because they feel more secure in their jobs. If the economy contracts and people lose jobs, consumer confidence decreases, leading to more saving and less spending.

GDP

Changes in GDP and evidence of a recession can have a powerful negative effect on marketing. Such conditions typically result in layoffs, increased unemployment, and reduced consumer confidence. These factors influence consumers' ability and willingness to buy products and contribute to nonprofit organizations.

Income Distribution

Shifting income distribution offers marketers new opportunities to satisfy consumer needs and wants at both higher and lower income levels. Many companies, such as Dollar General, have thrived by targeting consumers with modest incomes. Dollar General's marketing strategy includes offering low-income families food, health, and beauty products at reduced prices. On the other end of the spectrum, luxury brand Louis Vuitton has had success catering to high-income consumers. The company, which makes shoes, watches, accessories, and other premium items, has been one of the most successful luxury brands for years and has consistently increased its brand value. During the most recent recession, Louis Vuitton's marketing strategy involved raising prices and heightening its focus on quality. The result was additional sales to the firm's wealthiest clients and thus higher profits.*

Inflation

You may experiencing the impact of inflation soon. College tuition and fees have increased 440 percent since 1980. Meanwhile the average family's income has risen less than 150 percent.* Consequently, university marketing professionals are increasingly tasked with providing evidence of the value of higher education. Their strategy includes increasing career service staffs, offering more integrated course programs, and spending more time educating people about the financial benefits of a college degree.

Consumer Confidence

Marketers who can find strategic ways to help consumers feel confident about their purchases can improve performance during challenging times. For example, in 2009, Hyundai recognized that U.S. consumers were not confident in buying a new car as a consequence of their fear of losing their jobs. In response, Hyundai launched the Buyer Assurance program. The program allowed Hyundai buyers to return their car within 12 months, no questions asked, if they lost their job.* Hyundai identified and tapped into basic and powerful consumer fears to develop a strategy that sought to calm those fears and helped consumers feel more confident in purchasing. Marketers cannot dictate the state of the economic environment, but they must develop marketing strategies to put their firm in the best possible position for success, regardless of economic factors.

The Impact of Sociocultural Changes on Marketing Decisions

Our society and culture help shape our beliefs, values, and norms, which, in turn, define our tastes and purchasing habits. **Sociocultural** refers to the combination of social and cultural factors that affect individual development. One of the biggest sociocultural changes in the United States over the past half century has been the shift from a nation of primarily one-income families, where one spouse stays home to raise children, to one in which two-income families and single-parent households predominate. In fact, in 2016, only 69 percent of American households were composed of a married couple with children (as opposed to 87 percent in 1960). This change has led to a "money rich, time poor" society—that is, a nation with money to spend but little time to spend it.

The new sociocultural reality of busier families has created opportunities for firms to offer new kinds of value. Banks, for example, have expanded their offerings to include later evening hours and more services through ATMs. They've also launched additional online and mobile banking options that give busy consumers more opportunities to use the bank's services.

One of the most notable sociocultural changes is increased environmental awareness and concern, which started in the 1960s. In many places, pollution levels of lakes, streams, and even the air we breathe began to rise to dangerous levels. Today, companies produce green products in an effort to reduce our chemical and carbon footprints. For example, many products, from cleaners, to food containers, to reusable shopping bags and even clothing, are now designed for eco-friendliness.

Consumers are also increasingly concerned about nutrition and ingredients in their foods, prompting the arrival of new products, including gluten-free, non-GMO, and nut-free options for snack foods and other food items.

Another increasingly important sociocultural change relates to privacy and the use of big data to predict consumer needs and wants. Companies like Fitbit, Snapchat, Google, and almost any social media platform collect large amounts of data on consumer behavior. Many consumers are concerned that the collection and use of big data is leading to a reduction in privacy. Specific concerns focus on personal information that is gathered and then sold, or in some cases hacked, and used for reasons beyond what consumers realize. Marketers must factor privacy concerns of their customers into their decisions or risk alienating or losing customers in the future.

All of these sociocultural changes provide firms with opportunities to market products in new and different ways.

The Impact of Political/Legal Changes on Marketing Decisions

Political and legal environmental factors can greatly impact marketing strategy. The two important factors are interrelated, yet different. This section provides a brief overview of ways in which both political and legal change can affect how marketers pursue their goals.

Political

The political climate in the United States can change the direction of government policy quickly and impact how marketers position their products. Firms must understand how the changing political climate affects them and develop marketing strategies that allow them to succeed under various conditions.

Legal

The legal system represents another component of the external environment that affects how firms market their products. The legal environment within the United States continues to change, forcing marketing professionals to refine their strategies. The main federal agency that affects marketing is the Federal Trade Commission (FTC). The laws outlined in the remainder of this section are among those enforced by the FTC, which serves as the consumer protection agency for the United States. The FTC administers laws that serve two main functions:

1. **To ensure businesses compete fairly with each other.** For example, the Sherman Antitrust Act (1890) was passed to eliminate monopolies and guarantee competition. The Robinson-Patman Act (1936) refined prohibitions on selling the same product at different prices.

2. **To ensure businesses don't take advantage of consumers.** For example, the Fair Packaging and Labeling Act (1966) guarantees that products are labeled correctly. In 2009, banks and other financial institutions were required to change how they dealt with consumers following the passage of the Credit Card Accountability, Responsibility, and Disclosure (CARD) Act. The law banned unfair credit card rate increases and required that disclosures regarding minimum payments and interest rates be made in plain English to protect younger consumers.

Sherman Antitrust Act (1890)

Combats anticompetitive practices, reduces market domination by individual corporations, and preserves unfettered competition as the rule of trade.

Robinson-Patman Act (1936)

Prohibits firms from selling the same product at different prices in interstate commerce unless based on a cost difference or if the goods are not of similar quality.

Wheeler-Lea Amendment (1938)

Authorizes the Federal Trade Commission to restrict unfair or deceptive acts; also called the Advertising Act. Broadened the Federal Trade Commission's powers to include protection of consumers from false advertising practices.

Fair Packaging and Labeling Act (1966)

Applies to labels on many consumer products. It requires the label to state the identity of the product; the name and place of business of the manufacturer, packer, or distributor; and the net quantity of contents.

Telephone Consumer Protection Act (1991)

Limits commercial solicitation calls to between 8 a.m. and 9 p.m. and forces telemarketers to maintain a do-not-call list and honor any request to not be called again.

Credit Card Accountability, Responsibility, and Disclosure Act (2009)

Protects consumer rights and abolishes deceptive lending practices.

The Impact of Competitive Changes on Marketing Decisions

Marketers adapt marketing strategy based on actions of their competitors. Firms face two types of competition, both direct and indirect. **Direct competitors** are firms that compete with products designed around the same or similar characteristics. **Indirect competitors** are firms that compete with products that have different characteristics but serve a similar function. With both types of competition, changes in how competing products are sold are often the basis for a firm's marketing strategy. Thus, scanning the marketing environment for competitive changes is one prominent way that marketers can maintain or develop an advantage for their products.

For example, Lucy and Yak competes with direct competitors that also sell sustainable fashion. Lucy and Yak has to build relationships with customers in order to increase brand loyalty and top of mind awareness. Lucy and Yak also has to contend with indirect competitors. This can include other fashion and apparel brands that are maybe less sustainable and hand crafted. However, this also includes other ways that Lucy and Yak's target market spends their money. Some consumers might choose to buy experiences or other items instead of Lucy and Yak apparel.

In this sense, scanning what the competition is doing is considered a *lagging indicator,* as scanning of the competition usually takes place after changes have already occurred. Marketers that focus solely on following the competition will be unlikely to ever lead a market; rather, they will typically be distant followers trying to keep up with several market leaders.

The Impact of Technological Changes on Marketing Decisions

Of all the external factors, rapidly evolving technology arguably represents one of the most significant challenges, as well as one of the most significant opportunities, for marketing professionals. Technology influences how consumers satisfy their needs and wants, the basic concept underlying all marketing activities. For example, if you heard a song on the radio in the early 1990s and you wanted to buy it, you had a couple of options. You could buy the song as a single on compact disc (CD) for $3 to $5 or buy the artist's entire album on CD for $15 to $20, even though the album was filled with nine other songs you could care less about. In 2003, Apple's iPod and the iTunes Store changed the market by allowing consumers to purchase only the specific songs they liked for a mere $0.99. This technological advancement changed the way consumers purchased music forever.

In addition to affecting how consumers use products, technology changes the way firms promote their products. For example, the relatively recent adoption of mobile phones (the iPhone was introduced in 2007) by consumers has changed how marketers communicate. You're no doubt familiar with mobile applications that track consumer locations and electronic coupons such as those provided by Groupon that make it possible for marketers to know where a customer is and to communicate directly to him or her with an offer for that moment and location. For example, beacon technology is becoming a common way of reaching nearby customers. Beacons employ Bluetooth low-energy (BLE) wireless technology to pinpoint the location of customers in stores and other places and to deliver messages to their mobile devices. Once connected to a beacon, the retailer has a direct link to nearby consumers that allows coupons, videos, or other messages to be transmitted.

This direct channel of communication—through technology—to consumers who are literally "in the neighborhood" is transforming how marketers think about promotion and advertising for their firms. As you might expect, the close relationship that consumers have with their technology allows for nearly limitless applications for marketers.

Responsible Marketing

Mattel Makeover: Barbie Goes from Pink to Green

In June 2021, Mattel Corporation, the behemoth maker of toy brands like Masters of the Universe and the UNO card game, announced a major decision. In a historic move, it decided it would begin producing a new Barbie line from 90 percent ocean-bound plastic.

This is not the first time Barbie has been transformed. The doll has seen many makeovers in her over six decades of existence. However, what made this change different was the rationale behind it. The climate is talking, and consumers have listened and shared what they heard with the brands they purchase. In fact, post-pandemic, more members of society have shown a passion for brand authenticity, particularly concerning cultural and social issues.

When surveyed, 38 percent of respondents indicated "active engagement" within the last year in one of the following activities: attending a protest, signing a petition, or donating and campaigning for a cause. Many factors contribute to increased activism among consumers, including the growth of social media, which has connected like-minded consumers to common causes. In addition, the power of Millennial and Gen Z consumers—two demographics that rank climate change and protecting the environment as their top concern—is growing.

Savvy brands like Mattel are responding proactively and working to align themselves with their consumers' wants and wallets. The new Barbie collection, which costs from $9.99 to $19.99, is intended to "educate kids on the importance of sustainability in an easily digestible way." The company has set a goal of achieving 100 percent recycled, recyclable, or bio-based plastic materials across all of its products and packaging by 2030. Perhaps the progressive product line, called Barbie Loves the Ocean™, will align nicely with each of her dreamhouses.

Reflection Questions

1. Are purpose-driven brands that inspire environmental consciousness justified in charging customers a higher price point? Explain your response.

2. When does tying things like climate change and social activism into toys become too burdensome for kids, if ever? Explain your response.

3. The COVID-19 pandemic hit Millennials and Generation Zers especially hard. The Deloitte survey referenced in this case study shows that 30 percent of 25-to-30-year-olds surveyed reported either losing their job or being placed on unpaid leave. Is it ethical for marketers to target them without a demonstrated social responsibility track record? Explain your response.

Lesson 3-2
Global Marketing

Technology has made global markets easier to access, communicate with, and deliver to; thus, almost all large companies have expanded into the global marketplace. This lesson focuses on global marketing. We will explore how global markets offer new opportunities for growth but also present some difficult challenges.

By the end of this lesson you will be able to

- Define global marketing and why firms pursue it.
- Explain some of the unique considerations of global marketing research.
- Explain how demographic, economic, sociocultural, political/legal, competitive, and technological factors can present challenges to global marketing.
- Define the five major market entry strategies.
- Describe the alternative marketing mix strategies used in global marketing.
- Identify the major trade agreements, monetary unions, and organizations that impact the global marketing environment.

Amazon Delivers to the World

It's no secret that Amazon.com has changed the landscape of retail sales in the United States. Jeff Bezos started Amazon.com as an online bookseller in 1995. Since then, the company has evolved to become a retail giant that sells virtually every type of product imaginable. In addition, Amazon now distributes and produces streaming video and music content and makes its own digital devices like the Echo, Fire Tablet, Fire TV, and Kindle E-reader. With its 2017 acquisition of Whole Foods, it has entered the grocery retailing business in the United States, Canada, and the United Kingdom; and with its 2018 acquisition of Ring, it has entered the home security business in North America, Latin America, Europe, Africa, the Middle East, and Oceania.

Amazon.com has grown into a global company in an effort to expand its business in existing and developing international markets.

To understand Amazon's dominance, consider this: In 2020, Amazon.com sold more than $315 *billion* worth of goods, which means it was responsible for 40 percent of the total e-commerce sales in the United States. It is likely that you're a part of Amazon's customer network and that you frequently find a "smiling" Amazon box in your mailbox or at your door. What you may not know is that Amazon is working overtime to become the leader in e-commerce sales not only in the United States, but also across the globe.

By mid-2021, Amazon.com was officially operating in 16 global marketplaces, including the Americas, Europe, Asia-Pacific, North Africa, and the Middle East. Across these marketplaces, Amazon operates more than 110 fulfillment centers in the United States and 185 fulfillment centers globally and delivers to customers in more than 200 countries and territories. Outside of North America, Amazon's annual international revenue was more than $104 billion in 2020, up nearly 40 percent from the previous year. Thus, as product sales and growth of services like Prime have been stabilizing in the U.S. market, Amazon has looked across the globe to drive growth and expansion. Some recent highlights include:

- Amazon's Marketplace in India is experiencing aggressive growth, with mobile Amazon shopping having grown at a pace of 46 percent year over year. To address this growth, Amazon expanded its fulfillment capacity in India by more than 25 percent.

- Amazon recently introduced its Prime services in Mexico. In four of Mexico's largest cities, Amazon offers one-day delivery service, while the rest of Mexico can expect two-day delivery of Prime orders.

- Amazon recently expanded its presence in the Middle East by acquiring Souq.com, which was an e-commerce giant in the area. The company also launched in the United Arab Emirates for the first time in May 2019.

Amazon.com, like many businesses, has decided that to be a successful company, it must market its products and services globally. This growth into new markets has provided additional revenue and sales, while also presenting many new challenges related to the four Ps of marketing.

Why Firms Engage in Global Marketing

Global marketing is a marketing strategy that consciously addresses customers, markets, and competition throughout the world in an effort to sell more products in more markets. An increasingly globalized world opens up new opportunities for marketers to likewise develop and maintain a global vision.

There are many reasons why firms expand their marketing efforts globally. In the simplest terms, the main reason is to grow revenue from sales in some way. Other reasons why firms market globally include:

1. **Because they can:** Using the Internet, global marketing of products is a relatively simple proposition. In the past, only large companies could expand around the globe, but today almost all businesses can reach diverse markets. For small firms, global marketing is relatively simple because it is typically easier for small firms to react quickly to the unique demands of different markets. For example, a firm that sells custom bags can easily adapt its bags to meet consumer needs and wants in Cape Town, Boston, or Rio de Janeiro by making a specific "one-off" product for each.

2. **Larger markets:** More than 95 percent of the world's population lives outside the United States. That means that many markets larger than our domestic ones exist. In an effort to sell more of their products, most large firms cannot

avoid expanding into global markets, so they cross borders in search of new or growing markets abroad. Small firms that sell online can also reach these large, non-U.S. markets. Expanding markets can help firms reduce their costs in all markets based on larger sales volume.

3. **Increased profits:** The U.S. market has many competitors in almost all types of businesses. In general, higher competition means lower profits, as prices drop and costs increase to sell products. However, many foreign markets are not as well served by competitors, so firms can expand into markets where they can sell their products for higher prices at a lower cost, increasing profits.

4. **Necessity:** A big reason that firms market globally is that they have to, or they risk losing local market share. For example, if Apple did not compete in the global market, it would risk being overtaken by a larger company, such as Samsung, that could afford to underprice products in Apple's markets by using profits earned in larger global markets.

5. **Demand:** In many cases, firms market their products in international markets because customers demand that the products be made available. For example, Kenzai, a Web-based wellness and fitness program, started as a U.S. company but quickly grew internationally, as traveling businesspeople spread the word across the globe. In response, Kenzai has embraced global customers by adapting its program to accommodate regional preferences and food customs.

6. **Innovation:** Marketers must continually develop and improve products to remain relevant in the marketplace. Expanding into global markets often results in innovations in response to market differences, which later become product improvements across markets. For example, Nike has used its global presence to incorporate new technologies developed for foreign markets, such as the Nike FuelBand, into its domestic products.

Other conditions make global marketing possible, or even necessary, including foreign incentives, increasingly talented labor pools, the ability to learn new processes, and investments in foreign infrastructures to support growth. In short, for most businesses, "going global" is not a question anymore. Rather, the question is, "How do we most effectively globalize?" The remainder of this lesson addresses how firms increase their chances of success when marketing globally.

Establish Your "Why"

Consider Uniqlo, Japan's largest apparel retail chain. Uniqlo operates more than 1,000 stores globally in countries such as China, South Korea, the United States, and Russia.

"Made for all" is Uniqlo's slogan and philosophy. The company lives out this philosophy by offering clothing that is affordable and paying employees fairly.

"Tell the Truth, Not Stories"

The company lives out its truth. The firm's philosophy of "made for all people" is genuinely reflected in Uniqlo's business selection of global brand ambassadors.* For example, Uniqlo's sponsored athletes include Gordon Reid, the reigning Wimbledon wheelchair tennis champion and a Rio 2016 Paralympic Games gold medalist.

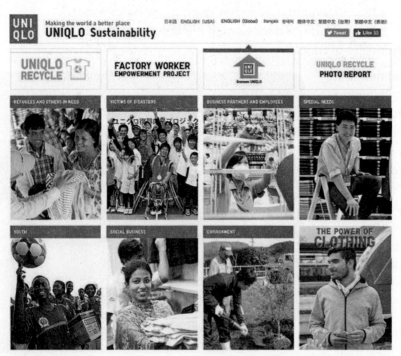

Be Good, Do Good

Uniqlo founder Tadashi Yanai has made philanthropy fashionable by building a global retail empire that gives back to local communities. For example, in partnership with the United Nations, Uniqlo Recycle has delivered 20.3 million

clothing items to refugees, evacuees, victims of disaster, expectant and nursing mothers, and others in need around the world since 2007.

Carefully Plan New Locations

Founded in Hiroshima, Japan, in 1984, Uniqlo today counts more than 1,900 stores (47 in the United States) across 18 markets. Every week, there's a Uniqlo store opening somewhere in the world. In addition to retail locations, Uniqlo places its research and development (R&D) centers in strategic locations: New York, London, Milan, Paris, and Tokyo.

Global Marketing Research

To better understand and learn more about global markets, some firms engage in global marketing research. Firms themselves can conduct global marketing research, but not all firms are either familiar with international data sources or equipped to use them even if they were.

Failure to understand new markets can cause firms to make huge mistakes. Many businesses have learned the hard way that even though something works in one country, it might not work in another country. For example, Clairol launched a product called the Mist Stick in Germany; however, the message was lost in translation as in German, the word "mist" means "manure."

A good way to overcome this deficiency is to use local marketing research companies, where available. These firms recognize potential problems with collecting primary data, such as higher costs, potential language barriers, and the lack of a technical infrastructure (e.g., working phone numbers or valid e-mail addresses) to reach respondents. Such companies are also familiar with the local culture and fluent in the language. This knowledge can help eliminate many of the mistakes companies might make as they seek to enter foreign markets.

Whether companies conduct their own research or hire an international research firm to do it, the key tasks remain the same:

- **Analyzing the global market:** To understand things like which markets are fastest growing overall, which are the largest markets for the firm's products, what trends in the various global markets relate to the firm's products, and what, if any, restrictions there might be on importing, a firm must analyze the global market. This type of information may be obtained through secondary sources, such as trade reports and websites, specific to the country being researched.

- **Acquiring specific information about the competition:** Such information includes the names of the competitors, what market share they hold, what promotional activities they engage in, and at what prices they sell their products. Additionally, potential market share and sales of products at particular price points should be estimated to ensure that the venture will be a profitable one.

The Challenges of Global Marketing

All marketers—whether marketing domestically or globally—must take into account demographic, economic, sociocultural, political/legal, competitive, and technological factors when they operate. International markets, however, present other factors and challenges that marketers must take into account. Multiply those factors by the almost 200 countries in the world and you can begin to understand the complexity and potential for error when operating in a global marketplace. This section illustrates the impact of these challenges on firms' global marketing efforts.

The COVID-19 pandemic created new challenges and shocked the global economy, splitting companies into two camps: those that are operating and thriving and those that have closed and have been brought to their knees. Companies such as Google, Facebook, Microsoft, Amazon, and Netflix all saw an increase in revenue in the first quarter of 2020. However, as the virus continues, even these companies and others that have had relatively few disruptions will begin to struggle. On the other side, firms in the entertainment, travel, retail, and food/hospitality industries have transitioned into survival mode. Disney and Marriott both saw a decrease in profits by over 90 percent in the first quarter 2020 compared to the previous year, and global firms such as Food First Global Restaurants, Virgin Australia, Aldo, and Hertz all filed for bankruptcy in the second quarter of 2020.

The Impact of a New Demographic Market

Marketers seeking government-provided demographic information about a potential new global market may find that what is available is less complete, up-to-date, or reliable than what is available for established markets. For example, well-established markets, such as the United States, Canada, and France, conduct a census every 5 to 10 years, whereas many emerging nations do not conduct an accurate census on a regular basis. In addition, not all countries collect or classify their data in the same way, making it all but impossible to compare characteristics across nations.

Even if marketing professionals obtain the information they seek, each country, including the United States, has regional differences that can make establishing quality market segments a difficult task. Thus, in many cases, marketers must invest in research to determine the demographic makeup and needs of a new market. Many research firms produce specialized reports for emerging markets, such as the ISI Emerging Markets, IBIS World Reports, and One Source Global Express, that are purchased by firms considering expansion into new markets.

The Impact of Economic Factors

Currency fluctuation refers to how the value of one country's currency changes in relation to the value of other currencies. Currency fluctuation can impact how firms market products internationally either positively or negatively. Consider, for example, the currency exchange rate between the U.S. dollar ($) and the EU currency, the euro (€). The currency **exchange rate** is the price of one country's currency in terms of another country's currency. On January 1, 2017, €1 was worth $1.05. On January 1, 2019, the same €1 was worth $1.15. As the value of the euro appreciated relative to the dollar, it reduced the spending power of U.S. consumers seeking to buy European products. In contrast, as the dollar depreciated, U.S. goods and services became more affordable to European consumers. This change in the currency was good for tourism businesses in the United States, as it would cost less for a European tourist to travel to the States. Conversely, U.S. travelers to Europe spent more for a hotel room than they did two years earlier.

A related topic, income distribution, is also important to marketers. **Income distribution** refers to how wealth is allocated across the population of a country. Income distribution often provides the most reliable picture of purchasing power. Marketers are particularly attracted to countries with a growing middle class because a nation's purchasing capability tends to increase as the proportion of middle-income households increases. For example, income growth in developing nations in Asia and Latin America is likely to stimulate world trade as more of their residents move into the middle class.

The Impact of Sociocultural Factors

Cultural fit relates to how well marketing efforts are designed to meet the needs of a specific market. One of the biggest mistakes domestic firms make when they attempt to take their business global is to believe that foreign consumers want exactly the same products that are sold in the United States. Burger King was widely criticized when it created an in-store ad for some European stores that showed a Hindu goddess atop a ham sandwich with the caption, "A snack that is sacred."18 Many of the nearly 1 billion Hindus throughout the world, most of whom are vegetarian, were offended and protested the use of the ad. Burger King eventually suspended the ad campaign. The negative attention and the potential long-term damage to Burger King's goal of expanding its market illustrate the importance of understanding cultural fit.

L'Oréal Advertising: Brazil

This L'Oréal ad is designed to reach a Brazilian audience. It focuses on long hair, which is a popular fashion trend in Brazil.

L'Oréal Advertising: France

This L'Oréal ad is designed to reach busy French women living in a fast-paced society. It focuses on a glowing foundation that lasts for up to 16 hours.

L'Oréal Advertising: China

This L'Oréal ad is designed to reach a Chinese audience. It focuses on a fair complexion, which is a popular fashion trend in China.

A growing concern for firms with overseas operations is consumer ethnocentrism. **Consumer ethnocentrism** refers to a belief by residents of a country that it is inappropriate or immoral to purchase foreign-made goods and services. This belief is on the rise in many developed nations, including the United States, France, Germany, and China. Consumer ethnocentrism is rarely grounded in fact, making the marketer's job even more difficult. For example, in 2003, following France's refusal to join the U.S. military operation in Iraq, many U.S. consumers refused to eat French fries, even though there was nothing French about the product. Proactive marketers across the country looked for a clever way to resolve the issue and, for a brief time, renamed their product "freedom fries." Analyzing cultural fit and overcoming consumer ethnocentrism are essential aspects of environmental scanning on a global scale that help firms create value for international consumers.

The Impact of Political/Legal Factors

International trade agreements, monetary unions, and organizations are political factors that can impact the environment in which a firm operates. They can affect how easy it is for firms to enter a foreign market, what the currency exchange rate is between countries, and even what competition firms will encounter in the domestic market. Trade agreements and monetary unions facilitate the exchange of money and products across borders. International organizations provide regulatory oversight to economic activity. You will learn more about how specific agreements impact global marketing later in this lesson.

Another political factor that affects marketers is currency manipulation. **Currency manipulation** occurs when the government of a country artificially controls the value of its currency relative to the currencies of other countries. The political aspect of this economic phenomenon is that currency manipulation can lead to one country having an advantage over another. For example, the world's largest country in terms of population, China, has been criticized for undervaluing its currency—that is, pricing it lower than it is actually worth. Many nations believe this gives China an advantage in selling exports because it can price its products cheaper than other countries' products. For marketers, concerns over currency manipulation isn't a political issue, but rather they must make sure they aren't caught on the wrong side of changes in the manipulation, which could have a negative effect on sales of their product or profits. For example, heavy equipment manufacturer Caterpillar has seen sales drop 20 percent when the Japanese yen dropped 20 percent in value. This change helped strengthen Komatsu, a competitor of Caterpillar, because Caterpillar's products became more expensive and Komatsu's became less expensive.

The Impact of Competitive Factors

Recently, the hospitality service website Airbnb.com launched operations in China. Airbnb expected strong competition and soon discovered that local companies were better able to connect with consumers looking for lodging. To combat this global competition, the company started localizing its operations and changed its website brand to be better suited to the Chinese. The new brand, Aibiying, which means "welcome each other with love," and the addition of payment through China's Alipay system, have helped Airbnb compete against the local competition. Overall, the changes seem to have helped, as recent estimates indicate that 8.6 million Chinese guests use Airbnb when traveling abroad, and that properties listed within China have increased by more than 1200 percent.

Airbnb changed its website brand to better suit the Chinese market. The new brand, Aibiying, means "welcome each other with love."

The Impact of Technological Factors

Today, technology enables even small businesses to reach consumers across the globe. Websites act as front doors to billions of potential customers, and social media help companies develop relationships with customers anywhere for very little cost. Tools such as Google Translate allow customers to view websites in their own language, making it easier to promote products in different countries. In addition, global shipping firms such as FedEx and UPS enable small manufacturers to ship their products to customers around the world and have those shipments tracked by both the buyer and the seller online. Understanding how technology impacts the global marketing environment benefits marketers as they attempt to meet the needs and wants of consumers in international markets.

Google Translate allows customers to view websites in their own language, making it easier to promote products in different countries.

Market Entry Strategies

The reality of globalization means a firm's strategic planning process must include a discussion about what, if any, international presence the firm wants to pursue. One of the most critical strategic decisions involves how to enter foreign markets. As part of developing a marketing plan that involves global marketing, the firm must choose from among the following five major **market entry strategies** for entering the international marketplace: *exporting, licensing, franchising, joint venture,* or *direct ownership,* as shown below. Let's look at the risks and rewards of each of these market entry strategies.

Low risk, lower potential return

High risk, higher potential return

Exporting

Typically, the least risky option for entering international markets is **exporting**—selling domestically produced products to foreign markets. Increasingly, firms of all sizes export their products to other countries. Large firms such as Cargill (a producer of food, agricultural, financial, and industrial goods and services) and ExxonMobil (an oil and gas company), two of the largest domestic exporters, ship tens of thousands of products annually in support of their various business units.

However, exporting is not just popular among large companies. Small companies account for 96 percent of all U.S. exporters. Online commerce platforms such as Etsy, eBay, and Amazon have decreased the barriers faced by small businesses going global. Social media tools allow small businesses to engage customers around the world in a way that was not possible a decade ago. Logistics firms such as FedEx and UPS also help increase export opportunities by providing small businesses with a quick, efficient way to deliver products almost anywhere in the world. These tools provide almost any small business in the United States with the opportunity to become an exporter.

Licensing

Licensing is a legal process in which one firm (the licensee) pays to use or distribute the resources—including products, trademarks, patents, intellectual property, or other proprietary knowledge—of another firm (the licensor). Licensing offers marketers the advantages of expanding the reach of their products quickly in a low-cost way. Licensing arrangements occur in domestic markets as well, but in a global context, the domestic licensor allows a foreign company to use its resources. In 2018, globally licensed products reported revenues of $280.3 billion.

The use of licensing to enter international markets has increased significantly in recent years due to several factors, including more regulation, rising research and development (R&D) costs, and shortened product life cycles. Licensing helps overcome some of these barriers because the licensee is typically locally owned and brings unique insight about its local regulations and what consumers expect in the local market. However, despite its growing popularity, licensing is typically a riskier option than exporting. Major risks include (1) that the licensor may be inadvertently creating a future competitor in the form of the licensee, (2) that the licensor shares information and the right to use its proprietary technology with the licensee, and (3) that the licensee could potentially misuse trademarks.

Franchising

You already may be familiar with franchising of U.S. companies such as McDonald's. **Franchising** is a contractual arrangement in which the franchisor (e.g., McDonald's) provides a franchisee (a local owner operator) the right to use its name and marketing and operational support in exchange for a fee and, typically, a share of the profits. International franchise agreements are the same as domestic agreements with the obvious exception that they must meet the commercial laws of the country in which the franchise exists.

Franchising is an attractive method of entering foreign markets because franchisees assume the majority of the capital costs and human resource issues. The franchisor provides knowledge and information about running the business, which increases the likelihood of success. The disadvantages of franchising include the risks of granting your name to a franchisee in a faraway place where direct oversight is difficult. In most cases, companies that sell franchises desire some oversight and control to ensure that product quality standards are maintained.

KFC has franchise locations in many countries (including Egypt, shown here) and territories.

Joint Ventures

A riskier option than exporting, licensing, or franchising is a joint venture. In a **joint venture,** a domestic firm partners with a foreign company to create a new entity and thus allows the domestic firm to enter the foreign company's market. The local partner shares equity in the new entity and provides the foreign entrant with valuable information about local consumers, suppliers, and the regulatory environment. Joint ventures work best when the partners' strategic goals align, their competitive goals diverge, and they are able to learn from each other without infringing on each other's proprietary skills. For example, Skechers and Luen Thai Enterprises (a South Korean company) recently announced a new joint venture called Skechers Korea Co., Ltd. From this venture, Skechers hopes to grow its brand in South Korea and have access to the resources and market knowledge of Luen Thai Enterprises.

Joint ventures come with inherent risk. Domestic and international firms often operate differently, which can lead to culture clashes. Joint ventures also can result in mistrust over proprietary knowledge, conflict over new investments, and disagreements about how to share revenue and profits. For example, a joint venture between Tiffany & Co. and Swiss watchmakers Swatch ended because Swatch felt Tiffany & Co. had been delaying the venture. At the end of a long legal battle, Tiffany & Co. was ordered to pay Swatch $450 million.

Direct Ownership

The riskiest method of entering an international market is **direct ownership,** which requires a domestic firm to actively manage an overseas facility. Direct ownership is a good strategic option when the firm sees substantial sales potential in the international market, very little political risk, and similarities between the foreign and domestic cultures. Still, maintaining 100 percent ownership of offices, plants, and facilities in a foreign country exposes the firm to significant risks. Direct ownership requires far more resources and commitment than any of the other options, and it can be difficult to manage local resources from afar. However, direct ownership provides the firm with more control over its intellectual property, advertising, pricing, and product distribution.

Marketers should diligently and thoroughly analyze the risks and rewards of each type of foreign entry as they develop their marketing plan. Regardless of which method a company ultimately pursues, as with all other strategic decisions, the strategy for entering international markets must align with the firm's objectives as defined in its mission statement and strategic planning process.

Alternative Marketing Mix Strategies for a Global Market

After establishing a global strategy, the next step is for firms to adjust their marketing mix for the new global market. In the following sections, we look at how product, price, place, and promotion decisions are made in a global setting.

Product

Any marketing strategy should consider the potential benefits and risks involved in bringing new products into international markets. Many multinational firms develop new products for international markets. Some of these products are standardized, whereas others are individualized for specific markets. For example, at McDonald's in India, you can get a Veg Pizza McPuff or a McPaneer Royale, which were created to appeal to the large vegetarian population in India. These types of product decisions are made based on the segmentation strategy that the firm is using, which can be *global* (one product in all markets), *multi-domestic* (a few products used in large market segments), or *adaptive* (completely new products in each market).

Cultural differences can also affect how product packaging is perceived. For example, mayonnaise is often sold in large two-pound bags in Chile because the average consumer there eats several pounds of mayonnaise each year, more than the average U.S. consumer.

Price

Pricing is a critical component of a successful global marketing strategy. Historically, companies have set prices for products sold internationally higher than the same products sold domestically. However, growing Internet access throughout the world has made global pricing more transparent and more competitive. The following are other issues marketers must take into account with regard to pricing:

- **Consumer willingness to pay:** A consumer's ability and attitude toward the price of a product when making a product decision is referred to as **consumer willingness to pay.** Marketers need to factor in how much their target customers can afford when setting prices. Of equal importance is a consumer's attitude toward price.

- **Gray market:** You have probably heard of the black market, which refers to the illegal buying and selling of products outside of sanctioned channels. A relative of the black market is the **gray market**, which consists of branded products sold through legal but unauthorized distribution channels. For example, sneakers are very popular on the gray market. Many sneaker companies manufacture their shoes at facilities owned by third parties. Sometimes these third parties will manufacture an excess and sell the excess on the gray market. This often occurs when the price of an item is significantly higher in one country than another. Individuals or groups buy new or used products for a lower price in a foreign country and import them legally back into the domestic market, where they sell them for less than the market price. Gray market goods cut into a firm's revenue and profits, leaving marketers looking for ways to control and repress such activity.

- **Tariffs:** Taxes on imports and exports between countries are called **tariffs**. For example, in 2019, the Trump Administration raised tariffs on a number of products entering the United States from around the world, resulting in higher prices for certain goods for U.S. consumers. Tariffs may also raise the price that foreign customers must pay for goods produced in the United States, negatively impacting a U.S. firm's ability to be price competitive in those markets. Marketers typically prefer targeting international markets with low tariffs or with which the United States has an international agreement, such as the United States-Mexico-Canada Agreement (USMCA), to lower tariffs.

- **Product dumping:** When a company sells its exports to another country at a lower price than it sells the same product in its domestic market, this is known as **product dumping.** For example, U.S. steelmakers complain that by selling steel at an unreasonable low price, foreign companies are negatively impacting American firms, leaving them with only two options: losing customers to the foreign producers or selling their steel at a loss. As they develop their international pricing strategy, companies must monitor how anti-dumping laws affect companies in the industry and calculate the potential impact of anti-dumping regulations on sales.

Place

Marketers entering new markets need to consider how they will get their products to that market. Modern international transportation has enabled companies to source materials and goods from low-cost producers wherever they are found. However, companies must analyze the costs involved in moving goods between countries and continents. The cost of international shipments generally makes up a much higher percentage of the product costs than domestic transportation due to the longer routes, use of multiple modes, and administrative costs. In addition, the complexities of international transportation—volatile fuel prices, congested ports, piracy, union strikes, and inland transportation delays—affect a company's ability to ship goods in a cost-effective and reliable manner. To effectively market goods in a global environment, companies must empower logistics professionals to navigate the maze of transportation choices and develop solutions that support the firm's marketing strategy.

Consumers from other cultures also often shop differently than U.S. consumers do. In the United States, having a store within driving distance may be acceptable (Americans typically have a car), but in other cultures where driving is not prevalent, a store that is far away may never be visited. In Japan, convenience is highly valued, which has led to a large number of unique and different vending machines. For example, Dole has developed a vending machine for the Japanese market that sells fresh bananas in places where they are needed, but previously have not been available, such as college dorms, apartment buildings, and offices.

Promotion

Global marketing efforts to promote can build on domestic brand awareness. Coca-Cola ranks at the top of the list as the most valuable global brand. The Atlanta-based company has had marketing success throughout the world, including in Mexico, where it was introduced in 1898. To expedite its entry into the Mexican market, Coca-Cola provided free refrigerators to restaurants to encourage distribution and trial. The strategy worked as increasing numbers of Mexican consumers tried the product and Coke became widely distributed even in the remotest parts of Mexico. Coca-Cola has used a similar strategy for success in many developed and emerging economies.

Promotion of U.S. brands has benefited across the globe from the growth of technology and social media. U.S. companies in general benefit from the collective success of U.S. tech companies such as Apple, Facebook, and Google due to the cachet of Silicon Valley in the global imagination. The Internet and the language associated with it are rooted in U.S brands and imagery, which increases marketers' success in building U.S. brands globally. Social networking sites such as Facebook and Twitter have served as important tools for promoting U.S. brands internationally.

However, not all brand messages translate well into other languages. Marketers need to be cautious when promoting in different languages, lest the true meaning of the campaign message gets lost in—or made wildly inaccurate by—the translation. The "Turn It Loose" campaign by the U.S. brewing company, Coors, for example, became "suffer from diarrhea" when it was translated for the Spanish market. In some cases, it is the imagery that gets mistranslated. When Procter & Gamble introduced Pampers diapers to the Japanese market, it retained its classic stork-delivering-a-baby image. Customers in Japan, however, did not understand this Western cultural reference and were therefore confused by the ad.

PRODUCT

McDonald's specializes its products for different markets. The McDonald's Maharaja Mac and Vegetable Burger with Cheese are menu items available in India. This is an example of how a firm's sensitivity to cultural differences will result in the modification of its product to recognize those differences.

PRICE

Many in the steel industry have accused China of dumping steel in the United States at artificially low prices. In response, tariffs have been proposed to help U.S. steel manufacturers compete.

PLACE

Would you buy bananas from a vending machine? Dole uses machines to market fresh bananas on college campuses, at train stations, and in offices in Japan.

PROMOTION

American companies must adapt their promotions for their locations abroad.

International Trade Agreements, Monetary Unions, and Organizations That Impact the Global Marketing Environment

International trade agreements, monetary unions, and organizations can impact the environment in which a firm operates, and likewise its marketing efforts. Let's look at the key entities that impact the global marketing environment.

United States-Mexico-Canada Agreement

The **United States-Mexico-Canada Agreement (USMCA)** entered into force on July 1, 2020, as an update to the North American Free Trade Agreement (NAFTA) that is mutually beneficial for North American workers, farmers, ranchers, and businesses. NAFTA's goal was to eliminate barriers to trade and investment among the three countries, and it brought the immediate elimination of tariffs on more than half of U.S. imports from Mexico and more than one-third of U.S. exports to Mexico. The new agreement maintains NAFTA's existing zero-tariff treatment, significantly expands U.S. access to Canada's dairy market, and includes a number of other important upgrades. The USMCA aims to create a more balanced environment for trade, support high-paying jobs for Americans, and grow the North American economy.

Dominican Republic–Central America Free Trade Agreement

The **Dominican Republic–Central America Free Trade Agreement (DR-CAFTA)** entered into force for the United States, El Salvador, Guatemala, Honduras, and Nicaragua in 2006, for the Dominican Republic in 2007, and for Costa Rica in 2009. Like the USMCA (previously NAFTA), the DR-CAFTA focuses on eliminating tariffs, reducing nontariff barriers, and facilitating investment among the member states. With the addition of the Dominican Republic, the trade group's largest economy, the region covered by DR-CAFTA is the second-largest Latin American export market for U.S. producers behind Mexico.

European Union

Agreements such as DR-CAFTA are designed to ease trade between nations. Entities such as the European Union go further by integrating countries to a much larger degree. The **European Union (EU)** was formed to create a single European market by reducing barriers to the free trade of goods, services, and finances among member countries. It is an economic, political, and monetary union of 27 European nations. In 2018, the EU generated an estimated 16 percent share of the global gross domestic product, making it the second-largest economy in the world. The EU is the largest exporter, the largest importer, and the biggest trading partner for several large countries, including China, India, and the United States. However, EU nations such as Greece, Spain, Portugal, and Italy have faced significant economic challenges in recent years. In addition, the result of the Brexit election in the United Kingdom in 2016 led to the UK's departure from the EU in 2020, which will affect the overall strength of the union because the UK has the second-largest economy in the EU.

World Trade Organization

The **World Trade Organization (WTO)** regulates trade among participating countries and helps importers and exporters conduct their business. In addition, the WTO provides a framework for negotiating and formalizing trade agreements and a dispute-resolution process aimed at enforcing participants' adherence to WTO agreements. The WTO, headquartered in Geneva, Switzerland, has 164 members, representing more than 98 percent of world trade, as well as observer nations seeking membership.

International Monetary Fund

The **International Monetary Fund (IMF)** works to foster international monetary cooperation, secure financial stability, facilitate international trade, promote high employment and sustainable economic growth, and reduce poverty around the world. Created in 1945, the IMF is governed by and accountable to the 189 countries that make up its near-global membership. The IMF was formed to promote international economic cooperation, trade, employment, and currency exchange rate stability, including by making resources available to member countries to help them manage their debts. Each country contributed to a pool that could be borrowed from, on a temporary basis, by countries with debt obligations they couldn't meet. The IMF was particularly important when it was first created because it helped stabilize the world's economic system following World War II. To this day, the IMF works to improve the economies of its member countries.

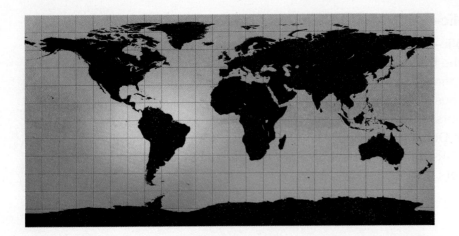

Responsible Marketing

Bribery and Doing Business Globally

Mateo is an attorney for a prominent international law firm. One of his top clients is an international food manufacturer, FOOD NOW, that is known for its integrity and commitment to excellence. Mateo is helping FOOD NOW build a manufacturing plant and distribution center in another country. FOOD NOW wants to build this facility in order to combat so-called "food deserts" in the area—places where people have a hard time accessing affordable and quality food—and to reach the country's residents faster and more efficiently than its current capabilities allow.

As an attorney, Mateo has sworn to uphold the law, and that includes not accepting or offering bribes or gifts in exchange for favors or services. Additionally, FOOD NOW requires that all of its employees and contract partners sign a formal agreement that their actions will reflect the highest ethical standards and behaviors.

As Mateo begins working in this new country, he quickly realizes that bribery and "gifts" are a common part of the cost of doing business there and that little progress will be made toward FOOD NOW'S new facility if bribes and gifts aren't given. However, both FOOD NOW and Mateo have contractual obligations that prohibit this type of transaction. If FOOD NOW wants to do business in this new country, it may have to change its policies.

Reflection Questions

1. If you were Mateo, what would you do in this situation?

2. Should Mateo adapt to the customs of the new country where he is working? Should he alert FOOD NOW? Explain your responses.

3. Should FOOD NOW change its policies? Why or why not?

Today's Marketplace: Test

1. A marketing strategy that consciously addresses customers, markets, and competition throughout the world is known as

    A. an import/export strategy.

    B. a global marketing strategy.

    C. a domestic market strategy.

    D. a foreign investment agreement.

    E. an international trade agreement.

2. Which of the following statements accurately describes global versus domestic marketing?

    A. Companies employ the same marketing strategies regardless of whether they are marketing globally or domestically.

    B. Expanding into a global market often spurs innovation as firms respond to market differences that later become product improvements for all markets.

    C. Most U.S. firms avoid engaging in global marketing efforts because the cultural differences make it too difficult to expand their brands.

    D. The costs of global marketing exceed the costs of domestic marketing.

    E. Global trends have less impact on small businesses than they do on large corporations.

3. A maker of high-quality smart phones risks losing domestic market share if it doesn't market globally like its main competitors.

    A. TRUE

    B. FALSE

4. U.S. companies often learn about consumer behavior in other regions of the world by

    A. taking advantage of the wealth of available secondary data.

    B. using local marketing research companies in other countries.

    C. applying data gathered from marketing research conducted in the United States.

    D. visiting the countries in which they seek to expand and familiarizing themselves with their consumers' buying behavior.

    E. relying on primary research data gathered by U.S-based marketing research firms.

5. Grapevine Corporation is interested in expanding its business globally and has begun the process of global marketing research. It considers several methods of gathering data. Which of the following is the weakest strategy it can employ?

    A. exclusively using data that is produced by domestic data collection firms

    B. using a local marketing research company to collect primary data

    C. using secondary data sources to obtain information regarding quotas and tariffs

    D. evaluating competitor pricing of similar products in global markets

    E. studying market share and price points to decide whether the venture would be profitable

6. If the value of the euro appreciates relative to the dollar, what impact would this have on a company seeking to expand globally?

    A. It would increase the spending power of American consumers seeking to buy European products.

    B. It would reduce the spending power of European consumers seeking to buy American products.

    C. It would reduce the spending power of American consumers seeking to buy European products.

D. The price of goods and services for European products would become evenly matched with American products.

E. The spending power of both American and European consumers would increase.

7. To gain the most reliable picture of a country's purchasing power, a company should study that country's

A. appreciating currency value.

B. income distribution.

C. level of gross domestic product.

D. stability levels of its currency fluctuation.

E. high level of consumer confidence.

8. A domestic firm believes that global consumers want exactly the same products as its U.S. consumers. This belief shows that the firm is most likely ignoring the importance of

A. consumer ethnocentrism.

B. demographics.

C. environmental scanning.

D. cultural fit.

E. consumer confidence.

9. Consumer ethnocentrism is defined as

A. a belief that the world's goods and services should be distributed fairly by all nations.

B. a belief by residents of a country that the goods and services they produce are superior to that of other countries.

C. a measure of how optimistic consumers are about the overall state of the economy and their own personal finances.

D. a measure of the amount of goods and services that can be purchased based on the currency exchange rate.

E. a belief by residents of a country that it is inappropriate or immoral to purchase foreign-made goods or services.

10. Technology in the global marketing environment is being used for

A. translation programs that allow customers to view websites in their own language.

B. advertising in both domestic and international social media.

C. joining free trade agreements to facilitate the import and export of products.

D. directly opening satellite facilities overseas.

E. outsourcing various functions such as customer service call centers.

11. Which of the following market entry strategies provides a firm with the least risk for entering the global market?

A. exporting

B. joint venture

C. licensing

D. direct ownership

E. franchising

12. Why has the use of licensing to enter international markets increased significantly in recent years?

A. the need to safeguard trademarks

B. rising research and development costs

C. less regulation

D. longer product life cycles

E. greater profit margins

13. For a fee, Sanjay can open a Subway sandwich shop in his hometown of Mumbai. The parent company has agreed to let Sanjay use its company name and will give him marketing and operational support. What type of arrangement is this?

A. licensing

B. joint venture

C. contract manufacturing

D. franchising

E. direct ownership

14. While traveling overseas, Martin drops and breaks his smartphone. He goes to a local mall and purchases a new one at a much lower price than it would have cost in the States. He decides to purchase several smartphones and brings them back to the States where he plans to sell them on eBay for less than the normal market price. The smartphones are considered to be

A. illegal imports.

B. black market goods.

C. contraband

D. smuggled goods.

E. gray market goods.

15. A company that sells its exports to another country at a lower price than it sells the same product in its domestic market is known as

A. gray market selling.

B. product dumping.

C. illegal importing.

D. exporting.

E. deceptive pricing.

16. A firm that makes a "place" decision in the global market should do which of the following?

A. Hire logistics professionals to navigate the maze of transportation choices.

B. Advance its brand awareness internationally.

C. Review import tariffs on products in all countries.

D. Understand how much consumers in certain markets are willing to pay.

E. Promote its name recognition on social networking sites.

17. Firms should keep consistent brand messages across both its domestic and global markets.

A. TRUE

B. FALSE

18. The North American Free Trade Agreement (NAFTA) was created with the goal of

A. allowing agricultural exports.

B. eliminating tariffs.

C. establishing trade embargoes.

D. establishing trade sanctions.

E. equalizing import/export fees

19. This international organization regulates trade among participating countries and helps importers and exporters conduct their business.

A. IMF

B. NATO

C. CAFTA

D. NAFTA

E. WTO

20. Among the reasons this was formed was to promote high employment and sustainable economic growth and reduce poverty worldwide.

A. IMF

B. EU

C. NAFTA

D. DR-CAFTA

E. WTO

4 Understanding Your Customer

What to Expect

When you hear about research, what is the first thing you think of? For many, its lab coats and clipboards. In marketing, research is used to understand customers — even you! Keep we reading to learn what marketing research truly means.

Chapter Topics:

- **4-1** Marketing Research
- **4-2** Consumer Behavior
- **4-3** Business-to-Business Marketing

Copyright © McGraw Hill Marco VDM/E+/Getty Images

Marketing Research

This lesson introduces you to marketing research, not only as a marketer *doing* research but as a consumer *being* researched. How would you feel about being experimented on? This question raises common ethical concerns with marketing research, and this lesson also discusses the ethics of marketing research in an increasingly digital world.

By the end of this lesson you will be able to

- Explain the importance of marketing research to a firm.
- Outline the steps of the marketing research process.
- Identify the different types of marketing research data.
- Describe the role of experiments in marketing research.
- Explain why the need to understand competitors as well as customers is an Important part of marketing research.
- Describe the main ethical issues in conducting marketing research.
- Explain the difference between marketing research and marketing analytics.

Marketing Analytics Implications

- Marketing research and marketing analytics work hand in hand, but they are not the same thing. Marketing research provides data to answer a specifically defined question. Marketing analytics explores the data to gain deeper insights about the data that collection alone cannot provide.
- Marketing analytics helps firms identify purchasing patterns; data mining is one method that firms use to explore and interpret data.
- Marketing analytics helps firms identify and interpret the relationships among multiple, complex sets of data; data modeling has become one of the fastest-growing methods for this type of data analysis.

LEGO: Building New Customers through Marketing Research

Believe it or not, marketing research has completely transformed countless companies, including LEGO. The LEGO philosophy is built on "systematic creativity—Lego bricks are all part of the Lego system, which essentially means that they can easily be combined in innumerable ways—and just as easily be dismantled. The more Lego bricks you have, the more fertile your creativity can become. The combination of a structured system, logic, and unlimited creativity encourages the child to learn through play in a wholly unique Lego fashion."

After reading this philosophy, you'll most likely think it can be applied to both boys and girls (and it can!), but for many years, LEGO catered to only one market segment: young boys. Girls comprised only 9 percent of LEGO's users. This provided the company an opportunity to increase sales by expanding its reach and empowering young girls in the process. But how would it appeal to this female audience? By conducting marketing research, of course!

LEGO used marketing research to understand how consumers interact with their products. As a result of marketing research, the company launched the LEGO Friends line focused on young girls.

LEGO conducted primary research in the form of focus groups, interviews, and observation to study the play habits of 4,500 girls and their mothers globally. Researchers discovered that boys and girls tended to play differently. For example, after boys finished building a castle, they tended to grab other item such as figurines and swords and use the castle simply as a backdrop for a battle. Girls, on the other hand, tended to want to play with the castle once it was built, and quickly discovered there was nothing to do with the castle because the inside was rather plain.

With these findings, LEGO launched the "LEGO Friends" line focused on young girls. The new line included a cupcake café, a giant treehouse, a supermarket, and other construction sets. The sets contained just as many pieces as other sets but offered girls the ability to assemble, play, and then continue assembling. While there were critics of the girl-focused line, it proved to be a success for the company. LEGO has seen a 15 percent average growth in sales since its launch.

Now that LEGO was focusing on both boys and girls, there was still one market that was underserved by LEGO: adults. Once again, LEGO conducted marketing research and discovered that adult LEGO users spend significantly more per year than average child users. One user even reported spending $50,000 in one year! This new information led the company to decide to sell more expensive sets, like the Star Wars Millennium Falcon, priced at $800 with 7,541 pieces.

The Importance of Marketing Research

Marketing research is the act of collecting, interpreting, and reporting information concerning a clearly defined marketing problem. When done properly, marketing research helps companies understand and satisfy the needs and wants of customers.

For example, LEGO's marketing research is aimed directly at its customers. LEGO:

- Collects information on its market segments by observing how children play and interact with the product.
- Interprets information by outlining the play process from start to finish for boys and girls.
- Reports its results by recommending the development of new products that match different types of play.

Marketing research has become more important, and more complicated, as markets continue to become globalized and product life cycles become shorter. In such a climate, companies need accurate information to reduce risk and make good decisions. This quest for information compels companies to spend billions of dollars each year on marketing research.

Patterns of consumer behavior that can change quickly, such as trends, styles, and preferences, also provide an incentive for firms to acquire fresh information through marketing research. Firms can no longer rely only on historical data to determine future marketing strategies. Now, they must generate timely information, interpret it quickly, and take action before the competition does. If they don't, they will be beaten before they can even begin.

Marketing research impacts almost every aspect of a company's business. We can see this impact clearly in terms of the four Ps: product, price, place, and promotion.

Product

Products need to be developed based on real customer needs and wants, not just the whims of marketing departments. Product developers in research and development departments must have an idea of what customers want before they can create new products, and finance departments must have good information concerning the viability of a product to approve expenditures on new product development.

Price

Pricing requires analysis of the size of the potential market and the effects of price changes on demand. This information comes from **demand analysis,** a type of research used to estimate how much customer demand there is for a particular product and to understand the factors driving that demand.

Place

Decisions regarding the place or distribution function must be made using **sales forecasting,** which is a form of research that estimates how much of a product will sell over a given period of time. Using this research, firms know how much product to hold in inventory at various points in the distribution network. We discuss sales forecasting in more depth later in the lesson.

Promotion

Promotional activities such as advertising must be evaluated based on their effectiveness. Firms use advertising effectiveness studies and sales tracking to gauge how well advertising and promotional campaigns are working. **Advertising**

effectiveness studies measure how well an advertising campaign meets marketing objectives such as increasing market share, generating consumer awareness of a product, or creating a favorable impression of the company's products. **Sales tracking** follows changes in sales during and after promotional programs to see how the marketing efforts affected the company's sales. Marketing professionals use the results of this research to adjust where and how they apply promotional efforts.

In addition to its impact on firms, marketing research matters to consumers like you. Consumers rely on companies to develop and market the products they need and want. Without marketing research, companies would mostly be guessing.

The Marketing Research Process

Marketing researchers use a variety of research techniques, and gather data from an assortment of sources, to provide the goods and services people truly desire. Regardless of the technique employed, firms follow five basic steps when they engage in the marketing research process.

STEP 1	STEP 2	STEP 3	STEP 4	STEP 5
Problem definition	Plan development	Data collection	Data analysis	Taking action

Step 1: Problem Definition

Problem definition is the first step in the marketing research process. Often firms know that they have a problem but cannot precisely pinpoint or clearly define the problem. Clarifying the exact nature of the problem prevents the firm from wasting time, money, and human resources chasing the wrong data and coming up with the wrong solutions.

To begin, a firm should set specific research objectives. As with overall marketing objectives, research objectives should be specific and measurable. They represent what the firm seeks to gain by conducting the research.

For example, LEGO started by defining a problem such as "Why are girls not interested in our products?" The genesis of any company comes from finding a better way to serve customer needs. For LEGO, focus on the initial problem definition guided the research process and ultimately led to a viable product solution. In other words, without clearly defining the problem, LEGO would have never found that girls tend to play with the product differently than boys and that they needed products to match their type of play to keep them interested.

Step 2: Plan Development

Plan development, often called *research design,* involves coming up with a plan for answering the research question or solving the research problem identified in the first step of the process. Plan development is about identifying what specific type of research will be used and what sampling methods will be employed.

Types of Research

How marketing researchers decide which type of research they need to do depends on the nature of the question or problem the firm has and how well the researchers understand it. There are three general types of marketing research:

1. **Exploratory research** is used when researchers need to *explore* something about which they do not have much information. Exploratory research seeks to discover new insights that will help the firm better understand the problem or consumer thoughts, needs, and behavior. This type of research usually involves a personal interaction between the researcher and the people being researched, often in the form of conversations, interviews, or observations.

2. **Descriptive research** is used when researchers have a general understanding of a problem or phenomena but need to *describe* it in greater detail in order to enable decision making. Descriptive research seeks to understand consumer behavior by answering the questions who, what, when, where, and how. (Note that descriptive research *cannot* answer "why" questions.) Examples of descriptive information include a consumer's attitude toward a product or company; a consumer's plans for purchasing a product; specific ways that consumers behave, such as whether they prefer to shop in person or online; and demographic information such as age, gender, and place of residence.

3. **Causal research** is used to understand the *cause-and-effect* relationships among variables. Causal research, also called *experimental research,* investigates how independent variables (the cause) impact a particular dependent variable (the effect).

Almost all causal research and most descriptive research starts with the development of a hypothesis. A **hypothesis** is an educated guess based on previous knowledge or research about the cause of the problem under investigation. For example, based on your life experience, you could *hypothesize* that people are more likely to wear shorts in the summer than in the winter. Although this is almost certainly true, we cannot actually prove that it is true without collecting evidence and using facts to empirically support our educated guess.

Marketers will establish specific data collection procedures, discussed in the next section, based on the general type of research the company decides is most appropriate to achieve the research objective. Companies need not limit themselves to one type of research and often use multiple approaches to help solve a problem. In many cases, marketers will begin with exploratory methods to gain a better-nuanced understanding of a marketplace phenomenon in order to make hypotheses about consumer behaviors, and then later use experimental methods to test whether their hypotheses are true.

Summary of Different Types of Research

Research Type	When Used	How Conducted	Type of Hypotheses
Exploratory	Typical when information is limited, such as when a firm enters a new market	Interviews and/or observation	Questions designed to gain broad understanding
Descriptive	For situations where specifics of a market are not well defined (i.e., who, what, when, where, how)	Surveys and/or focus groups	Multiple and specific questions to gain specific understanding
Causal	Used in situations where clarifying what caused an action to happen, such as "why are sales increasing at only some stores?"	Experiments, often in a store setting	Questions that assess why something happens

Sampling

It would be impossible—from both a budget and time perspective—for marketing professionals to obtain feedback from all the members of their target market. Instead, they must rely on sampling. **Sampling** is the process of selecting a subset of the population that is representative of the whole target population. Feedback gathered from this sample can then be generalized back to the entire target market. How researchers conduct sampling is critical to the **validity** of the research findings—how well the data measure what the researcher intended them to measure.

Sampling can be broken down into two basic types:

- **Probability sampling** ensures that every person in the target population has a chance of being selected, and the probability of each person being selected is known. The most common example of probability sampling is **simple random sampling,** where everyone in the target population has an equal chance of being selected. Simple random sampling is the equivalent of drawing names from a hat; every name in the hat has an equal chance of being chosen.

- **Nonprobability sampling,** on the other hand, does not attempt to ensure that every member of the target population has a chance of being selected. Nonprobability sampling contains an element of judgment in which the researcher narrows the target population by some criteria before selecting participants. Examples include **quota sampling,** in which the firm chooses a certain number of participants based on selection criteria such as demographics (e.g., race, age, or gender), and **snowball sampling,** in which a firm selects participants based on the referral of other participants who know they have some knowledge of the subject in question.

While probability sampling enables researchers to generalize findings from a portion of a target population, nonprobability sampling can generate findings that may be more appropriate to the research question.

Sampling is a valuable way to gather feedback about a large target market when obtaining feedback from every member of that target market is not feasible. In this photo, an Earth Balance representative is handing out free soy milk samples to consumers at a local market, and then asking them to complete a quick product survey—the link for which is found on the postcards displayed on the table.

Step 3: Data Collection

The third step of the marketing research process, data collection, begins with a decision about data. Researchers must ask themselves: Can we acquire the data we need to answer our questions from someone else, or do we need to collect the data ourselves? This section focuses on methods used when researchers must go out and collect the data themselves, and a later section discusses the advantages and disadvantages of using data already collected by someone else.

Primary data collection is when researchers collect data specifically for the research problem at hand; this can be either qualitative or quantitative in nature. Qualitative research studies the *qualities* of things, whereas quantitative research examines the *quantities* of things. **Qualitative research** is characterized by in-depth, open-ended examination of a small sample size, like in-depth interviews or focus groups, and is used for exploratory and descriptive research. **Quantitative research** is characterized by asking a smaller number of specific and measurable questions to a significantly larger sample size and is used for descriptive or causal research. The figure below shows the relationships among research objectives and the collection methods and types.

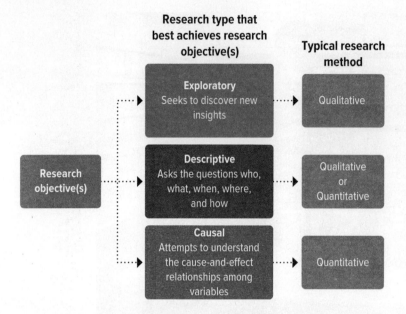

The Relationships among Research Objectives and Collection Options

Qualitative Research

As noted above, qualitative research includes exploratory types of research such as interviews, focus groups, and observation:

- In an **interview,** the researcher typically works with one participant at a time. During the interview, the researcher asks open-ended questions about how the individual perceives, uses, or feels about various products or brands. Interviews can be useful in figuring out what or how people think, but they have the limitation of being very time-consuming and must be conducted by experienced researchers who know how to properly ask and follow up on questions in a way that elicits rich and useful information from participants.

- **Focus groups** are conducted by a moderator and involve interviewing a small number of people (usually 8–12) at a time, where participants interact with one another in a spontaneous way as they discuss a particular topic or concept. The interactive nature of this setting lends itself to drawing out opinions and generating insights into a marketing question, much like one-on-one interviews but with the added advantage of group interactions. However, focus groups are expensive to perform and must be led by experienced moderators that are able to direct group discussions and manage interpersonal dynamics.

Qualitative methods like interviews and focus groups can provide researchers a great deal of insight, but they don't always allow researchers to draw generalized conclusions about the larger consumer population. To collect the necessary data to achieve their research objective, companies often turn to quantitative research.

Quantitative Research

Quantitative research includes surveys, experiments, and mathematical modeling:

- **Surveys** or questionnaires pose a sequence of questions to **respondents.** They provide a time-tested method for obtaining answers to who, what, where, why, and how types of questions and can be used to collect a wide variety of data. In addition, they help determine consumer attitudes, intended behavior, and the motivations behind behavior. Surveys often employ multiple-choice questions, making them appropriate for gathering feedback from a large number of participants. They can be administered by mail, at shopping malls, on the telephone, or online.

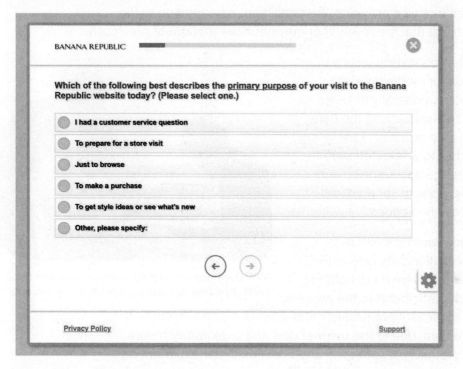

Banana Republic conducts survey research of its customers to find out many things, including their primary purpose for visiting its website.

- **Experiments** are procedures undertaken to test a hypothesis. They allow researchers to control the research setting so that they can examine causal relationships among variables. Researchers can see how a change in an independent variable (for example, price or product package color) might cause changes in another dependent variable (for example, sales or customer preferences). Experiments traditionally take place in a simulated environment known as a laboratory, but researchers can also perform **field experiments** in natural settings like stores or malls. While field settings offer an element of realism, they are also more difficult to control, which can lessen the validity of the experiment.

- Another type of causal research involves **mathematical modeling,** in which equations are used to model the relationships among variables. Statisticians have developed fairly reliable models for predicting certain elements of market behavior under normal market conditions, like how much consumer demand for a given item would go up or down based on a price change.

Step 4: Data Analysis

The purpose of data analysis is to convert the data collected in Step 3 into information the firm can use to answer the question or solve the marketing problem originally identified. If a research hypothesis was developed to be tested, analysis of the data should allow marketers to accept or reject the hypothesis.

Analyzing Qualitative Data

Researchers can gain substantial insights from qualitative research. Qualitative data gathering and analysis can give researchers ideas about the subject that can later be tested through quantitative research. Because there are no predetermined sets of responses (e.g., multiple-choice answers), the participant can open up to the interviewer and cover things that the researcher may not have thought of previously.

Because qualitative research usually results in great quantities of textual or media data, the qualitative analysis process generally involves a systematic approach to summarizing the data called coding. **Coding** is the process of assigning a word, phrase, or number to a selected portion of data so that it can later be easily sorted and summarized. The words, phrases, or numbers assigned to the data are called *codes*.

NVIVO software is used by researchers to sort through qualitative data and assign codes that help analyze the data.

Data analysis, particularly when applied to qualitative data, can be challenging. Due to the immediate and personal involvement of both the researcher and research participants in the process, results may be difficult to measure objectively and without bias. Qualitative data, due to the open-ended and exploratory nature, often requires the interpretation of subtleties and nuances that inexperienced researchers may miss.

Analyzing Quantitative Data

Quantitative methods for collecting and analyzing data can often be done quickly, at relatively low cost, and can help researchers describe large groups of customers and activities and understand cause-and-effect patterns in consumer behaviors.

Quantitative analysis almost always involves the use of **statistical analysis,** which is the mathematical classification, organization, presentation, and interpretation of numerical data. Simple analysis can be done in common programs like Microsoft Excel, whereas more advanced statistical modeling requires expensive and dedicated software like IBM's SPSS. There are two main types of statistics: descriptive and inferential.

Descriptive statistics are used to *describe* characteristics of the research data and study sample. Descriptive statistics, called "descriptives" for short, speak only to the properties of the data on hand and cannot be used to make claims about the larger population from which the sample was drawn.

Descriptives are the most common type of statistics we see in our daily lives, and they are the most straightforward way to tell a story about data. Anytime you see a percentage, a ratio, a bar graph, or a pie chart, that's a descriptive statistic. Descriptives can detail one variable at a time or multiple variables at a time. For example, your GPA uses one number to tell a story about your academic performance overall, across the dozens of classes you've taken. But statisticians could dig a bit deeper to introduce a second variable and compare your GPA for required versus elective courses.

Keep in mind that although descriptives allow us to summarize large quantities of data with a small amount of information, just like any type of summary, we lose a bit of detail along the way. For example, by looking at a GPA of 3.46 we would have no way of knowing whether a given student consistently earned B+ grades in all of his classes, or whether she was almost a straight A student that received a single D last semester.

Inferential statistics, on the other hand, are used to make *inferences* about a large group of people from a smaller sample. The purpose of inferential statistics is to make generalizable conclusions about a population by studying a small group from that population. Inferential statistics almost always include two or more variables.

Some inferential statistics are used to explore *differences* between two groups of people. For example, we might want to explore the differences between how men and women perform on an exam (studying two groups of people at the same time), or see if there is a difference in exam performance between the midterm and the final (studying the same group of people at two different times).

Other types of inferential statistics examine *relationships* among variables. For example, we might want to see if there was any correlation between time spent studying and exam grade performance.

As with qualitative data, there are drawbacks to quantitative analysis. While statistical analysis might give marketing professionals insight into consumer behavior, whether the results should be accepted depends on how well the sample represents the general population and how well the statistics were conducted. Remember that statistical results are only as good as the data you have, and the data are only as good as the methods you used to collect them and the sample you collected them from. As we mentioned earlier, validity concerns how well the data measure what the researcher intended them to measure. If you have ever come across a statistic in the newspaper or in an advertisement and thought "that doesn't seem right" or "yeah, but . . ." that's calling the validity of the results into question. Often, validity is called into question because of poorly worded questions that can be interpreted in several ways.

The following table lists some of the advantages and disadvantages of qualitative and quantitative research methods.

Advantages and Disadvantages of Qualitative and Quantitative Research

Research Method Type	Advantages	Disadvantages
Qualitative	• Uncovers details concerning the motivations behind behaviors • Is not limited to a predetermined set of responses • Can be a good way to start research into a marketing problem • Can be very flexible in approach • Can be used to generate marketing ideas	• Results may be difficult to measure objectively • Research can take longer than quantitative methods • Potential for researcher bias • Individual participants may not represent general target market • Small sample size
Quantitative	• Results may be generalizable to a larger population • Some methods can be conducted quickly and inexpensively • Analysis of data can be faster than in qualitative research • Can conduct causal studies that indicate why behaviors occur • Can be cost-effective • Often convenient for respondent	• May be limited by researchers' questions • Response rates can be very low • Difficult to determine nonresponse bias • Possible respondent self-selection bias • Participant resistance to giving sensitive information

Step 5: Taking Action

The culmination of the marketing research process is a formal, written report to decision makers. The report typically includes a summary of the findings and recommended actions to address the problem. Often, an oral report on the project is presented along with a written one.

Research report findings should be presented in a clear and understandable manner and include appropriate visual data, such as figures and tables, to support the findings and recommendations. The research report should allow the marketing manager to solve the marketing problem or provide answers to the marketing manager's questions.

Both reports should communicate any limitations of the research. Limitations could include a variety of things, including inadequate sample sizes or samples that do not adequately represent the population under study. It is important that marketers honestly discuss the limitations of the research. Such limitations should be considered before the firm makes any final decisions based on the research.

Step 1: Problem Definition

Redbox has a problem. Rental of DVDs at its self-serve kiosks have been steadily declining. This problem helped Redbox to define a problem and set the research objective: can Redbox compete in the online streaming market?

Step 2: Plan Development

To study this objective, Redbox was able to determine that its 27 million customers (enrolled in its loyalty program) would be a suitable test group and that it needed to study consumer preferences for online streaming.

Step 3: Data Collection

Redbox conducted primary research through surveys and focus groups comprised of current disc renters. These primary data were used to uncover unique needs and wants of Redbox customers.

Step 4: Data Analysis

The results of Redbox's research uncovered a market need for a non-subscription content streaming service that focuses on new release movies (more titles than competitors) and lower rental process (starting at $3.99) for a longer duration (2 days).

Step 5: Taking Action

The report of the research findings helped Redbox to design and launch its new streaming service. The new service focuses on the needs of 27 million customers and also hopes to attract new customers.

Marketing Research Data

Data are facts or measurements of things or events. *Qualitative data* might take the form of interview transcripts, video recordings of focus groups, or field notes from an in-store observation of shoppers. *Quantitative data* may be a collection of filled-out questionnaires or a spreadsheet with thousands of rows and columns of responses to an online survey. Data can also be presented in the form of a graph for easier digestion.

Data are the building blocks of research. Without data, researchers would simply be guessing when making marketing decisions about strategy, product design, pricing, distribution, promotions, and advertising. Likewise, without data, we cannot tell a story about our sample, but data by themselves do not tell the story. Researchers need to analyze data to help tell the story and turn the data into information.

Information is the result of formatting or structuring data to *explain* a given phenomenon, or to *define the relationship* between two or more variables. Information is the result of data analysis. Qualitative information from interviews may take the form of a theory about customer behavior or insights about why certain customers make the choices they do when they shop. Quantitative data can be presented using charts or diagrams and can provide information about customer behavior such as who a store's customers are, when they shop, what they buy, and how they move through the store. Unlike data, information can support and enable marketing decision making.

Survey engines such as Survey Analytics help researchers collect data through its survey platform; it can also automatically generate a bar or circle graph, like the ones shown here, out of the complex data it collects, making the data easier for marketers to digest and interpret.

Primary versus Secondary Data

As discussed previously, all research endeavors must begin with researchers asking themselves about the nature and type of data they need to answer their questions. Sometimes questions can be answered using existing data, but sometimes they cannot, and researchers must go out to collect new data to answer their particular question.

Primary data are data that are collected specifically for the research problem at hand. For example, LEGO conducted MRI scans of children's brains as they played to understand how their brains reacted. In doing so, the company collected primary data because the data were collected specifically for the purpose of this research.

Secondary data are data that have been collected for purposes other than answering the firm's particular research question. For example, if LEGO wanted to dig into its data to see what the top three selling products are for children under age 10, this would be a secondary use of the company's primary data because the data were collected for another purpose.

As with most decisions, using secondary data requires the marketing researcher to make a trade-off: On one hand, secondary data are much less expensive and easier to access than primary data, but on the other hand, secondary data collected for another purpose may not shed light on the specific nuances of a firm's particular problem.

Primary data collection may be necessary if secondary information cannot adequately answer the research question. Using primary data requires marketing researchers to invest far more time and money into their research endeavors, with respect to both collecting and analyzing data, but this can result in far more firm- or problem-specific information that allows for a more nuanced understanding of customers and their behavior.

(You're probably curious about the results of this research. The MRI scans showed that children had the highest level of engagement in skill-based play and from gaining the satisfaction of completing the building process. This research finding led LEGO to rank its sets based on skill level.)

Differences between Primary and Secondary Data

	Primary Data	Secondary Data
Collection Method Examples	• Focus groups • Surveys • Observations • Data gathered by equipment (e.g., video) • In-depth personal interviews	• Literature reviews • Online electronic searches • Company records • Marketing information systems • Private research companies • Boundary spanners (e.g., salespersons)
Advantages	• Pertain only to firm's research • May provide insight into why and how consumers make choices	• Less expensive (often free) • Information typically readily accessible
Disadvantages	• More expensive • May be difficult to enlist customer participation • May take excessive amount of time to collect	• Data may not be relevant • Data may not be accurate • Data may have been altered • Data may contain bias
Examples of Use	• To understand what motivates consumers • To determine the effect of variables (e.g., price) on product choice • To gain feedback on company's existing and proposed products	• To gather macroeconomic data • To gather socioeconomic data • To obtain information about competitors • To gain insight into international cultures and markets

Sources of Primary and Secondary Data

Primary data can be qualitative or quantitative in nature and can take the form of interviews, focus groups, observations, video recordings, questionnaires, surveys, or experiments.

Secondary data can come from internal or external sources. *Internal secondary data* are collected by the company and can include things such as sales by product, information about individual purchases from loyalty cards, previous research reports, accounting records, and market information from the sales force. Companies often build large internal databases in which to store such data.

External secondary data can come from many sources. Governments compile a lot of data and make them available to the general public. U.S. government agencies such as the Economics and Statistics Administration and the Census Bureau provide a great deal of useful secondary data in various publicly available reports. The Census Bureau, for example, provides geographic and demographic data about U.S. citizens for free. Other sources of secondary data, some free and some paid, include trade associations, academic journals, business periodicals, and commercial online databases. The Internet is a rich source of secondary information.

Sources of Primary and Secondary Data

Primary Data Sources	Secondary Data Sources
Observations	Previous research
Audio or video recordings	Literature reviews
Interviews	Online electronic searches
Correspondence	Company records
Focus groups	Marketing information systems
Case studies	Private research companies
Questionnaires	Government reports
Online surveys	Academic journals
Experiments	Periodicals and mass media
	Historical information

Experimental Research in Marketing Research

Earlier in this lesson, we briefly discussed marketing research experiments. As you will recall, experiments are procedures undertaken to test a hypothesis. Just like traditional scientists, marketing researchers are *social* scientists, and they conduct research experiments in much the same manner, and for much the same purpose.

Experiments are a type of causal research: the purpose is to allow researchers to control and then manipulate the research setting so that the researchers can examine causal relationships among variables. In a normal marketplace setting, there are lots of things going on at once, so it makes testing cause-and-effect relationships difficult. By carefully and intentionally manipulating—in the literal sense of skillful control and adjustment, not the colloquial sense of being unscrupulous—certain elements of the research setting, researchers are able to test cause-and-effect relationships in a much more definitive way.

Independent and Dependent Variables

Usually experimental research means investigating how a change in an **independent variable** might cause changes in one or more **dependent variables.** Dependent variables are the ones being tested and measured. A change in the dependent variable *depends* on something changing with another variable in the study. Independent variables, on the other hand, are the variables that *cause* the change in the dependent variables.

Some independent variables can be *controlled* for but not changed or manipulated specifically by researchers. Examples of this type of independent variable would be age, height, income, grade point average, weather, or time of day. But other types of independent variables can be intentionally *manipulated* by researchers, such as price, package color, store temperature, number of staff working, or product placement.

Research Considerations

In addition to considering how to study a problem (e.g., survey, experiment, etc.), marketers also need to determine who is being researched in different experimental conditions. **Experimental conditions** are the set of inputs (independent variables) presented to different groups of participants. In an experiment, typically only one or two inputs will be changed in each condition. This tight control over each condition allows researchers to identify what is causing something to happen in one group versus not happen in another. Those assigned to conditions are called **participants** because they are *participating* in the experimental conditions. Participants are typically selected using a defined selection process and randomly assigned to one of the different experimental conditions.

For example, we might be interested to research if the new low-calorie ice cream, Halo Top, contributes to weight loss. Halo Top is a low-sugar, high-protein ice cream that lists the calorie count for each pint prominently on the label. To test this question, we could set up three experimental conditions.

- **First Condition:** Participants eat only Halo Top ice cream (in specific quantities and with all other food intake being equal).
- **Second Condition:** Participants eat a higher-calorie ice cream (in the same specific quantities and with all other food intake being equal).
- **Control Condition:** Participants eat no ice cream (with all other food intake being equal).

This would be a very difficult experiment to conduct because it is hard to control exactly how much a person eats, especially outside of a laboratory. However, if this test could be conducted, we would expect to see differences in weight gain/ loss between the groups. Marketers analyze these differences to determine, with all other things being equal, what happens when one thing is changed (in this hypothetical experiment, type of ice cream). The results of this experiment would seem obvious, but in many cases, determining exactly what is causing something to happen can be complex. In cases where it is important to test what is causing change (e.g., pharmaceutical trials, tire blowouts, etc.), experiments may be the only way to sort out a complex problem.

Experiments traditionally take place in a simulated environment: a laboratory—for social scientists as well as traditional scientists—because researchers can have more control over the research setting, environment, and extraneous variables. An **extraneous variable** is anything that could influence the results of an experiment that the researchers are not intentionally studying. For example, if one of our Halo Top participants were to eat a few extra protein bars each day during the experiment, the experiment's results would be skewed by the extraneous variable (protein bars).

For marketers studying consumer behavior, it often makes more sense, both practically and financially, to study customers in a realistic marketplace environment. As we discussed earlier, experiments where researchers control or manipulate variables but that are conducted in natural settings are called *field experiments*. Although field settings offer an element of realism, they are also more difficult to control and more extraneous variables can possibly influence the outcomes.

Deciding whether to conduct experiments in a laboratory or in the field is certainly a question of resources, but it is importantly a question of validity. As you'll recall, validity is the extent to which an experiment measures and tests what it is supposed to measure and test. There are two types of validity to consider, which highlight the trade-offs a researcher must make when choosing between laboratory and field experiments.

The first type considers the validity of the **experimental manipulation** itself. **Internal validity** is the extent to which changes in the outcome variable were actually caused by manipulations of the independent variable conditions. The higher the internal validity, the more faith we can have in the cause-and-effect relationship demonstrated. The second type of validity to consider is **external validity,** which is the extent to which the results of the experiment can be generalized beyond the study sample of subjects. The higher the external validity, the more likely it is that the cause-and-effect relationship demonstrated by the experiment will be found across similar real-world settings. Field experiments tend to be lower in internal validity but higher in external validity, whereas laboratory experiments tend to be higher in internal validity but lower in external validity.

For example, in the Halo Top study, if the experiment were conducted in a laboratory setting, it would be easier to control what the participants ate. Exact portions could be weighed and measured. Other variables could also be closely controlled, such as exercise and sleep, which might affect weight

loss. The problem is, not many people would want to live in a controlled lab setting for a long period of time.

Conversely, the study could be conducted as a field experiment. In this setting, participants would record their eating, exercise, and sleep. However, it is pretty easy to imagine that errors would affect the study results. The lab experiment is most effective, but unrealistic, and the field experiment is most realistic, but least effective. In most cases, there is no great solution, so these trade-offs are considered by marketing researchers and the best solution is followed, based on the objectives of the experiment.

In addition to considering how to study a problem, marketers also need to determine who to study in different experimental conditions. If all of these individuals participated in the Halo Top experiment in a laboratory setting, it would be easier to control what they ate and therefore observe the differences in weight gain/loss among them more accurately. The problem is, not many people want to live in a controlled lab environment for extended periods of time.

The Importance of Competitors and Consumers in Marketing Research

The purpose of marketing research is to help marketing managers make better decisions. Often when we talk about marketing research, we discuss it in the context of the consumer marketplace. Why people make the purchase decisions they do, and when, how, and where they make those decisions are common questions marketers ask. Marketers also often want to understand what their customers want and how things such as product features and prices influence their decision making. When done properly, marketing research helps companies understand and then satisfy the needs and wants of customers, which creates value for all.

Another important component of marketing research involves gathering data about what competitors are doing in the marketplace. **Competitive intelligence** involves gathering data about what strategies direct and indirect competitors are pursuing in terms of new product development and the marketing mix. Such information can provide a firm with foreknowledge of a competitor's upcoming promotions or products, allowing it to respond in a way that blunts the effects of the competitor's actions.

A company can obtain information about another company's activities and plans in a number of ways, including conferences, trade shows, social media sites, competitors' suppliers, distributors, retailers, and competitors' customers. Competitor websites and the websites of government agencies such as the Securities and Exchange Commission (SEC) and the U.S. Patent and Trademark Office may contain financial and new product information about competitors. Firms also can obtain information for a fee from a number of companies that collect data and make them accessible.

There are also unethical ways of obtaining competitive intelligence. Bribing competitor employees for information and hiring people to tap phone lines or place surveillance cameras on the premises of rival companies are both examples of unethical and illegal ways of getting competitive information. Firms should scrupulously avoid these activities for legal reasons as well as for the damage they can do to a company's image and reputation.

Trade shows, such as the annual Consumer Electronics Show (CES) in Las Vegas, are a great place for marketers to gather competitive market intelligence.

Marketing Research Ethics

Whether conducting research for a domestic or international market, firms must always consider the ethical implications of gathering data on customers and competitors. Such considerations have become even more essential in a world of rapidly changing technology.

Privacy in a Digital Age

The increasingly powerful computer capabilities available to companies allow for the collection, storage, and analysis of data from millions of consumers. Online platforms like Twitter that are used by millions of consumers present firms with almost unlimited access to personal data. But companies must be careful not to go too far in collecting information of a sensitive nature. There is enough concern in the marketing industry about consumer privacy that many companies now have chief privacy officers who serve as watchdogs, guarding against unethical practices in their company's collection of consumer data.

At issue is the willful intrusion on the privacy of individuals. Consider mobile ad network companies that use unique identifiers embedded in smartphones to collect information about consumer preferences as they move from one app to another. To do this, such companies have worked around efforts by Apple to protect the privacy of iPhone and iPad users. The companies say they need personal data from users or they will lose millions of dollars of revenue from the firms that hire them, such as Mazda and Nike. Although Mazda and Nike don't participate in the actual tracking of data, they could potentially use what's collected to target customers based on geographic or demographic profiles. Meanwhile, the Federal Trade Commission (FTC) is evaluating mobile tracking technology as part of its ongoing mission to protect consumer privacy. It is trying to determine how far a company can go in tracking personal information before it invades customer privacy and becomes unethical.

Using Data Appropriately

Another ethical issue in marketing research is the misuse of research methods and findings. Marketing research firms may be compelled by their clients to return findings favorable to the client or to arrive at a conclusion predetermined by the client. Or marketing research firms may have employees who report false data, do not follow the directions for conducting the research, or claim credit for surveys that never were conducted.

Organizations such as the American Marketing Association, the Marketing Research Association, the International Chamber of Commerce (ICC), and ESOMAR, concerned with the reliability of research results, have established ethical standards for conducting research. The table below lists the key components of the ICC/ESOMAR standards. Such standards are important to the industry because they help gain the trust of consumers. Without that trust, individuals will be less likely to participate in marketing research. Expectations of privacy, or other ethical concerns, may vary depending on geography. By developing a global set of marketing research expectations, firms can ensure that they are adhering to appropriate marketing research standards, regardless of location.

Key Fundamentals of the ICC/ESOMAR International Code on Market, Opinion and Social Research and Data Analytics

1. Market researchers shall conform to all relevant national and international laws.
2. Market researchers shall behave ethically and shall not do anything which might damage the reputation of market research.
3. Market researchers shall take special care when carrying out research among children and young people.
4. Users' cooperation is voluntary and must be based on adequate, and not misleading, information about the general purpose and nature of the project when their agreement to participate is being obtained and all such statements shall be honored.
5. The rights of users as private individuals shall be respected by market researchers and they shall not be harmed or adversely affected as the direct result of cooperating in a market research project.
6. Market researchers shall never allow personal data they collect in a market research project to be used for any purpose other than market research.
7. Market researchers shall ensure that projects and activities are designed, carried out, reported and documented accurately, transparently and objectively.
8. Market researchers shall conform to the accepted principles of fair competition.

This Code was drafted in English and the English text is the definitive version. As the Code and ESOMAR guidelines are updated on a regular basis, please refer to www.esomar.org for the latest English text.

Clearly, it is in the best interest of companies to conduct marketing research in the most ethical manner possible. If they don't, consumers may refuse to participate in research studies, provide personal information either online or face-to-face, visit company websites, or order products online. Additionally, ethical behavior on the part of companies will make it unnecessary for the government to increase regulation of marketing research practices.

Marketing Analytics: An Insightful Look into Marketing Research Data

Differentiating between Marketing Research and Marketing Analytics

Marketing research is an iterative process that gathers data to answer a clearly defined research question; the end result provides clear outcomes and deliverables. Marketing analytics focuses on the data collected by marketing researchers and *how* those data can be used. These data typically are stored in the firm's data repository. The analytics team then examines the complete set of data gathered by the firm and may reuse them for further study, for example, to identify the effectiveness of distribution channels.

Using Data Mining to Identify Purchasing Patterns

Marketing analytics uses several methods to explore data collected through marketing research; data mining is one popular, growing method that firms use to identify and interpret purchasing behaviors. For example, several years ago Target sent a teen girl in Minneapolis coupons for maternity clothes, baby clothes, and baby furniture, which provoked her father to go to his local Target demanding to speak to the manager, to whom he angrily accused the company of trying to encourage his daughter to get pregnant. It wasn't until his daughter later informed him that she was, in fact, pregnant that the father backed down and apologized for his behavior. Although the local store manager had no idea why the Target coupons were sent to the teen, Target's marketing team knew it was because of the marketing analytics they conducted on the daughter's purchasing behavior. According to Target's statistician Andrew Pole, Target assigns every customer who visits its site a Guest ID number, which is tied to that guest's personal information, including name, credit card, or e-mail address. This information is stored in a data repository that statisticians like Pole then analyze to identify buying patterns. The teen's buying patterns aligned with that of expectant parents. These customers would purchase certain sets of products together (e.g., lotions, hand sanitizers, and vitamin supplements), and then a few months later, diapers. Target made an educated guess that this teen was also expecting and thus sent her coupons and advertisements of items an expectant parent would buy.

Statistical Modeling of Complex and Voluminous Data

Marketing research and marketing analytics work hand in hand; marketing analytics requires marketing research to provide data for further analysis and marketing analytics evaluates those data to provide deeper insight into the defined research question, in some cases, even raising additional questions for continued or additional study.

Marketing analytics have come to the forefront of marketing due to the availability of multiple large and complex data sets. Analytics assists firms in identifying relationships among these multiple data sets—relationships that are not evident from looking at marketing research studies or data alone. The data provided by online sites such as Google and Facebook, and even the firm's own website, provide a wealth of information about consumers' preferences and shopping behavior. However, these data must be organized and studied in order for them to be of any use to the firm. Data mining and data modeling are two growing fields within data science. Statistical models are used to organize and review the data. Once relationships between variables are found, then researchers must interpret the meaning of those relationships. In the Target example, if the data showed that customers who purchased items related to expectant parents also showed higher instances of purchasing blue bath mats, one may conclude that expectant parents really like blue bath mats. However, sometimes the relationships between data points are not relevant or meaningful. A more meaningful interpretation may be illustrated by the Connect Marketing Analytics exercise related to this section. In the exercise, you are asked to examine the differences between men and women in their preference for online learning. Marketing analytics can use these data to potentially validate a relationship relating to the differences in general online purchasing behavior between men and women.

Responsible Marketing

Is Marketing Research Worth the Investment?

Companies rely on marketing research to streamline their marketing efforts and effectively reach their target markets. Dove is an example of a company that continues to capitalize on its marketing research strategies and deliver advertisements that are both culturally and emotionally relevant.

Dove's "Campaign for Real Beauty" catapulted Dove from a creamy soap brand to a giant in the beauty industry. Data have shown that 68 percent of women feel the media supports unrealistic standards of beauty. Dove paid attention to the data, and in the early 2000s the company devised a campaign that focused on using real women, not models. This was a drastically different approach from other beauty companies that typically used models and air-brushing.

Over the past 20 years, Dove has continued to promote the "Campaign for Real Beauty," which has proven to be a very successful long-term strategy for the brand. In 2007 Nielsen discovered that every $1 Dove spent on advertising resulted in $4.42 in sales, which validated the investment Dove makes in marketing research.

Reflection Questions

1. What types of primary and secondary data would a company like Dove collect for its marketing campaigns?

2. How can companies use data mining to further their marketing research?

Lesson 4-2

Consumer Behavior

This lesson concerns the foundations of *consumer behavior,* which is rooted in the psychology of *why, when, where,* and *how* customers purchase products. To address these questions more specifically, marketers delve into customer buying behaviors. We all make decisions for different reasons, yet we follow a very similar process to do so. Understanding the process people use to make decisions and how this process functions in different settings is one of the more interesting and useful aspects of marketing—and the focus of this lesson.

By the end of this lesson you will be able to

- Describe the stages in the consumer purchase decision process.
- Identify the influences on consumer behavior.
- Distinguish among three variations of the consumer purchase decision process: routine, limited, and extended problem solving.
- Explain how data can be used to understand the consumer decision-making process.

Marketing Analytics Implications

- Many of the models used in consumer behavior were brought over from psychology. Models such as the consumer decision-making process were developed and studied in both psychology and marketing. The data from these studies are reviewed to ensure that these models, often used to predict consumer behavior, are still accurate.

- Because consumers are influenced by a wide variety of factors, marketers must constantly study not only the data gathered from the marketing environment, but also the data gathered from the economic, demographic, political/legal, technological, competitive, and sociocultural environments in which consumers live. All these data combined help marketers determine how changes in these environments are impacting consumers.

- Marketers have developed a series of tools to help structure and assist consumers in making complex decisions.

"Why Do I Buy?": The Psychological Basis of How Consumers Make Buying Decisions

The study of consumer behavior asks *why?* Why do we buy some things and not others? Why do we buy things that we didn't intend to buy? Why do people see products differently? Why does the setting of the store change how we value a product? Asking "why?" is a powerful component of studying consumer behavior topics.

Marketers strive to understand you—the customer—to help firms better adapt their marketing mix in order to create value for you.

During the onset of the COVID-19 pandemic in 2020, consumer behavior changed dramatically. Many retailers were selling out of necessity items like toilet paper and rice, but consumers "panic shopped" other items as well. Because they couldn't get to a gym, consumers ordered weights and gym equipment for their homes, causing up to a three-month delay for some products. Pet owners rushed to order grooming supplies and nail clippers while parents of young children stocked up on bicycles, puzzles, and workbooks to make the time at home more bearable.

Consumer behaviors are likely to experience lasting changes caused by the pandemic. Consumers once reluctant to shop online or utilize curbside pickup are embracing the new shopping experience. At the same time, physical retail locations are unlikely to vanish as many consumers find happiness at the chance to touch and feel products again.

Behaviors have also changed for things like self-care, mental health, and physical wellness. Consumers are spending increased time reading and taking up hobbies— all leading to a shift in spending habits focused on experiences rather than material possessions.

Because researchers have no comparison for the pandemic, understanding how consumers will behave in the post-pandemic world is impossible. What they do know is that periods of crisis often reshape consumers' values and individual psyches and that most brands will need to update their pre-pandemic personas and communication tactics.

As many consumers feared shopping in public, online ordering and delivery services became increasingly important to both consumers and businesses as they responded to changing consumer behaviors.

The Consumer Purchase Decision Process

Consumer behavior is the way in which individuals and organizations make decisions to spend their available resources, such as time or money. Firms that understand these principles of *why, when, where,* and *how* individuals and organizations make their purchases are often able to sell more products more profitably than firms that do not.

Most consumers go through a common decision-making process that involves the following five stages:

1. Problem Recognition
2. Information Search
3. Evaluation of Alternatives
4. Purchase
5. Post-Purchase Evaluation

Stage 1: Problem Recognition

Problem Recognition Question:

What need do I have to satisfy?

Answer:

I am hungry, and I need to eat something.

Stage 2: Information Search

Information Search Question:

What products are available to meet my need?

Answer:

I had pizza at the place down the street and it was good, but let me check online for other choices.

Stage 3: Evaluation of Alternatives

Evaluation of Alternatives Question:

What product will best satisfy my need?

Answer:

I am low on money so I should go for the cheapest option.

Stage 4: Purchase

Purchase Question:

Where should I purchase from and how much should I pay?

Answer:

I will go with the place down the street because it charges only $10.50 for a large pizza.

Stage 5: Post-Purchase Evaluation

Post-Purchase Evaluation Question:

Am I satisfied with my purchase?

Answer:

It was cheap, but it was so greasy I feel sick. I have to remember not to order there again.

Note that this response exemplifies *cognitive dissonance* because the consumer is feeling regret over the purchase.

Problem Recognition

The buying process begins when consumers recognize they have a need to satisfy. This is called the **problem recognition** stage.

Imagine leaving class to find that high winds had blown one of the oldest trees on campus onto your car. You need your car to get to school, work, and social events. Because your car is destroyed, you immediately recognize that you need a new mode of transportation. In this case, due to a lack of public transportation and the distance you must travel to meet your day-to-day obligations, you need to buy a new car.

Alternately, consider the panic shopping induced by COVID-19. All of a sudden, consumers felt the need to buy large amounts of toilet paper.

Sometimes problem recognition is easy, such as when you get a flat tire. In other cases, problem recognition is more difficult, such as deciding when to begin to plan for retirement; many don't realize they need to do so until it is too late.

From a marketing perspective, it's important to keep in mind the following two issues related to problem recognition:

1. **Marketers must understand all aspects of consumers' problems, even those that are less obvious, to create products that improve or enhance consumers' lives.** For example, if marketing professionals don't know what problem, beyond the need for transportation, you want to solve by purchasing a new car, they are not likely to develop strategies that will resonate with you. Are you looking for added prestige, or do you want to spend less on monthly payments? Marketers also must recognize that consumers might be buying the same car to solve different problems.

2. **Marketers must remember that if consumers are not aware of a problem or do not recognize a need, they are unlikely to engage in any of the subsequent steps of the buying process.** For example, Whole Foods may offer free samples of a new organic coffee brand to help consumers realize that they have a problem (in this case, the problem would be that they don't own this brand of coffee that tastes better than the leading brands). For Whole Foods, helping consumers recognize how new products solve unknown needs is an important way to encourage evaluation of alternatives and potential purchase of new products.

Information Search

Once consumers recognize a problem, they begin the **information search** part of the decision process. They seek information to help them make the best possible decision about whether to purchase a product to address their problem. Consumers expend effort searching for information based on how important they consider the purchase. Consumers search for both external information and internal information.

External Information Search

When consumers seek information beyond their personal knowledge and experience to support them in their buying decision, they are engaging in an **external information search.** Marketers can help consumers fill in their knowledge gaps through advertisements and product websites. Many firms use social media to empower consumers' external information searches. For example, Ford uses Facebook, Twitter, YouTube, Flickr, and Scribd to communicate information and to deepen relationships with customers.

Consumers' friends and family serve as perhaps the most important sources of external information. Recall again the example of buying a new car and what those in your life might say about different vehicle brands or models. If your parents or friends tell you about a negative experience they had with a certain car model, their opinions probably carry more weight than those of other external sources. Reviews of a product or service by other customers also carry a great deal of weight. How often have you decided not to stream a movie, visit a restaurant, or stay at a hotel based on what others have said in reviews? The power of these personal external information sources highlights why marketers must establish good relationships with *all* customers. It's impossible to predict how one consumer's experience might influence the buying decision of another potential customer.

When buying a new car, booking a hotel room, or choosing a restaurant, consumers increasingly use sites such as Yelp and TripAdvisor to read other consumers' evaluations prior to making a purchase.

Internal Information Search

Not all purchases require consumers to search for information externally. For frequently purchased items such as shampoo or toothpaste, internal information often provides a sufficient basis for making a decision. In an **internal information search,** consumers use their past experiences with items from the same brand or product class as sources of information. For example, you can easily remember your favorite soft drink or vacation destination, which will likely influence what you drink with lunch today or where you go for spring break next year.

In our car example, your past experience with automobiles plays a significant role in your new car purchase. If you have had a great experience driving a Ford Escape or Toyota Camry, for example, you may decide to buy a newer model of that same car. Alternatively, if you have had a bad experience with a specific car, brand, or dealership, you may quickly eliminate those automobiles from contention.

Evaluation of Alternatives

Once consumers have acquired information, they can use it to evaluate different alternatives, typically with a focus on identifying the benefits associated with each product. Consumers' **evaluative criteria** consist of attributes that they consider important about a certain product. Returning to our car-purchase example, you would probably consider certain characteristics such as the price, warranty, safety features, or fuel economy more important than others. Car marketers work hard to convince you that the benefits of their car, truck, or SUV reflect the criteria that matter to you.

Marketing professionals must not only emphasize the benefits of their good or service but also use strategies to ensure potential buyers view those benefits as important. For example, Whole Foods sells its food products at premium prices, so Whole Foods focuses its marketing efforts on highlighting how its program of working with local vendors is good for the local economy and helps the global environment.

Purchase

After evaluating the alternatives, customers will most likely buy, or *purchase,* a product. At this point, they have several decisions to make. For example, once you have decided on the car you want, you have to decide where to buy it. The price, salesperson, and your experience with a specific dealership can directly affect this decision. In addition, items such as financing terms and lower interest rates at one dealership versus another can affect your decision. If you decide to lease a car rather than buy one, you would make that decision during this step.

Marketers do everything they can to help you make a decision to purchase their product, yet in the end, it comes down to many small decisions. Thus, an effective marketing strategy seeks to minimize any small decisions that could delay or derail the actual decision to purchase. Companies like Lowe's are very aware that consumers often get bogged down in details related to delivery, installation, and cost before deciding to make a purchase. For this reason, Lowe's offers a variety of services, such as design, delivery, installation, and financing. Lowe's main reason for doing this is not explicitly to earn profit from selling services, but rather to remove any roadblocks from a consumer making a purchase at its store versus a competitor like Home Depot. Because profits start with sales, providing goods and services that make it easy for consumers to make a purchase decision is one of the most important activities that marketers perform.

Post-Purchase Evaluation

Consumers' **post-purchase evaluation** occurs after the sale is complete. Consumers will consider their feelings and perceptions of the process and product, which can result in either positive or negative assessments. Post-purchase assessments are important because these evaluations about the purchase will likely impact how consumers ultimately feel about the purchase and whether or not they will purchase the product again.

Three main things are evaluated after purchase: customer satisfaction, cognitive dissonance, and loyalty. Customer satisfaction is the first evaluation.

Customer satisfaction is a state that is achieved when companies meet the needs and expectations customers have for their goods or services. If the product purchased meets expectations (for instance, it works well, it performs as described, and the buying experience was as expected), consumers will most likely be satisfied with the purchase.

In many cases, consumers experience **cognitive dissonance,** which is the mental conflict that occurs when consumers acquire new information that contradicts their beliefs or assumptions about the purchase. Cognitive dissonance, also known as *buyer's regret,* often arises when consumers begin to wonder if they made the right purchase decision. Cognitive dissonance after making a purchase can arise for numerous reasons. For example, perhaps you discover that the car you just bought doesn't get the gas mileage you'd expected or you find out someone you know bought the same car for a lower price. Marketers do various things to reduce the level of dissonance felt by consumers. A car company might offer an extended warranty, a price-match guarantee, or a no-questions-asked return policy. Reducing cognitive dissonance is important to keeping customers satisfied.

By helping consumers feel better about their purchase, marketers increase the likelihood that those consumers will remain satisfied to the point of becoming loyal to the product. **Loyalty** is an accrued satisfaction over time that results in repeat purchases. In addition, loyalty can increase consumer actions, such as positive word of mouth, reviews, and/or advocacy for the product.

Because the post-purchase period has such potential to affect long-term results, many companies work just as hard *after* the sale as they do *during* the sales process to keep the customer satisfied.

Influences on Consumer Behavior

Numerous factors affect the consumer decision-making process at every stage. These include psychological, situational and personal, sociocultural, and involvement influences. An effective marketing strategy must take these factors into account. The following sections provide an overview  of the different influences and how marketers incorporate them into their marketing activities and strategies.

Psychological Influences on Consumer Behavior

Through the course of the consumer decision-making process, consumers engage in certain psychological processes: they develop *attitudes* about a product's meaning to their lives, they *learn* about the product, and they must be *motivated* to purchase the product. Marketers need to understand each of these psychological processes as they develop strategies to reach consumers.

Attitude

An **attitude** is a consumer's overall evaluation of a product, which involves general feelings of like or dislike. Attitudes can significantly affect consumer behavior. For example, many consumers have strong positive feelings about Whole Foods because they appreciate the high-quality products it offers. Conversely, many consumers have strong negative feelings about Whole Foods, terming it "Whole Paycheck" because it charges premium prices for its products. These diverse positions represent different attitudes.

Attitudes can be both an obstacle and an advantage to a marketer. Choosing to discount or ignore consumers' attitudes toward a particular good or service when developing a marketing strategy guarantees the campaign will enjoy only limited success. In contrast, perceptive marketers leverage their understanding of attitudes to predict the behavior of consumers and develop a marketing strategy that reflects that behavior.

Learning

Learning refers to the modification of behavior that occurs over time due to experiences and other external stimuli. Consumer learning may or may not result from things marketers do; however, almost all consumer behavior is learned. Marketers can influence consumer learning, and, by doing so, impact consumer decisions and strengthen consumer relationships, but they must first understand the basic learning process.

Learning typically begins with a stimulus that encourages consumers to act to reduce a need or want, followed by a response, which attempts to satisfy that need or want. Marketers can provide cues through things like advertisements that encourage a consumer to satisfy a need or want using the firm's good or service. Reinforcement of the learning process occurs when the response reduces the need. For example, eating a pizza reduces hunger and an immediate need for food. Consistent reinforcement satisfied by a particular type of pizza (Sicilian, plain, veggie, pepperoni) can lead the consumer to make the same purchase decision repeatedly as matter of habit. Think about it, when you're hungry, do you tend to put a lot of thought into what pizza to buy or where to buy it?

Marketers can capitalize on consumer learning by designing marketing strategies that promote reinforcement. Bounty has promoted its paper towels using the slogan "the quicker picker upper" for more than 40 years. Through repetition in promotion, the company hopes to influence consumer learning by associating Bounty with the idea of cleaning up spills quickly. The strategy has worked: Bounty remains the market share leader in paper towels and continues to grow in sales volume even as other paper towel brands have faced years of declines.

Learning is important in the consumer decision-making process. When a consumer learns that a product or service satisfies a need and solves a problem, that product or service moves up to the top of the information search the next time the consumer recognizes a problem.

Motivation

A third psychological process that affects consumer behavior is motivation. **Motivation** is the inward drive people have to get what they need or want. Marketers spend billions of dollars on research to understand how they can motivate people to buy their products. One of the most well-known models for understanding consumer motivation was developed by Abraham Maslow in the mid-1900s. He theorized that humans have various types of needs, from simple needs such as water and sleep to complex needs such as love and self-esteem. Maslow's hierarchy of needs illustrates the belief that people are motivated to meet their basic needs before fulfilling higher-level needs.

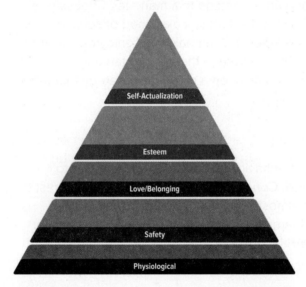

Maslow's Hierarchy of Needs has five levels organized in a pyramid shape, with levels ordered from bottom to top.

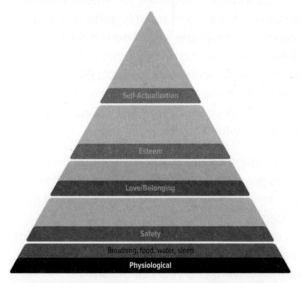

A consumer who lacks food, safety, love, and esteem would consider food his or her greatest need and, according to Maslow, would seek to fulfill that need before any of the others. Marketers of food, bottled water, and medicine are often focused on meeting the **physiological needs** of their target customers. This level represents the most basic of human needs.

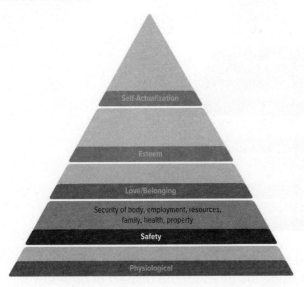

Safety can take different forms, including physical safety and economic safety. For example, the absence of economic safety—due to economic crisis or lack of job opportunities—leads consumers to want job security, savings accounts, insurance policies, and reasonable disability accommodations. Marketers match products to these types of needs because they know consumers will be motivated to buy them.

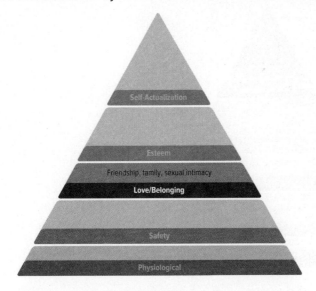

The third level of human needs, after physiological and safety needs are fulfilled, involves **love and belonging**. Love has become big business for marketers. The online dating site eHarmony was founded to serve this need by establishing its brand as the site for the serious relationship seeker.

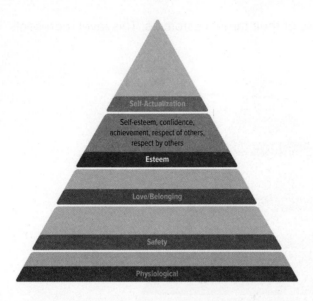

Esteem is the need all humans have to be respected by others as well as by themselves. Lower esteem needs include the need for the respect of others. Jewelry stores and luxury car makers often target their marketing at consumers with lower esteem needs or those looking to increase their status or prestige. Higher esteem needs include the need for self-respect, competence, and mastery. For example, makers of foreign language education software market their products as a way for consumers to fulfill a lifelong dream of speaking a new language.

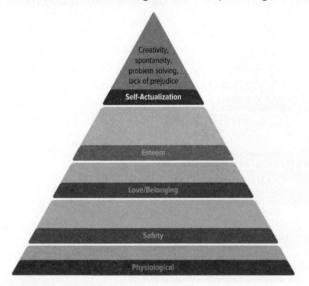

Maslow describes the top tier of the hierarchy as the aspiration to become everything that one is capable of becoming. **Self-actualization** pertains to what a consumer's full potential is and the need to realize that potential. For example, one individual may have a strong desire to become an ideal parent, another may want to become a superior athlete, and another may want to excel at painting, photography, or inventing.

Creativity,
spontaneity,
problem solving,
lack of prejudice

Self-Actualization

Self-esteem, confidence,
achievement, respect of others,
respect by others

Esteem

Friendship, family, sexual intimacy

Love/Belonging

Security of body, employment, resources,
family, health, property

Safety

Breathing, food, water, sleep

Physiological

Situational and Personal Influences on Consumer Behavior

Situational and personal influences on the consumer decision-making process include factors such as time, personality, lifestyle, values, and surroundings.

Time

Consumers value their time greatly, and time considerations often affect their purchasing decisions. Companies throughout the world understand this about their customers, so they will design goods and services accordingly. For example, banks have created apps that allow banking to occur without a customer ever needing to step foot inside an actual bank (think mobile check deposits). Reducing the time a consumer spends interacting with a firm is vital in almost all consumer settings.

Time can also impact what consumers ultimately pay for a good or service. Consumers are often willing to pay more for products if the placement of those products saves them time. For example, though a consumer may realize that milk, eggs, or bread are significantly more expensive at a small, local grocery than they might be at a larger chain supermarket farther away, that consumer might be willing to pay more for these items for the convenience of shopping closer to home and checking out more quickly. By placing products in a more accessible location, marketers can often increase their profits on individual items while still providing value to their customers.

Immediacy of the need plays a role in how fast the buying process occurs. If a product is needed now, such as a replacement dishwasher, the time to elaborate on choices before making a decision will be shortened. However, if a new dishwasher is needed for the remodeling of a kitchen that is occurring in a few months, more time is available to elaborate on the decision. Often consumers are forced to skip steps in the buying process in response to an emergency or immediate need.

Personality

Personality is the set of distinctive characteristics that lead an individual to respond in a consistent way to certain situations. A consumer's personality might include traits that make them confident, personable, deferential, adaptable, or dominant. Personality strongly influences a consumer's decision to purchase products. For example, shopping at Whole Foods might demonstrate a consumer's passion for locally grown organic products, whereas shopping at Walmart might represent an individual's commitment to frugal spending. Marketing professionals strive to identify personality traits that distinguish large groups of people from each other and then design strategies that best appeal to those consumers.

Lifestyle

Lifestyle is a person's typical way of life as expressed by his or her activities, interests, and opinions. For example, Whole Foods's overall marketing message focuses on exceptional products that cater to individuals that appreciate healthy living. Lifestyle characteristics are often easier for a firm to understand and measure than personality traits because it is relatively easy to observe how consumers express themselves in social and cultural settings. Marketers can potentially reach their targeted consumers by sponsoring events related to interests or activities those consumers are passionate about, or through social media advertising.

Values

Think about how your personal value system influences how you live your life. It is likely that your value system also corresponds to your buying behavior. **Values** reflect a consumer's belief that a specific behavior is socially or personally preferable to another behavior. Personal values, which include everything from a consumer's religious beliefs to a belief in self-responsibility, can impact the decision-making process. For example, many consumers look for goods and services that embrace sustainability—that is, products intended to benefit the environment, society, and the economy. Firms take into account consumer values to design specific products, such as BPA-free water bottles or LED bulbs. Marketers then message the products as eco-friendly to appeal to customers' sense of environmental responsibility.

DON'T BUY THIS JACKET

In this ad by clothing company Patagonia, people are told not to buy its apparel. This message was a counterintuitive attempt to appeal to consumers' concerns about environmental responsibility and sustainability.

Surroundings

Often, factors in our surroundings influence our decision-making process. Have you ever been in a grocery store that smelled bad? If you have, chances are pretty likely that you might have questioned the quality of the food products at that grocery store. Or, if you've ever had to choose between two different fitness centers, you might decide to join the one that features more modern equipment and has a more open and friendly workout space. Crowd size is another factor that can influence consumer behavior; if the crowd at the mall on the day of a huge sale is too large, it might deter you from shopping there that day. Conversely, if you notice a restaurant with just a few people dining in it during the busiest part of the day, you might be less likely to eat there.

Sociocultural Influences on Consumer Behavior

Sociocultural factors develop from a consumer's relationships with others and can significantly impact his or her buying behavior. Sociocultural influences that impact the consumer decision-making process include family members, reference groups, and opinion leaders.

Family Influences

Family members are one of the greatest influences on consumer behavior. The level of influence can vary across families and can evolve as a family ages and new members join the family through marriage or birth. The composition of families has changed greatly in recent decades to include more single parents and same-sex households, which can impact consumer decisions in different ways important to marketers. For some cultures, family is the primary influence. It can be a central theme for marketers seeking to target Hispanic market segments in both the United States and throughout the world.

Children often greatly influence a household's purchase decisions, particularly when it comes to grocery shopping and dining out. Marketers at McDonald's, Sonic, and Burger King spend a significant amount of money advertising to young consumers, giving away toys and books with their kids' meals. McDonald's has dedicated an entire website, **www.happymeal.com**, to marketing to children through games and technology. These promotions aimed at children can enhance restaurant traffic and revenues. However, marketing food to children has become a controversial topic for firms. Some suggest that marketing to children is unethical because children are impressionable and can be easily manipulated by marketing messages. Thus, marketers must practice caution when marketing to children so as not to take advantage of them.

Few situational factors have a more significant impact on consumer behavior than the family life cycle. The **family life cycle** describes the distinct family-related phases that an individual progresses through over the course of his or her life time. There are six stages that an individual might go through as part of his or her family life cycle:

1. Unmarried

2. Married with no kids

3. Married with small children or tweens

4. Married with teens

5. Married without dependent children

6. Unmarried survivor

Each stage impacts consumer behavior, and therefore a firm's marketing strategy to that consumer. Companies like State Farm Insurance actively promote their ability to service consumers throughout the entire family life cycle. They promote products and services that help customers insure their first car, secure family life insurance, and invest for retirement. In addition to family, consumers typically belong to or come into contact with various other social groups—schools, workplaces, churches, and volunteer groups—that can influence their purchase decisions.

1. Single

Single

This stage is characterized by **individual spending** (such as fitness, fashion, dating, travel, hobbies, etc.).

2. Married

Married

This stage is characterized by personal **couple's** spending (travel, self-improvement, work clothing, fine dining, education, etc.).

3. Young Family

Young Family

This stage is characterized by **family spending** on infants and toddlers (such as diapers, baby food, clothing, safety equipment, daycare, etc.) and tweens (sports equipment, summer camps, fast food, clothing, etc.).

4. Family with Teens

Family with Teens

This stage is characterized by spending on **teens** (such as college funds, cars, fashion, travel, technology, etc.).

5. Empty Nest

Empty Nest

This stage is characterized by **future spending** (such as travel, retirement planning, first homes, cars, etc.).

6. Survivor

Survivor

This stage is characterized by **freedom spending** (such as travel, hobbies, grandchildren) and **health** (doctor fees, insurance, medication, etc.).

Reference groups can provide consumers with a new perspective on how to live their lives. When you accept your first job, how will you know what to wear on your first day at work? You might recall what you saw people around the office wearing when you interviewed. In such a case, your coworkers serve as your reference group. A **reference group** is made up of the people to whom a consumer compares himself or herself. Marketers should understand that the more public the purchase decision, the more impact reference groups are likely to have. For example, reference groups tend to significantly influence a consumer's clothing purchases.

Firms typically focus on three consumer reference groups when developing a marketing strategy: **membership reference groups, aspirational reference groups,** and **dissociative reference groups.**

Membership Reference Group

A membership reference group is the group to which a consumer actually belongs. Membership groups could include school clubs, fraternities and sororities, and the workplace.

Marketers who understand the influence other members of these groups have on consumers can target products that would be ideal for the group members. For example, a local bank might market itself as the official bank of your university and offer a debit card featuring the school's logo.

Aspirational Reference Group

An aspirational reference group refers to the individuals a consumer would like to emulate. For example, professional athletes represent an ideal for many people. Serena Williams, who was ranked world No. 1 in singles by the Women's Tennis Association from 2002 to 2017, is one such iconic sports figure. She is featured in an inspirational Gatorade ad speaking to her newborn daughter, and in so doing, delivering a greater message to all young girls interested in sports, or aspiring to be an athlete: "Whether your bond is by blood or by ball, whether she shares the color of your skin or the color of your jersey, you'll find sisters in sweat."*

Dissociative Reference Group

Dissociative reference groups include people that an individual would *not* like to emulate. Teenagers and young adults tend to actively dissociate themselves from groups they view as "uncool" or in which their parents might express interest. But dissociative reference groups can play a role in marketing to all consumers. Marketers for mouthwash and certain types of chewing gum encourage consumers to use their products as a way to avoid being associated with those who have bad breath.

Opinion Leaders

Individuals who exert an unequal amount of influence on the decisions of others because they are considered knowledgeable about particular products are called **opinion leaders.** Opinion leaders range from Michael Jordan endorsing a pair of Nike shoes to Rachael Ray promoting a specific type of cooking utensil.

Opinion leaders are not just celebrities; social media allow small numbers of consumers to influence the consumer decisions of a much larger group. More marketers are trying to tap into the power of these social media opinion leaders through rewards and other benefits. For example, credit card companies offer special rewards to customers who have the potential to influence others; airlines give these same types of opinion leaders free flights in an effort to encourage them to use their influence on behalf of the company's products.

Involvement Influences on Consumer Behavior

How consumers make choices is influenced by their level of involvement in the decision process. **Involvement** is the personal, financial, and social significance of the decision being made. The study of high and low involvement focuses on how consumers choose which alternative to purchase and is important for firms to understand as they develop strategies to sell their products.

Characteristics of Purchasing Decisions

	Low-Involvement Purchase	High-Involvement Purchase
	Usually inexpensive products that pose a low risk to the consumer if the purchase ends up being a mistake.	More significant purchases that carry a greater risk to the consumer if the purchase ends up being a mistake.
Price	inexpensive	expensive
Preparation	requires little forethought	requires research
Frequency	frequently purchased	rarely purchased
Risk	limited risk	high risk

Copyright © McGraw Hill (l)MarioGuti/iStock/Getty Images. (r)Africa Studio/Shutterstock

Low-Involvement Buying Decisions

Most likely, you have made an impulse purchase sometime in the past month. Impulse buying is purchasing a product with no planning or forethought. Buying gum in a grocery store checkout line or a new cap that you notice as you walk through a mall are examples of impulse buying. Impulse purchases usually occur with low-involvement products. **Low-involvement products** are inexpensive products that can be purchased without much forethought and with some frequency.

Consumers often do not recognize their desire for a low-involvement product until they are in the store, which influences the strategic decisions for marketing these items. In-store promotion, for example, is a very useful tool for marketing low-involvement products. Unique packaging or special displays help to capture the consumer's attention and quickly explain the product's purpose and benefits. Marketing strategies for low-involvement products include Kellogg's signage at Walmart stores explaining the relatively low cost of eating breakfast at home versus at a restaurant, or a promotional sign by Sara Lee in a grocery store aisle that focuses on a promotion for its "soft and smooth bread." Tactics like low-tech cardboard displays found at the end of aisles can potentially drive more impulse purchases than temporary price reductions.

High-Involvement Buying Decisions

High-involvement products include more significant purchases that carry a greater risk to consumers if they fail. Two common examples of high-involvement purchases are a car and a house. Companies that market high-involvement products must provide potential consumers with extensive and helpful information as they go through the decision-making process. An informative advertisement can outline to the consumer the major benefits of a specific product purchase. Residential brokerage firm Coldwell Banker provides a wealth of information about the homes it is attempting to sell to help potential buyers understand more about the house itself as well as about local schools, financing options, and moving services.

Variations of the Consumer Purchase Decision Process

Marketers must be aware that there are three variations of the consumer buying process. All start with recognition of a need, but follow a slightly different path based on if the consumer purchase is the result of a *routine, limited,* or an *extended* problem-solving decision. For regularly purchased products, the consumer will not go through all the steps of the decision process. These products are typically low-involvement purchases. On the other extreme, for first-time purchased products, the consumer will engage in extended problem solving and most likely go through all the steps. These purchases require more involvement from the customer. The differences between these variations are highlighted in the table that follows.

Variations of the Consumer Buying Process

	Routine Problem Solving	Limited Problem Solving	Extended Problem Solving
Example	**Toothpaste**	**New Cell Phone**	**New Car**
Frequency of Purchase	Regularly	Occasionally	Infrequently or never
Experience in Purchasing	Experienced	Some experience	None or little experience
Level of Involvement	Usually low level	Mid-level	High level
Coverage of Decision-Making Stages:			
Problem Recognition	Yes	Yes	Yes
Information Search	No	Limited amount	Yes
Evaluation of Alternatives	No	Limited amount	Yes
Purchase	Yes	Yes	Yes
Post-Purchase Evaluation	Limited amount	Yes	Yes

Marketing Analytics: Using Data to Understand Consumer Behavior

Using Research to Develop Predictive Models of Consumer Behavior

The models used in consumer behavior were developed primarily in psychology and were based on theories of how people think and behave. In marketing we are primarily interested in how consumers think and behave in a *marketplace*. Firms use data to convert these theories into workable models that could be used to predict how consumers will respond to market offerings. Often to evaluate these models fully, ongoing data collection must be conducted. For example, consumer satisfaction measures are often created from survey data. Consider the surveys you might find at the end of a receipt from a store or restaurant. The data from these surveys inform the consumer's post-purchase evaluation stage and give firms the information they need to ensure they are fully satisfying a consumer's needs.

Understanding Influences on Consumers

What motivates a consumer to choose one product over another is a complex process that involves both physical and psychological states. These states are influenced by the broader environments in which consumers live. To truly understand these influences, marketers must collect and analyze data from these different environments, including the economic, demographic, political/legal, technological, competitive, and sociocultural environments.

Improvements in technology have changed how consumers interact with the market—for example, data show that more and more consumers are shopping online versus at brick-and-mortar stores. Economic data trackers, such as the Consumer Confidence Index, allow marketers to better understand how consumers are feeling about making big purchases. Marketers must stay on top of changes in the political/legal environment as well, and consider the impact these changes have not only on consumer behavior, but also on consumer access. For example, in some states consumers are able to order wine from online merchants, whereas in other states this is against the law. Regulations can be written that prohibit or limit consumer's access to such products, so marketers must take this into consideration when making marketing decisions. The physical environment, such as global climate change, is also influencing consumer behavior. Data reveal a growing interest in the purchase of more sustainable products—from automobiles to food to clothing and even travel. Using data that track the changes in these environments allows marketers to make predictions about what will happen next for consumers. It also adds a new dimension to their strategic planning.

Models of Consumer Decision Making

Not only can marketers use consumer decision-making models to understand consumers, but consumers can also use these models to help *them* make marketplace decisions. For routine or known problem solving, most consumers use an established method of decision making; they buy whatever they bought before (based on previous experience) to solve their market problem. In such cases, they often do not evaluate other alternatives. This form of decision making is called **noncompensatory decision making.**

For novel problems that involve some level of risk, consumers will consider multiple decision-making criteria and a more complex decision-making process. This form of decision making is called **compensatory decision making.** In this model, consumers assign **weights,** measures of importance or preference, to the various criteria they've predetermined. Assigning weights—for instance, on a scale of 1 to 5, with 5 being most important and 1 being least important—is a way for consumers to evaluate how different product options compare overall to each other.

The weights that consumers assign to the same set of criteria will vary with each consumer. Let's consider the criteria involved in purchasing a car: price, comfort, fuel economy, and style. Consumer A may rate style as more important than price, whereas Consumer B may rate price as the most important criterion of all. The sum (or score) of these weights, known as the **weighted preference,** will be used to make a purchase decision. After completing this exercise, some consumers may find that the product they thought they originally wanted is now, in fact, no longer the same product they end up buying. For example, a couple planning a wedding may have decided they wanted to marry in a specific location because it holds special meaning for them. They have three other venues in mind as well. They evaluate all their options using the compensatory decision-making model, considering location, cost, ease of access for guests, and catering options. After assigning weights to each of these criteria and calculating the weighted scores, the couple realizes that their original choice of venue is not the best option for them after all.

Responsible Marketing

Patagonia and Consumer Decision Making: Sustainability and Beyond

During the consumer purchase decision process, certain psychological, situational, personal, sociocultural, and involvement influences can affect a consumer's decision to purchase. What mood is the customer in? What values does the customer have that affect shopping behavior? What is the state of the economy? With so many options available to consumers and so many factors affecting purchasing decisions, it can be difficult for companies to convert sales. One company that has grown its customer base and continues to attract customers with shared values is Patagonia.

The text references Patagonia and its unusual campaign strategy in 2011, which told people to not buy its apparel. This campaign was designed to bring attention to the company's position on sustainability (ironically, Patagonia sales increased by 30 percent that year). In subsequent years, Patagonia has amplified its mission and promotion of sustainability. The company resells gently worn items on its website and encourages customers to turn in used products for credit. This is in keeping with its customers' values; a 2017 survey indicated that 69 percent of Patagonia customers said they were concerned where the products come from.

While Patagonia has increased awareness of sustainability, COVID-19 has greatly impacted consumer behavior. Euromonitor reported that "73% of professionals believe sustainability efforts are critical to success"; however, with many customers forced to order merchandise online during the pandemic, 51 percent reported an increased use of plastics. How Patagonia will deal with these and other challenges to sustainability in the future remains to be seen.

Reflection Questions

1. Consider the five steps in the consumer purchase decision process. What are considerations a Patagonia customer might make in each step?

2. What are some potential psychological, situational, personal, sociocultural, or involvement influences on Patagonia customers in their decision-making process?

3. How has COVID-19 impacted consumer behavior and how may it affect Patagonia's message going forward?

Lesson 4-3

Business-to-Business Marketing

This lesson will explain how businesses that sell to other businesses market their products and services. Marketing to business customers involves some similar and many different types of activities than those used in B2C marketing. For example, as a consumer, you may be familiar with Champion products. However, have you considered the importance of the retail relationships that Champion has to develop to have a successful brand?

By the end of this lesson you will be able to

- Describe the buyer–seller relationship.
- Describe the different types of buying situations.
- Recognize the major forms of B2B e-marketing.
- Outline the steps in the B2B buying process.
- Define buying centers, how they influence organizational purchasing, and the roles of their members.

Champion: Becoming "Cool" Again

Have you ever heard of Champion? Originally called the Knickerbocker Knitting Company, Champion was founded as a family business in 1919. It started out selling athletic uniforms for the University of Michigan and gained traction with its durability and comfort. Growing in style and notoriety, Champion obtained partnerships with the National Collegiate Athletic Association (NCAA) in the 1960s, the National Football League (NFL) in the 1970s, and the National Basketball Association (NBA) in the 1990s.

After years of partnerships with collegiate and professional sports, demand for the brand waned. Consumer demand shifted from what turned into the "frugal dad" look to new brands such as Lululemon and Under Armour.

So how did Champion become trendy again? For starters, in 2006 the Sara Lee Corporation spun off HanesBrands to contain all of its clothing brands, including Champion, when it no longer had time for "niche businesses" and became fully invested in the food industry. Champion started partnering with trendy brands such as Supreme, UNDEFEATED, and Vetements with the mindset that if it wanted to become "cool" again, it had to partner with trending brands.

As Champion began making its comeback, vintage clothing and athleisure were also becoming popular. Thus, products designed around these trends were created for maximum hype and designer brand partnerships. Currently focused more on pop culture than sports, Champion clothing can be found in trendy clothing stores such as Zumiez, Urban Outfitters, PacSun, and more.

So who is HanesBrands, and how does it get Champion products into the hands of consumers? HanesBrandsb2b.com is the business-to-business website that business partners can order from directly. This wholesale website puts business customers in control of their order. Hanes makes it easy for businesses big or small to order with no minimum order quantity, receive fast and affordable shipping, chooses from drop shipping options, and have access to bestselling brands.

In March 2020, HanesBrands teamed up with Amazon offering the C9 Champion performance athletic wear as part of a multiyear partnership. More than 100 styles from C9 Champion will be available via Amazon's online store, providing consumers with a new channel to access the brand. HanesBrands signed a multiyear agreement for the C9 Champion Athleticwear brand to be sold exclusively through Amazon.

Factors Affecting Business-to-Business Marketing

As noted in the introduction to this lesson, **business-to-business (B2B) marketing** consists of marketing to organizations that acquire goods and services in the production of other goods and services that are then sold or supplied to another business or consumer groups.

B2B marketers face many of the same challenges B2C marketers do, as both work to connect or build relationships with their customers. However, unlike in B2C marketing, where marketing is targeted at individual consumers, in B2B marketing, marketing is targeted at **professional purchasers or buyers**—employees of companies who make purchase decisions in the best interest of their organizations. Because of this, B2B marketing is based on long-term relationships that are often maintained among multiple people from the selling firm and a group of people from the buying center of a purchasing company. In addition, B2B buyers and sellers are motivated by derived demand for products rather than individual decision making based on personal needs and wants. Let's look at each of these factors in a bit more detail.

Professional Purchasing

Businesses typically buy things through professional purchasing managers who are experienced in the policies and procedures necessary to make a large deal. For example, a professional buyer at Old Navy will be responsible for purchasing the clothing styles and accessories that will eventually be featured in Old Navy's stores. Because businesses strategically plan well into the future, the B2B purchasing process is often far longer than the consumer decision-making process. In addition, the B2B buying process requires standardized procedures, such as a request for proposal (RFP) and contract negotiations, which are not typically found in consumer buying. Finally, the strategic nature of business purchasing removes much of the emotion and personal interest from purchase decisions. For example, Sysco is a large wholesale distributor of food and beverage products. A buyer–seller relationship at Sysco may involve discussion of the sale/purchase of 200 cases of tomato soup for a large catering event. It should be easy to see how the buyer of a large quantity of soup will be more strategic and less emotionally invested in the purchase than he or she would be in a personal decision to buy a new iPhone or concert tickets.

Importance of the Buyer–Seller Relationship

In addition to buyers being professionals, business marketers typically deal with far fewer buyers than consumer marketers. Usually, each of these customers is larger and more essential to the firm's success than a consumer because there are fewer business buyers. For example, the potential demand for pizza is almost unlimited in the United States. However, the demand for large-scale pizza ovens is confined to medium and large pizza establishments, many of which belong to Pizza Hut, Little Caesars, and other major national chains.

Because there are fewer buyers, B2B marketers feel even more pressure to make sure they offer high-quality products to and establish good business relationships with their customers. A B2B **buyer–seller relationship** is a connection between a firm and/or its employees intended to result in mutually beneficial outcomes. B2B buyer–seller relationships are critical because a bad relationship with an individual pizza consumer might cost the local Pizza Hut $20 per week, whereas a bad B2B relationship with Pizza Hut might cost a pizza oven maker its entire annual profit or even its future.

Derived Demand

The need for business goods comes from demand for consumer goods. **Derived demand** occurs when demand for one product occurs because of demand for a related product. Even though Sysco, for example, does not sell directly to consumers, the success of its business depends on the buying patterns of individual consumers. Derived demand also provides an important reason to develop mutually beneficial relationships with B2B partners.

Imagine a scenario in which Sysco has a contract as the only provider of food to campus dining at your school. Because Sysco is the only supplier, it has the option to charge far more than is necessary for the goods it sells, thereby maximizing profit. However, if Sysco chooses to charge your university higher prices, the university, in turn, will have to pass on those cost increases to students like you. Once the price of campus dining gets too high, you and other students will simply find other places to eat. In turn, overall student demand for cafeteria meals will decrease, thereby decreasing the university's need for Sysco's products.

This scenario illustrates why marketers must take a strategic view of business relationships and understand all of the potential impacts their actions can have on derived demand.

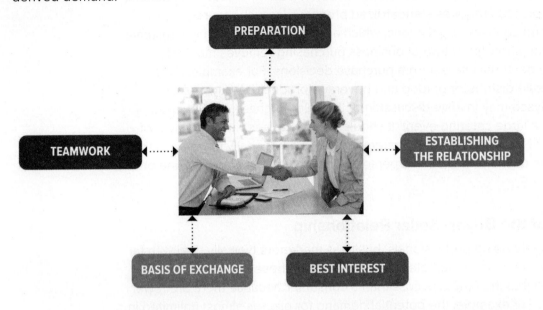

The Different Types of Buyer–Seller Relationships

B2B marketing professionals focus on and maintain relationships with buyers who work in different business markets. In addition, within each market buyer–seller relationships can vary. Let's discuss different business markets and the continuum of business relationships.

Reseller Markets

Resellers include retailers and wholesalers that buy finished goods and resell them for a profit. A retailer, such as a clothing or grocery store, sells mainly to end-user consumers like you. A **wholesaler** is a firm that sells goods to anyone other than an end-user consumer. Sysco is a perfect example of a wholesaler. Wholesalers frequently purchase a large quantity of a good (e.g., 200,000 pounds of hamburger meat) at a low cost and then sell off smaller quantities of the good (e.g., 20-pound cases of hamburger meat sold to restaurants) at a higher per-unit price. Thus, wholesalers are often called intermediaries because they don't produce products, they just process them.

Government Markets

Government markets include thousands of federal, state, and local entities that purchase everything from heavy equipment to paperclips. The U.S. government is one of the world's largest customers, spending billions of dollars a year. Marketing goods and services to the U.S. government requires adherence to certain policies, procedures, and documentation obligations. Because the public holds the government accountable for its purchases, complex buying procedures are often used to ensure that purchases meet the necessary requirements. Firms must be detail oriented and complete extensive documentation to succeed at marketing to federal agencies and departments. For example, Mississippi-based Gulf Coast Produce spent considerable time and resources winning a government contract to provide millions of dollars in fruits and vegetables to the military. This complicated and often slow process has made some marketers, especially in small businesses, reluctant to bid on government business. Of the 20 million small businesses in the United States, only about 500,000 of them have completed the documentation necessary to be eligible to sell to the government. However, government markets can be highly lucrative for smart marketers and organizations.

Institutional Markets

Institutional markets represent a wide variety of organizations, including hospitals, schools, churches, and nonprofit organizations. Institutional markets can vary widely in their buying practices. For example, a megachurch with thousands of members and a multimillion-dollar budget will likely have a buying manager for firms to work with, whereas marketing to a new church with a small congregation might simply require speaking with the pastor. These diverse buying situations pose unique challenges for institutional marketers. They must develop flexible, customized solutions that meet the needs of differently sized organizations. Educating institutional customers about how specific goods and services can make their organizations more efficient or effective is a firm's best tool for selling products in this type of market. For example, marketers for a medical technology firm could show a hospital how their customized technology solutions can reduce costs for the hospital while improving patient care.

Reseller Market

Nordstrom is an example of the reseller market. Nordstrom processes products as they come in through a distribution center, and resells them by such brands as Madewell, Patagonia, and Clinique to the end consumer.

Government Market

The government is a huge user of food products. The Armed Forces, federal cafeterias, and even the White House all purchase food on behalf of the government. Sellers to government commissaries (foodservices) must comply with complex and specific regulations to do business with the government.

Institutional Market

School lunch is just one example of an institutional market for food products. Though many schools and institutions are government run, they are a category of their own. Schools, hospitals, and nursing homes are regulated by different organizations and standards that require sellers to comply with specific guidelines and policies.

Continuum of Buyer–Seller Relationships

B2B relationships are often described as being somewhere on a continuum, which means they vary but fall within a range of two different types:

- **Transactional relationships:** On one end of the continuum, relationships are transactional, which means they are based on low prices, commodity products, and are not usually stable in the long term. The pure **transactional relationship** exists where both parties protect their own interests and where partners do things for each other purely as an exchange.

  ▶ Many organizations, for example, purchase office supplies like printer paper, staples, paper clips, and pens. While the organization may have a relationship with the provider they work with, many of these purchases are just transactional. Organizations mostly look for the best price for these types of products.

- **Collaborative relationships:** At the other end of the continuum, relationships are collaborative, which means that both parties in the relationship are highly engaged and dependent on working with each other for a long time. The **collaborative relationship** occurs when both parties share resources (e.g., financial risk, knowledge, and employees) in an effort to attain a common goal that provides beneficial outcomes to both parties.

  ▶ It is important for Champion to work closely with its retail partners. The success of the retailer and the success of Champion are often tied together. Champion is dependent on its retail partners to advance and build the Champion brand. As retailers are often the final interface with consumers and advocates for the brands they sell, Champion must provide its retail partners with information and materials to educate the consumers they serve.

Most B2B relationships fall somewhere between transactional and collaborative. In many cases, one firm may act as a transactional partner for some products and a collaborative partner for others. Thus, selling firms must evaluate the relationships they maintain with their buyers and invest their resources accordingly.

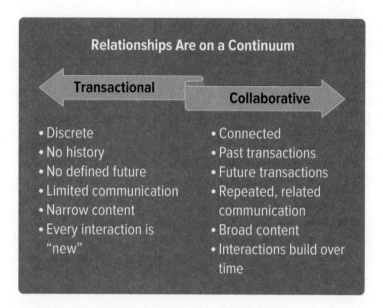

B2B Relationship Continuum. This continuum shows how B2B relationships vary but fall within a range of two different types: transactional and collaborative.

Buying Situations

Although the types of B2B customers can vary, the buying situations for each are often quite similar. Marketers can classify B2B buying situations into three general categories.

A **new buy** involves a business customer purchasing a product for the first time. For example, let's say that Dell is looking to market its personal computers to a college that has not previously bought from the company. Because the college has little or no experience with Dell, its decision process will likely be extensive, with the college requiring a significant amount of information and negotiation. From a marketing standpoint, Dell's reputation for meeting specifications and providing high-quality service to its current business and college customers could prove to be a critical factor in selling to the college for the first time.

A **straight rebuy** occurs when a business customer signals its satisfaction by agreeing to purchase the same product at the same price. B2B marketers prefer the straight rebuy outcome to any other because straight rebuys normally do not require any additional design modifications or contract negotiations. Another major advantage of a straight rebuy is that the customer typically does not look for competing bids from other companies. To revisit the Dell example, in a straight rebuy scenario, Dell should work hard to produce high-quality computers at a competitive price and offer great service that makes the college feel good about its purchase decision. Marketers who make it as easy as possible for customers to do business with their firm increase the likelihood that the customers will perceive value and develop loyalty.

A **modified rebuy** occurs when the customers' needs change or they are not completely satisfied with the product they purchased. In our Dell example, the college might want Dell to modify its computers to add additional features, lower its prices, or reduce delivery times to get new products to the school. Modified rebuys provide marketers with both positive and negative feedback. By buying from Dell again, the college signals that it is pleased with at least certain parts of its purchase experience. However, modified rebuys can also be negative if the college asks Dell to reduce its price or modify design characteristics to a point where the agreement no longer earns Dell a profit. If Dell agrees to terms that cause the company to lose money, its long-term health as an organization could be in jeopardy. Thus, the ability to walk away from potential modified rebuys provides marketers with an important tool that should be used to negotiate a positive solution that works for both the buyer and seller.

Regardless of what type of business customer B2B marketers are selling to or what type of buying situation they're in, B2B marketing should seek to create, communicate, and deliver value to customers in a way that is ultimately profitable, just as marketers would with individual consumers.

New Buy

For Sysco, *new buys* will come from new customers or by expansion of what an existing customer purchases. Acquiring new buys takes much of a salesperson's time and effort. However, once a new buy is established, it becomes a *rebuy,* which takes less effort and is the basis of a seller's recurring business portfolio. Thus, much time is spent developing new buys with new or existing customers.

Straight Rebuy

For *straight rebuy* situations, like weekly ordering at a restaurant to replenish inventories, Sysco foodservice provides an online and mobile application. Because straight rebuys require the least amount of effort for both the buyer and seller, e-commerce is an effective way to accomplish them without extra hassle.

Modified Rebuy

A *modified rebuy* for a Sysco salesperson may require that specialists in menu planning or meal preparation be consulted to resolve any issues with the original product order. Sysco invests heavily in test kitchens and services to help restaurant chefs find the right solution and taste portfolio for their customers.

The Forms of B2B E-Marketing

B2B marketers use several digital marketing tools to reach their customers, many of which are very similar to consumer marketing, and many that are not. Major B2B e-marketing tools include websites, mobile applications, e-commerce platforms, social media, and other forms of communication.

Websites

B2B website marketing is similar to consumer website marketing. B2B companies are expected to have a quality website that contains useful information concerning the company's products, services, and contacts. B2B marketing websites do pay special attention to a few additional things, often not included in consumer websites. For example, B2B websites may contain highly technical information and specifications concerning the selling firm's products. Inclusion of deeper levels of information and specifics of a product is useful to professional buyers who require more specific information than is typically available to consumers. Champion is a good example of a company that provides its retailers with a great deal of information about its prodcuts. This helps the retailers sell Champion products better.

Mobile Applications

In addition to websites, B2B sellers offer mobile applications that help their customers have real-time access to their accounts and orders on a mobile platform. Integration of a supplier's mobile technology into a buyer's supply chain flow is a good way to keep everyone in the process (e.g., purchasing, logistics, accounting, and manufacturing) informed. For example, Sysco's mobile application allows customers to access inventory levels in real time, track delivery status, review accounting statements, and even message with delivery drivers. Having a value-adding, time-saving application in a B2B setting is becoming a way that selling firms can create competitive advantage over those vying for a customer's business.

E-Commerce Platforms

E-commerce allows buyers and sellers to conduct exchange through a digital channel. For B2B firms, e-commerce is most closely associated with standard rebuy situations and transactional relationships. B2B sellers have traditionally facilitated exchange through a salesperson who meets with a buyer on a regular basis in an effort to collect orders and communicate them to the selling firm. In many cases, e-commerce eliminates this function of salespeople. However, most salespeople look at this change as positive because it allows them to focus on more important customer needs, such as customized solutions and new buys, or modified rebuys, while less complex orders process electronically. One important consideration for B2B e-commerce is security. This is similar to consumer e-commerce, but B2B e-commerce security is concerned with cyberattack and competitor security. To reduce security risks, most B2B e-commerce sites are accessed through permission of the selling firm that is granted only to customers.

Social Media

B2B firms use social media to keep their customers informed and to learn about their customers. In many B2B settings, social media have become an effective way for customers to interact with each other as well as interact with selling firms. The use of forums, blogs, and groups allows engagement to occur around a firm or product. This type of engagement, especially B2B customer-to-customer engagement, is helping firms get more immediate feedback on their products and customers to learn new ways to use a product or to solve a problem. For example, Sysco Foodservice has a LinkedIn page that has more than 90,000 followers, which allows Sysco to disseminate new information about the company quickly.

Other Forms of Communication

New forms of e-marketing are helping B2B marketers be more in touch with their customers than ever. **Enablement tools** are applications that streamline buyer–seller engagement in an effort to improve the customer's experience. Customer relationship management (CRM) systems are the basis of enablement tools. CRM allows sellers to manage their relationships with customers more effectively by closely tracking the activities and needs of customers. Enablement tools use CRM information to automate communications and follow up on customer needs. Today, tasks that once were time-consuming sales and marketing burdens are handled through enablement tools. From proactively predicting customer orders to delivery reminders, enablement tools are changing how and when B2B firms communicate with their customers in an effort to keep customers highly engaged.

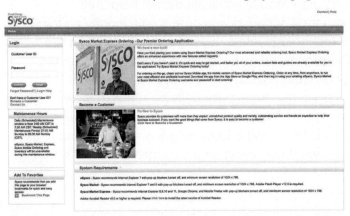

The B2B Buying Process

The B2B buying process involves research and strategic planning. As information and access to specific seller details have become more available to B2B buyers, the timing of when a seller becomes involved has changed. Sellers used to be involved in the strategic process early, as a conduit to needed information (specifications, prices, availability, etc.). Today, sellers often become involved in the process much later, with some firms estimating as late as 70 percent of the way through the process. This change in where and whom the information comes from is reflected in the process that professional buyers follow to make B2B purchase decisions, which we show in the figure below and discuss next.

The B2B Buying Process

STEP 1	STEP 2	STEP 3	STEP 4	STEP 5	STEP 6	STEP 7
A need is recognized	The need is described and quantified	Potential suppliers are researched	Requests for proposal are sought	Proposals are evaluated and suppliers are selected	The order is placed	A post-purchase review is conducted

Step 1: A Need Is Recognized

At some level of the organization, a need is recognized. Need recognition can come from any functional area, for example, new product development, operations, and/or information technology. New product development typically generates needs as a new product is developed, such as a new plastic formulation, a new type of product label, or an innovative sealing process. Operations often realize needs in the area of manufacturing machinery, plant safety, or energy efficiency. Information technology often identifies needs such as new software, hardware upgrades, or security enhancements. All of these needs eventually are communicated to the department in charge of purchasing, where the procurement process starts.

Step 2: The Need Is Described and Quantified

The buying center, or group of people brought together to help make the buying decision, considers the need that has been communicated and develops parameters of what needs to be purchased. In some cases, this is easy (rebuy) because a good deal is known about what is needed. In other cases, this is difficult (new buy) because what is needed is totally new and different from what is normally purchased by the firm. Specifications of the needed product are determined, such as features, amount required, where it is needed, and method of delivery. Technical or complex products require users of the new product to define the product's technical specifications. When the details are well defined, the type of product will be clear—does this product already exist or does it need to be custom made?

Step 3: Potential Suppliers Are Researched

At this stage, purchasers seek information about the product they need and the vendors that can supply the product. Online research allows most of this stage to be completed without contacting potential suppliers. Purchasers may also investigate industry forum websites and blogs of industry experts. Professional buyers often play a key role when it comes to deciding which vendors are the most qualified. Some strategic questions are also addressed in this stage, such as: Is the supplier reliable and financially stable? Where is the supplier located and will logistics be an issue? Does the supplier have enough inventory or capacity to deliver what is needed? Typically, at the end of this stage only a few qualified vendors are identified as probable suppliers.

Step 4: Requests for Proposal Are Sought

The next step is to get supplier quotes for supplying the needed product. It is important to note that to this point in the process, many purchasers will not have contacted a seller. In the past, suppliers would have been consulted much earlier in the process, but often they are now first contacted when most specifications are already decided. A **request for proposal (RFP)** is a formal listing of specifications for a needed product that is sent to a supplier firm asked to bid on supplying the product. The RFP clearly states what is needed and asks the supplier to submit a proposal for the business that includes specifics on factors such as quality, price, financing, delivery, and after-sales service. Most RFPs also include a strict deadline that indicates when competing proposals will be considered. Because in many cases this is the first time the supplier hears about the need, good sales and marketing professionals go deeper than just providing what is asked for in the RFP. Rather, selling firms use the RFP to investigate the buyer's problems and needs and how to adapt their offers to solve those problems and needs.

Step 5: Proposals Are Evaluated and Suppliers Are Selected

During this stage, the supplier responses to RFPs are reviewed and ultimately one supplier is chosen. Companies evaluate supplier responses differently, but all use some formula to evaluate what value will be realized from each supplier's offer. Many things can indicate value differently, such as high levels of service or quality that one firm offers but others do not, or innovative solutions. In the end, the lowest-priced bid may not be the one chosen, as many variables are considered in determining the best value.

Step 6: The Order Is Placed

Once a supplier has been selected, the agreement is finalized. Finalizing an agreement can be simple or complex depending on the buying situation. For example, a restaurant that decides to buy taco shells from Sysco likely just needs to complete some simple paperwork. However, a government agency that decides to purchase a missile system from a defense contractor likely involves a team of lawyers to finalize the details of the sale.

Step 7: A Post-Purchase Review Is Conducted

Once a purchase is complete, buyers conduct a review of how well the order was fulfilled. The review can be simple—did the order arrive on time and correctly?—or it can be a complex and formal process. Regardless, "getting it right" is important for suppliers that desire to continue supplying the needs of a customer.

The Influence of Buying Centers and the Roles of Their Members

Many buying decisions are complex and require decisions by multiple members of a buying center. A B2B **buying center** is a group of people responsible for strategically obtaining products needed by the firm. The buying center is everyone involved in any way in the B2B buying process. This includes the purchasing manager who places the order, the manager or executives who approve the purchase, and anyone who recognizes a purchase need. Buying center roles are informal and are not built into the company's corporate hierarchy. It is important that sellers be able to determine all the decision makers in the buying process. In an engineering firm, for example, the engineers may be the ones making buying decisions; in a goods production firm it may be product managers, and in a small company almost everyone may be involved.

Because many people are involved in buying centers, it is important to understand the roles they may play in purchase decisions. Several different general buying center roles exist. Let's discuss the role of each.

Users

Users are the individuals who will use the product once it is acquired. Users often make specifications about what the product needs and start the buying process.

Gatekeepers

Gatekeepers control the flow of information into the company that all other users review in making a purchasing decision. A personal assistant can be a gatekeeper by deciding who gets appointments with buyers and deciders.

Deciders

A **decider** is the person who chooses the good or service that the company is going to buy. The decider may not actually make the purchase but decides what will be purchased. This role can be filled by many people ranging from the CEO of the company in a large purchase decision, to the head engineer on a project, to an individual office manager buying office supplies.

Influencers

Influencers affect the buying decision by giving opinions or setting buying specifications. IT personnel and those with special knowledge on projects are often important influencers.

Buyers

Buyers are those who submit the purchase to the salesperson. This role is often more formal, such as a purchasing manager. Many firms also have specialized buyers, trained to help negotiate for the best prices of materials for the firm.

The buying center can be different across firms, and even across decisions within firms. Thus, not every role is present in all purchasing decisions. However, understanding who is performing which role, and which roles are a factor in decision making, is critical to firms that wish to effectively win a sale of their product.

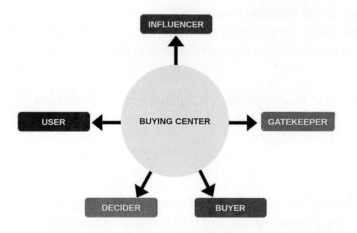

Responsible Marketing

Transparency and Communication in B2B Marketing

B2B marketing occurs when the customer is another business or organization. Business customers behave differently than traditional consumers, and are often engaged in significant purchase decisions that impact their ability to perform their own business or organizational functions.

We could explore a variety of responsible marketing challenges in the B2B context, but one of the most prevalent involves pricing. Because of the unique structure of B2B relationships, pricing can change by customer in a way that's different from consumers.

Consider the following fictional scenario:

Tom is a sales associate for a top robotic equipment manufacturing company that makes robotic arms for production lines for software companies. The product Tom sells is very specialized and there is little competition, yet a lot of demand, in the market.

One of Tom's main clients is Gary. Gary has been a customer for 15 years and has been a lucrative customer. Gary and Tom are also good friends. Gary tells Tom that his business suffered during COVID-19 and asked for a discount on a big order of 12 new robotic arms. Tom agreed, valuing the relationship with Gary, and offered a 20% discount off the normal asking price.

Phil is a potential new customer of Tom. Tom just read an article that Phil's firm just received $100 million in investment funding. Tom believes that Phil has a lot of money for the robotic arms and could be a great customer long term. In order to compensate for the lost revenue with Gary, Tom decides to charge Phil 20% over the normal asking price for the robotic arms. Phil, not knowing any better, agrees to the price.

Reflection Questions

1. Do you feel that Tom's behaviors are ethical or unethical? Explain your answer.

2. Should Tom charge both customers the same price?

Understanding Your Customer: Test

1. When customers look beyond their personal knowledge and experience to help them in buying something, they are engaging in

   A. ritual consumption.

   B. external information search.

   C. cognitive dissonance.

   D. business-to-business marketing.

   E. self-actualization.

2. Robin, a 28-year-old fashion designer, is very finicky about the brands that she consumes as a customer. She believes in buying brands that she has previously used. When Robin walks into Brown and Grey, a supermarket, she is most likely to engage in _____ to buy toothpaste and mouthwash.

   A. cognitive dissonance

   B. an evaluation of alternatives

   C. an internal information search

   D. impulse buying

   E. business-to-business marketing

3. Tom recently bought a pair of headphones, XOB 10, from Blue for $99. Within a month, Tom's friend, Leo, bought a brand new pair of headphones XGD 20, released by Soundz, online at a reduced price of $86. Tom tried the XGD 20 and found it to be significantly better the XOB 10. In this scenario, Tom is most likely to experience

   A. cognitive dissonance.

   B. self-actualization.

   C. problem recognition.

   D. external information search.

   E. criteria evaluation.

4. Consumer learning typically begins with

   A. knowing that a particular product exists.

   B. a response which attempts to satisfy a need or want.

   C. reinforcement that a particular product satisfies a need or want.

   D. cues that discourage the use of a particular product.

   E. a stimulus that encourages consumers to act to reduce a need or want.

5. According to Maslow's hierarchy of needs, once their physical needs have been satisfied, consumers' _____ take precedence.

   A. need for belonging

   B. safety needs

   C. esteem needs

   D. self-actualization needs

   E. need for food

6. Which of the following exemplifies impulse buying in a supermarket?

   A. Picking up a candy bar while waiting in line at the checkout counter

   B. Remembering to buy toothpaste while browsing the groceries sections

   C. Buying an reasonably priced watch from the accessories sections

   D. Picking up a bottle of wine to go with meat

   E. Opting for a different brand of tissues from the usual brand

7. Jerome has a dish full of dirty dishes and has run out of his favorite dishwashing liquid, so he heads to the corner market to pick up a new bottle. This type of consumer purchase exemplifies which variation of the consumer buying process?

   A. extended problem solving

   B. motivated purchase

   C. high-involvement purchase

   D. routine problem solving

   E. limited problem solving

8. The soles of Asha's Asics running shoes are wearing thin, so she needs to purchase a new pair. She knows the style she has worn the last two years has been updated, so she plans to purchase the same brand, but in the updated style. This type of purchase is considered a limited problem-solving purchase.

   A. TRUE

   B. FALSE

9. Hassan is getting ready to purchase his first home. His involvement throughout the buying process will likely be mid-level because this purchase is considered an extend problem-solving purchase.

   A. TRUE

   B. FALSE

10. In order to ensure that firms sell more products more profitably, marketers need only to understand *why* individuals and organizations make their purchases.

    A. TRUE

    B. FALSE

11. In which of the following would primary data be *most* helpful to a firm?

    A. determining whether price cuts on store-branded products impact what products consumers buy

    B. determining the competition's prices and inventory

    C. understanding the international market for its products

    D. identifying which in areas to open a second location

    E. determining interest rates before deciding whether to expand

12. A toy company is testing kites to see which one can be flown the highest. In this example, which of the following would be considered a dependent variable?

    A. the height the kite can be flown

    B. the temperature at the time the kite is flown

    C. the wind conditions at the time the kite is flown

    D. the type of kite being flown

    E. the person flying the kite

13. Antoine wants to know if his main rival in organic baked goods is increasing its prices. He talks to his rival's distributors, suppliers, and some of his rival's customers to get this information. Antoine is engaging in

    A. data mining.

    B. trending.

    C. user research.

    D. usage studies.

    E. competitive intelligence.

14. The marketing research industry relies on ethical standards to help gain the trust of consumers. Establishing trust

    A. increases individuals' willingness to participate in research.

    B. ensures accurate research results.

    C. informs companies when to cease research.

    D. decreases the cost of research.

    E. increases the cost of research.

15. The Blooms & Blossoms company has been selling its garden supplies to retail outlets for years. The company has decided it wants to tap into new markets, so it is looking to sell its garden supplies to county institutions for planting gardens in public parks. If the county institutions purchase the Blooms & Blossoms garden supplies, it would be an example of a(n)

    A. modified rebuy.

    B. first buy.

    C. new buy.

    D. straight rebuy.

    E. institutional buy.

16. Fenton orders the same office stationery each month for the same price from the same supplier. This is an example of a

    A. modified rebuy.

    B. quick buy.

    C. continual buy.

    D. new buy.

    E. straight rebuy.

17. A B2B seller can use mobile applications to create a competitive advantage by integrating its mobile technology into the its own supply chain.

    A. TRUE

    B. FALSE

18. E-commerce is *most* likely to be associated with which type of B2B buying situation?

    A. modified rebuy

    B. collaborative relationship

    C. new buy

    D. transactional relationship

    E. relationship marketing

19. This application is a type of enablement tool.

    A. delivery reminder

    B. CRM system

    C. e-commerce platform

    D. social media presence

    E. mobile application

20. Someone who is acting as an influencer in the buying center may be best illustrated by which of the following scenarios?

    A. The firm's purchasing agent has been asked to get the price of office supplies down before committing to the purchase.

    B. The company's CEO has said that the head engineer will choose what surveying equipment the company will buy.

    C. All information flow from potential suppliers is filtered through the executive assistant who makes appointments and distributes what he has received to different members of the buying center.

    D. The members of the call center have stated that the desks that are to be purchased for them must be adjustable so that they can have the option of standing.

    E. While discussing an upgrade to the company's computer software, the head of IT discussed current trends in computing that she would like the company to consider.

5 Segmentation, Targeting, and Positioning

What To Expect

How much do you know about your customers? Your ability to understand who you are marketing to can help you be a success. In this chapter we will look at techniques to make the most out of what information you have.

Chapter Topics

- **5-1** Segmentation
- **5-2** Targeting
- **5-3** Positioning

Lesson 5-1

Segmentation

This lesson introduces you to the concept of *market segmentation*. As the Gap Inc. example illustrates, before a product can be positioned to meet customer needs, it is critical to separate a larger market into well-defined market segments. Each segment is defined relative to specific customer or market characteristics deemed vital to a firm's success.

By the end of this lesson you will be able to

- Define market segmentation.
- Explain three key ways in which market segmentation plays a role in an organization's success.
- Explain the four methods of market segmentation.
- Explain the criteria for successful market segmentation.
- Explain how marketers use data to identify and refine consumer segments within a market.

Marketing Analytics Implications

- All product markets are characterized by *heterogeneous* demand, which means different consumers are seeking different things from the same product or service. Market segmentation is the process of finding homogeneous groups within the market. This requires a lot of data to do correctly.
- For any marketing activity to be successful firms *must* identify their market segments. Firms use data to understand the needs and wants of different consumers in a given market, categorizing each according to the four main bases of segmentation discussed in this lesson: demographic, geographic, psychographic, and behavioral. Because firms cannot meet the needs of all consumers in a market, they must use data to understand those needs.
- Just because a segment is identified does not mean it is appropriate to pursue. Data are used to evaluate and determine the most appropriate segments to target later.

Gap Inc.: Finding a Segment for Every Style

As consumers, we have access to millions, if not billions, of products and services. One of the most difficult aspects of marketing is understanding which groups of customers will buy which products. As you have learned, a target market is a relatively homogeneous group of consumers that will respond favorably to a particular product or service. The next few lessons (segmentation, targeting, and positioning) describe how a firm identifies and responds to its target market.

Consider global retail conglomerate Gap Inc. Gap was founded in 1969 by Doris and Donald Fisher. Their initial goal was to help make it easier for consumers to buy a pair of jeans that fit. Their first store, located at 1950 Ocean Avenue in San

Francisco, sold only men's Levi's jeans as well as records and tapes. By 1974, Gap started selling its own branded merchandise, and by 1976, the company had gone public. Today Gap employs over 130,000 people worldwide, operates in over 90 countries, and has almost 4,000 stores. Gap Inc. operates not only Gap stores but also Athleta, Old Navy, Banana Republic, and Intermix. Gap also recently announced a collaboration with Kanye West for a new clothing line called Yeezy Gap. In order to not compete with itself, Gap uses segmentation methods to identify unique target audiences for each of the brands it operates.

Our Brands

Consumers have different needs when it comes to clothing and accessories. Gap Inc. has a portfolio of offerings to meet the different needs of multiple segments.

Defining Market Segmentation

Gap's roots involve helping customers find the best-fitting pair of jeans. As Gap Inc. grew and added new stores, it had to identify and reach out to new audiences to support this growth. This growth strategy begins with market segmentation. **Market segmentation** is the process of dividing a larger market into smaller groups, or market segments, based on shared characteristics, such as an enjoyment of high-quality burgers. **Market segments** are the similar groups of consumers that result from the segmentation process.

Market segmentation

The process of dividing a larger market into smaller groups, based on meaningfully shared characteristics.

Market segmentation helps firms identify the needs and wants of customers.

Targeting

Evaluation of market segments to determine segments that present the most opportunity to maximize sales. The segments selected are the firm's target markets.

Targeting markets helps firms design specific marketing strategies.

Positioning

Consumers compare products and brands based on benefits. The features that lead to benefits desired by target markets are used to develop products and brands.

Market positioning helps firms decide how to allocate their marketing resources.

The Segmentation, Targeting, and Positioning Process

Red Wing Shoes began 110 years ago. It focused on designing sturdy shoes intended for hardy industries such as mining, logging, and farming. Workers in these fields needed durable shoes that would stand up to the elements. Although not always fashion forward, these boots are known for their quality and the heritage brand promise of Red Wing. When Red Wing was first built, the company did not try to target every person who wore shoes. Rather, it targeted an audience who needed quality, performance shoes that could stand up to the environment. Similarly, Five Guys did not try to meet the needs of every person who eats food. Instead, it developed its products and brand (*image*) in an effort to best position it to the customers in the selected target markets (everyone *who will eat high-quality burgers*). For Five Guys, this strategy has proven successful and has made it the fastest-growing restaurant chain in North America.

The Red Wing footwear brand focuses on durability against the elements.

The Importance of Market Segmentation

Market segmentation plays an important role in the success of almost every organization in the United States and throughout the world. To identify their market segments, firms start by conducting marketing research based on a market's specific needs and wants. For example, Five Guys's marketers identified that a large segment of customers were willing to pay more and wait longer for tasty, fresh, high-quality burgers than customers of other fast-food restaurants.

Thoughtful market segmentation allows Five Guys Burgers and Fries to enjoy success serving only a limited number of menu choices, which it knows will satisfy its customers.

Market segmentation is important. But what specific characteristics are used to define a market segment? For Five Guys, quality and variety preferences are important. For competitors of Five Guys, these importance preferences may differ. McDonald's may emphasize low price and consistency, whereas BurgerFi may focus on burger variety and beer preference. Because every firm is different, marketers consider numerous methods (called *bases*) to segment markets. The four main segmentation bases are demographic, geographic, psychographic, and behavioral.

Define Needs and Wants

Market segmentation helps firms define the needs and wants of the customers that are most interested in buying the firm's products. Five Guys customers typically pay more and wait longer than they would at other fast-food restaurants. The restaurant's target market values taste, freshness, and quality and is therefore willing to pay a premium for a more "gourmet" hamburger experience.

Define Specific Marketing Strategies

Market segmentation helps firms design specific marketing strategies that appeal to the characteristics of specific segments. This allows firms to increase revenues by gaining a much larger share of the market segments they target. For example, the marketing department at Five Guys understands its customers' needs and wants; it develops promotional campaigns and advertisements that focus on fresh, high-quality products. It chooses this strategy over lowering prices or offering quicker service, which are not high priorities for its targeted segment. Five Guys has capitalized on this, even using signs that urge people to seek their burgers elsewhere if they're in a hurry. Five Guys focuses on preparing the best possible burger for customers rather than simply trying to get everyone through the line.

Allocate Marketing Resources

Market segmentation helps firms decide how to allocate their marketing resources in a way that maximizes profit. By understanding the needs and wants of its market segment, Five Guys has been able to funnel resources toward more profitable markets, such as in areas where hamburgers and fries are purchased more often than not, and in areas with higher family income levels, such as northern Virginia and Dallas, Texas.

The Four Bases of Market Segmentation

Marketers use **segmentation bases,** which are characteristics of consumers that influence their buying behavior, to divide the market into segments. The four broad bases are: demographic, geographic, psychographic, and behavioral—and specific variables within each that can be used to segment the market. These bases help firms develop customer profiles that highlight the similarities within segments and the dissimilarities across segments. They are discussed in the sections that follow.

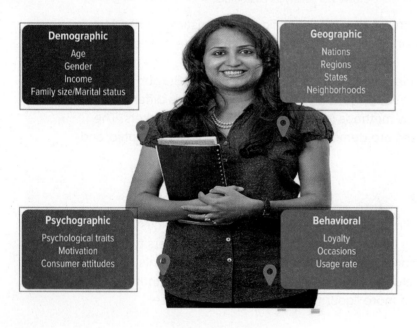

Demographic Segmentation

Companies divide markets using characteristics such as age, gender, income, education, marital status, and family size to achieve **demographic segmentation.** Age and gender are the most commonly used demographic variables because they are often the easiest to identify. However, for most firms, demographic segmentation involves more than just age or gender. Firms can find more advanced demographic details in a number of places, for example, the U.S. Census Bureau, which is a free primary source. By searching its website (**www.census.gov**), firms can find details such as the net worth and asset ownership of households segmented by race, education, age, and occupation. Research firms that package census and other demographic data in user friendly formats also compile more advanced demographic information. The relative low cost and ease of access of segmenting by demographics makes it a popular and effective option for small and large businesses.

Review the demographics of these two individuals. What does this information tell you as the owner of a business? How would you market to each?

Age: 16
Gender: Male
Income: $3,500
Occupation: Intern at a tech firm
Marital Status: Single
Race: African American

Age: 26
Gender: Female
Income: $69,500
Occupation: Journalist
Marital Status: Single
Race: Pacific Islander

Age

Age is an especially valuable segmentation tool because it is relatively easy to determine and provides a lot of information about each segment. As indicated in the following table, typical age segments include Generation Z, Millennial, Generation X, Baby Boomer, Silent, and Greatest. Age segmentation works because variations in customer preferences tend to follow age progression, leading to different needs and wants for younger, middle, and older segments. For example, research shows that older Americans spend significantly more than younger Americans in areas such as food, housing, and health care because they typically have more money and time to spend it. This is important because the population of the United States is experiencing an age bubble, providing marketers with expanding market segments of older Americans that represent billions of dollars in potential sales. Firms can use this knowledge to develop marketing strategies for products that appeal to this age group, for example, anti-aging products and healthy foods.

Businesses that segment based on age define markets according to differences within each age range. For example, McDonald's developed a strategy focused on children whose appetites have outgrown Happy Meals®. Its marketing research found that children between ages 8 and 10 had growing appetites and no longer wanted to order a "little kids meal." In response, the company introduced Mighty Kids Meals®. McDonald's marketed these meals to a slightly older segment as meals with expanded menu choices and larger portions. Thus, it was able to take an existing product and repackage it to appeal to kids of a variety of ages. This move helped drive additional revenues for the company.

The Generations Defined

	Born	Age in 2020	U.S. Population in 2020 (in millions)
Generation Z	1997 to 2012	8 to 23	67.06
The Millennial Generation	1981 to 1996	24 to 39	72.26
Generation X	1965 to 1980	40 to 55	64.95
The Baby Boom Generation	1946 to 1964	56 to 74	70.68
The Silent Generation	1928 to 1945	75 to 92	21.78
The Greatest Generation	Before 1928	92 and up	1.33

Sources: Pew Research Center, https://www.pewresearch.org/fact-tank/2019/01/17/where-millennials-end-and-generation-z-begins/; https://www.statista.com/statistics/797321/us-population-by-generation/.

The U.S. population split between females and males is approximately even, making *gender* a simple, yet valuable segmentation variable for products ranging from clothing to soft drinks to medications. For example, marketers for home improvement store Lowe's recognized that females were becoming an increasing percentage of its overall customer base through gender segmentation. Using its knowledge of this shift from its primarily male customer base, Lowe's decided to appeal to its growing female segment by marketing a new line of Martha Stewart products and other home decor items.

As perceptions, roles, and definitions of gender evolve, so do marketing strategies. Marketers are also consciously shifting their marketing dollars away from purely male- or female-oriented marketing to try to appeal to both genders. Marketers for companies such as Target, for example, have responded by reducing gender-specific signage and product suggestions in its stores. In general, marketers are trying to determine what new segments may emerge and how to market to them thoughtfully. In their efforts to recognize these new segments, they must take care to strategize with caution so as not to offend either the new or existing segments, and to ensure that they are working toward adapting to changes in gender segmentation in a positive and effective way.

Income

Income affects consumers' ability to buy goods and services and provides marketers with a valuable segmentation tool. It can be considered at various levels, including household gross income, household disposable income, or individual income. For example, the demand for wealth management and financial planning services in the United States has risen dramatically in recent decades. Firms such as Merrill Lynch pursue American households that are considered "mass affluent," that is, those that have between $100,000 and $250,000 in investable assets. Mass-affluent audiences are one of the fastest-growing segments in the United States and many need the type of financial planning that Merrill Lynch offers.

By segmenting the market in this way, Merrill Lynch can better assess what needs exist in each income-based segment and tailor its global investment advice and professional money management services to best meet those needs. Think also about luxury products such as certain brands of automobiles, handbags, jewelry, or fashion products. Companies that produce these products will most likely consider income as an important factor when engaging in segmentation.

Marital Status and Family Size

Marital status and *family size* can be helpful demographic segmentation tools. These life-stage demographics are typically good indicators of common characteristics based on life events that many people experience. For example, a company may segment based on marital status and family size to determine what markets exist for items such as car seats, diapers, and crossover vehicles. Jewelry stores can use marital status to determine the size of segments that are likely to need engagement rings versus those that need anniversary jewelry. In recent years, the size of families living under one roof has increased as more college-age students move back home and more adult couples take care of their older parents. Marketers at major U.S. homebuilders such as Lennar appeal to these larger family units by building and promoting houses that come with additional rooms or garage apartments with separate entrances.

Decisions

Review these questions and identify the demographic variable addressed.

1. McDonald's marketing research found that children between ages 8 and 10 had growing appetites and didn't want to order a "little kids meal." In response, the company introduced Mighty Kids Meals, which offer this slightly older segment an expanded menu and larger portions.

   A. Age

   B. Gender

   C. Income

   D. Family Size and Marital Status

2. Marketers for the home improvement store Lowe's recognized that women were becoming an increasingly significant part of their customer base, but that they were largely being ignored by Lowe's promotional strategies. In the effort to redirect these strategies toward women, Lowe's introduced a new line of Martha Stewart products.

   A. Age

   B. Gender

   C. Income

   D. Family Size and Marital Status

3. Merrill Lynch's strategy to focus on the mass affluents allowed the firm to tailor its global investment advice and professional money management services to best meet the needs of these targeted consumers.

A. Age

B. Gender

C. Income

D. Family Size and Marital Status

4. In recent years, the size of families living under one roof has increased as more college-age students move back home and more adult couples take care of their elderly parents. Marketers at major U.S. homebuilders such as Lennar hope to appeal to these larger family units by building and promoting houses that come with additional rooms or garage apartments with a separate entrance

A. Age

B. Gender

C. Income

D. Family Size and Marital Status

Correct Answers: 1. A; 2. B; 3. C; 4. D

Geographic Segmentation

The value consumers place on a product can vary greatly by region. For this reason, marketers may find it helpful to segment on the basis of geography. **Geographic segmentation** divides markets into groups such as nations, regions, states, and neighborhoods. Marketers pay special attention to local variations in the types of goods and services offered in different geographic regions. A Walmart located near the beach in Southern California, for instance, might sell surfboards, but a similarly sized Walmart store near the mountains of Denver, Colorado, likely will not. Walmart understands the importance of geographic differences and, based on the results of geographic segmentation, empowers local managers to stock products that are most appropriate for their local community. Variables in geographic segmentation include market size and customer convenience.

market size

The size of a market is an important geographic segmentation tool. IKEA marketers prefer to locate new stores in areas where at least 2 million people live within a 60-mile range. This type of geographic segmentation requires information about the entire market, beyond just a city or county. The U.S. Census Bureau divides cities and urbanized areas into Metropolitan Statistical Areas (MSAs), which are free-standing areas with a core urban population of at least 50,000.

The 10 Largest MSAs in the United States

Rank	Core City	Metro Area Population	Metropolitan Statistical Area	Region
1	**New York City**	19,979,477	New York-Newark-Jersey City, NY-NJ-PA	Northeast
2	**Los Angeles**	13,291,486	Los Angeles-Long Beach-Anaheim, CA	West
3	**Chicago**	9,498,716	Chicago-Naperville-Elgin, IL-IN-WI	Midwest
4	**Dallas**	7,539,711	Dallas-Fort Worth-Arlington, TX	South
5	**Houston**	6,997,384	Houston-The Woodlands-Sugar Land, TX	South
6	**Washington, D.C.**	6,249,950	Washington-Arlington-Alexandria, DC-VA-MD-WV	South
7	**Miami**	6,198,752	Miami-Fort Lauderdale-West Palm Beach, FL	South
8	**Philadelphia**	6,096,372	Philadelphia-Camden-Wilmington, PA-NJ-DE-MD	Northeast
9	**Atlanta**	5,949,951	Atlanta-Sandy Springs-Roswell, GA	South
10	**Boston**	4,875,390	Boston-Cambridge-Newton, MA-NH	Northeast

Source: U.S. Census Bureau, "Top 10 Most Populous Metropolitan Areas: 2018," https://www.census.gov/newsroom/press-releases/2019/estimates-county-metro.html#table6.

IKEA marketers generally target their stores to the largest MSAs in an attempt to reach the volume of consumers they desire. The table above shows the 10 largest MSAs in the United States and all of the cities and areas that are included in each. The figure below shows the demographic details of Washington, D.C.

Demographic Details of the Population of Washington, D.C.

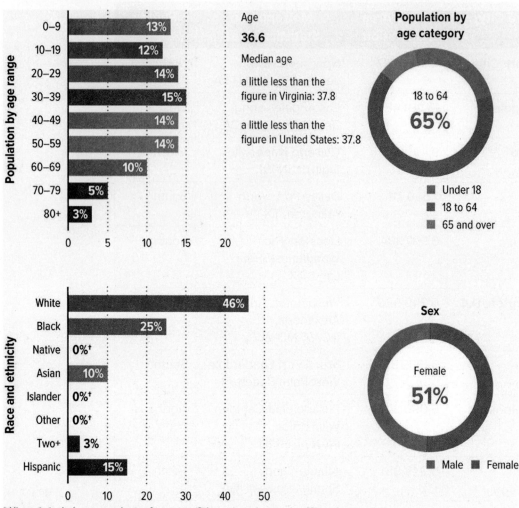

Population by age range
- 0–9: 13%
- 10–19: 12%
- 20–29: 14%
- 30–39: 15%
- 40–49: 14%
- 50–59: 14%
- 60–69: 10%
- 70–79: 5%
- 80+: 3%

(scale: 0, 5, 10, 15, 20)

Age

36.6

Median age

a little less than the figure in Virginia: 37.8

a little less than the figure in United States: 37.8

Population by age category

18 to 64
65%

- Under 18
- 18 to 64
- 65 and over

Race and ethnicity
- White: 46%
- Black: 25%
- Native: 0%†
- Asian: 10%
- Islander: 0%†
- Other: 0%†
- Two+: 3%
- Hispanic: 15%

(scale: 0, 10, 20, 30, 40, 50)

Sex

Female
51%

- Male
- Female

* Hispanic includes respondents of any race. Other categories are non-Hispanic.

Source: Census Reporter, "Washington-Arlington-Alexandria, DC-VA-MD-WV Metro Area," https://censusreporter.org/profiles/31000US47900-washington-arlington-alexandria-dc-va-md-wv-metro-area/.

Customer Convenience

Segmenting by geography also allows marketers to capitalize on *convenience to the customer.* Cracker Barrel Old Country Store® has developed a successful restaurant and retail business across the United States by locating its stores in convenient geographic locations, such as along interstate highways. This decision to focus on locational convenience is a major component of Cracker Barrel's marketing strategy. Consumers have almost come to expect a Cracker Barrel at major exits, especially in the southern United States. Travelers make up approximately one-third of the Cracker Barrel guest base, so statistics like miles driven and the percentage of people who travel mostly by automobile affect Cracker Barrel's revenue. Cracker Barrel marketers have built on the need for traveler convenience by implementing an operating platform that focuses on getting customers through the door and eating in 14 minutes for those guests in a hurry. Through geographic segmentation Cracker Barrel can identify customers in the travel segment and appeal to their specific wants and needs.

Gas stations, banks, and restaurant and retail businesses such as Cracker Barrel Old Country Store® make customer convenience based on geographic segmentation a central part of their marketing story.

Psychographic Segmentation

The science of using psychology and demographics to segment consumers is called **psychographic segmentation.** While demographic characteristics give us multiple ways to segment markets, psychographic characteristics help us understand what influences those segments' purchasing behaviors—for example, certain psychological or personality traits. Psychographic segmentation allows marketing professionals to create a more meaningful profile of market segments by focusing on how those psychological traits intersect with demographic characteristics. When marketers segment based on psychographics, they divide the market into groups according to the reason consumers purchase products. For example, consumers that purchase a new Mercedes automobile do so for a variety of reasons. One segment of consumers may buy a Mercedes for the status that a luxury car will provide them. A different segment may buy the car because of its superior safety features. Mercedes salespeople who highlight specific features of their product, based on an understanding of individual consumer motivations, increase the likelihood of a sale because the features they highlight are more likely to match the customer's specific needs and wants.

Lifestyle

Many firms, including Kraft, have successfully segmented by lifestyle in recent years. **Lifestyle segmentation** divides people into groups based on their opinions and the interests and activities they pursue. Through lifestyle segmentation, Kraft Foods was able to identify a segment of consumers that, because of busy schedules, wanted something that would make mealtime planning easier. Kraft used this lifestyle information to develop an iPhone app called "iFood Assistant," which helps those too busy to plan or make a home-cooked meal simplify mealtime preparation.

Review these examples of different lifestyles:

Fitness Lifestyle

Student Lifestyle

Green Lifestyle

VALS Framework

Perhaps the most commonly used psychographic segmentation tool is the **VALS™ framework,** which stands for Values and Lifestyles Framework. VALS classifies adults age 18 years and older into eight psychographic segments—Innovators, Thinkers, Believers, Achievers, Strivers, Experiencers, Makers, and Survivors— according to how they respond to a set of attitudinal and demographic questions. VALS measures two dimensions: primary motivation and resources. How these dimensions interact explains why different consumer groups exhibit different behaviors, and why different consumer groups can exhibit the same behaviors but for different reasons.

VALS identifies three primary motivations: ideals, achievement, and self-expression. The availability of emotional and psychological resources, such as self-confidence, as well as key demographics such as income and education, affect each group's ability to act on its primary motivation. Though segmenting by psychometric frameworks like VALS allows firms to understand consumer behavior better, psychographic segmentation can be more difficult and expensive than demographic or geographic segmentation. Thus, marketers need to carefully evaluate which type of segmentation is truly needed to successfully guide their products.

Psychographic Segments of the VALS Framework

	Ideals			Achievement		Self-Expression		
Psychographic Segments	Innovators	Thinkers	Believers	Achievers	Strivers	Experiencers	Makers	Survivors
Psychological Descriptors	Sophisticated In Charge Curious	Informed Reflective Content	Literal Loyal Moralistic	Goal-Oriented Brand-Conscious Conventional	Contemporary Imitative Style Conscious	Trend-Seeking Impulsive Variety-Seeking	Responsible Practical Self-Sufficient	Nostalgic Constrained Cautious

To see where you fit in the VALS framework, complete the VALS survey by visiting this link: http://www.strategicbusinessinsights.com/vals/presurvey.shtml.

Behavioral Segmentation

Segmentation by behavior involves categorizing customers based on what they *actually do* or *how they act* toward products. **Behavioral segmentation** variables include usage rate, loyalty, and occasions (e.g., a wedding or business trip). For example, marketers at Netflix might segment the market into groups of users and nonusers. Next, they may further segment current Netflix users based on their streaming plan (basic, standard, premium). In the case of Netflix, very loyal customers form the firm's profit core and should be treated accordingly. Many firms subscribe to the **80/20 rule,** which suggests that 20 percent of very loyal customers account for 80 percent of their total demand. If a firm can identify its loyal users, it is in a better position to create an effective marketing strategy to reach those consumers who contribute most to the firm's success. This information can also be used to identify low-use segments that may be ripe for conversion to more loyal categories.

When done well, behavioral segmentation helps marketers clearly understand the benefits sought by different consumer segments. However, behavioral segmentation is often the most difficult of the four bases to use. The marketing research required to track and understand how consumers behave with a certain product is very expensive and time-consuming. Firms must weigh the benefits of such segmentation against the costs associated with obtaining the necessary information.

Criteria for Effective Market Segmentation

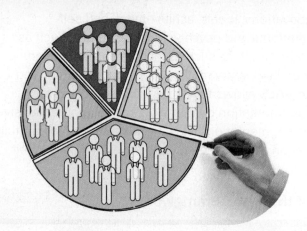

Although there are multiple ways to segment markets, none of them are guaranteed to prove helpful to marketers. Simply dividing a larger group of consumers or businesses into smaller ones serves no purpose unless doing so improves how the firm markets its goods and services. To be effective, segmentation should create market segments that rate favorably on five important criteria. How segments rate on these criteria is important because marketers use the results of market segmentation studies to choose target markets and position products. Thus, defining segments correctly has implications for future firm performance.

1. **Substantial.** The segments must be large enough for the firm to make a profit by serving them. Segments that are not substantial should not be considered as potential target markets, whereas very substantial markets should be considered as possible target markets because they hold promise of greater rewards (sales, profits, market share, etc.).

2. **Measurable.** The size and purchasing power of the segment should be clearly identified. Many successful marketers believe that if you cannot measure something you cannot manage it.

3. **Differentiable.** Dividing the market into segments does no good if all the segments respond the same to different marketing strategies. Many marketers make this mistake. For example, segmenting students for a specific marketing class by gender or age would not provide a textbook firm any value because everyone in the class would be required to buy the book. Thus, allocating resources to both genders and all ages would be an inefficient use of marketing activities and funds.

4. **Accessible.** Marketers must be able to reach and serve the segment. If the firm doesn't have the size, financial capital, expertise, or government permits to serve a certain market segment, all of the other criteria are irrelevant. Thus, just because a market segment exists doesn't mean that it is accessible to the firm that would like to serve it. This criterion should be considered early in the development phase to help guide resource allocation or withdrawal from a project.

5. **Actionable.** Marketers should be able to develop strategies that can attract certain market segments to their firms' goods and services. A firm should be reasonably certain that its marketing mix can inform consumers about the product, how it adds value to the consumer, and ultimately how to purchase it. The allocation of promotional resources should also be considered in the context of how actionable a segment is. If more than a normal amount of resources is required to motivate the segment, then the long-term viability of that segment needs to be carefully evaluated.

Once a marketer has determined that a potential market segment meets these five criteria, the marketer can move on to the next step: choosing a strategy for targeting that market.

Substantial

Athleta could decide to segment markets based on just people that lived in a certain geographical area. However, this market might not be substantial enough to support Athleta's aggressive business model.

Measurable

The management of Athleta recently added masks to their prodcut asortment. Whether this addition will result in long-term success will be determined by whether the estimates of the number of consumers willing to purchase masks from Athleta are correct.

Differentiable

Athleta knows it can differentiate its segments based on those who just quality workout and athleisure wear and those who really love the customer service and experience.

Accessible

Because of Athleta's global supply chain and infrastructure, most audiences are accessible to Athleta and can receive their products and services.

Actionable

For Athleta, adding masks to its product assortment is actionable. Masks are complementary to Athleta's core products and can be advertised in conjunction with existing marketing efforts.

Marketing Analytics: Using Data to Identify and Refine Market Segments

Defining Market Segments

Different consumers can purchase the same product for different reasons, that is, the perceived value of that product changes. Some consumers may purchase a cup of coffee to warm their hands, whereas others may buy it for the caffeine. Still others may purchase a cup of coffee as a force of habit, as part of a social activity with friends, or simply because they enjoy the taste of it. All of these consumers are coffee drinkers, but the *value* the coffee provides to each group is *different*. Understanding this difference allows marketers to tailor their marketing mix to each consumer group. To better understand these differences, marketers can research what motivates consumer purchasing in a number of ways, including via a combination of surveys and in-store observation. Some of these motivations are based on physical needs, psychological needs, or social needs and wants.

Using External Data

Firms can purchase or utilize external data and combine it with their own internal data to refine their segmentation strategies. The most common forms of external data that marketers collect are demographic and psychographic.

Demographic data are the easiest to collect. An important source of demographic data is the government. Every 10 years in the United States the federal government conducts a census. The data gathered from this census are public and are a treasure trove for marketers. Let's consider a fictional firm for a minute—one that creates skincare products for young mothers. The firm's internal data show that the customer base is primarily women between the ages of 23 and 30 who have children younger than age 5. The firm can supplement this internal data with external census data to identify where this demographic group lives. It can then use these data to identify potential new store locations or potential new customers to whom it can send marketing materials.

Psychographic data provide valuable information for marketers as well; if marketers want to understand what influences a consumer's purchasing behavior, for example, lifestyle or personality, then marketers turn to psychographic data, often provided by such syndicated data providers as Nielsen through Claritas, or other data companies. Nielsen's primary mission is to connect clients to audiences "to fuel the media industry with the most accurate understanding of what people listen to and watch." Marketers listening to Nielsen's Database podcast series can learn such information as which devices are growing in popularity and how Millennials are driving that trend. For a company like our skincare product firm catering specifically to Millennial mothers, this information can be especially valuable, particularly in terms of how it decides to promote its product or expand its market segment. By understanding not just who these consumers are but also their purchasing motivations, marketers can tailor their marketing mix to best meet the needs of the consumers.

Determining Segment Appropriateness

Just because a segment exists does not mean the firm should pursue it. Often segments are not going to be profitable for the firm for a variety of reasons. Thus the identification of the segments within a market is simply a first step. Data then must be collected from that segment and evaluated to determine whether or not that segment is suitable for further development as a target market.

Let's consider again the firm that creates skincare products for young mothers. One of the products this firm offers is an organic face wash made from all natural ingredients with no additives or preservatives. The firm notices that the census data show a heavy concentration of young mothers living in an area the firm has not yet explored. The demographic variables match those identified by the firm's internal data—between ages 23 and 30 with children all younger than age 5. But before the firm decides to promote its product line in this area, it needs more data to show whether there is even a demand or a need for organic face wash among this unexplored segment. Perhaps there is, but further data show that the market need for this product is already satisfied by a competing brand that offers the same product at lesser costs. Or, perhaps there isn't because further data show this segment uses generic brands and the need or desire for an organic face wash is just not there. These additional data can save the company a lot of unnecessary time, effort, and money.

Responsible Marketing

Segmentation: The Perils of Reaching the Wrong Audiences

As you learned in the lesson, market segmentation is the process of dividing a larger market into smaller groups based on meaningfully shared characteristics. Companies like Vans might target customers who have a shared passion for skateboarding, while companies like Patagonia might target customers who value conservation. Segmentation is a critical part of understanding the needs and wants of a firm's customers. However, sometimes segmentation can backfire.

Firms often use data to drive their marketing campaigns to target and reach specific audiences. However, when those data are interpreted only at face value, they can be misleading. The Internet abounds with examples of firms that send "congratulatory" messages to people who have been identified as new moms or moms-to-be. Often these e-mails or other promotional messages are sent because a customer has engaged in some kind of behavior (e.g., purchased formula or diapers) that triggered a firm's system to identify them as a new mom or mom-to-be. However, this can be devastating if the recipient of this promotional message is not actually a new mom or a mom-to-be. For example, someone who desperately wants a baby may be identified as a mom-to-be while not actually expecting. This could be very upsetting to customers who receive this kind of message.

While using segmentation effectively is critical to business, it is equally important that segments be identified correctly. Furthermore, firms must be intentional in how they communicate with customers and not make assumptions that someone identifies a certain way based on their previous purchase behaviors or attitudes.

Reflection Questions

1. Can you think of other examples where segmentation and targeting have gone wrong?

2. How do you feel when someone makes assumptions about you based on external variables like your age or gender or what you have purchased in the past?

Lesson 5-2
Targeting

This lesson focuses on targeting (or target marketing), the second step of the segmentation, targeting, and positioning process. Targeting occurs after the segmentation process and is the point at which a firm defines which segments will be the focus of the final step of the process, positioning.

As the Gap Inc. example illustrates, a firm doesn't need to serve every possible customer segment; instead, it should concentrate on specific target markets that will respond favorably to its products and services.

By the end of this lesson you will be able to

- Define target marketing and its importance in establishing a firm's target marketing strategies.
- Describe how a firm chooses which consumer group(s) to pursue with its marketing efforts.
- Differentiate among the key strategies for targeting a market.
- Explain how marketers use data to identify target market segments and develop a targeting strategy.

Marketing Analytics Implications

- Not all segments are chosen to be target markets. Only segments that data reveal will be responsive to the firm's marketing efforts are chosen as target markets.
- Target market segments are analyzed continuously to ensure that the segment remains viable in terms of size, competitive intensity, and overall fit with the firm's strategic mission.
- In addition to selecting and evaluating the target market, a firm must use market data to decide how many target markets to serve.

Gap Inc.: Creating Value for Different Target Markets

The days of one-size-fits-all mass marketing are over. Companies that try to appeal to everyone are no longer successful because they don't use their resources efficiently and don't engage in a target marketing strategy.

Look at Gap Inc.'s progress since it started in 1969. Initially, Gap offered only jeans. It began with a target market who was interested in finding affordable jeans that fit right. Eventually, Gap began to grow its product offerings to encourage its customers to buy more. Over time, Gap acquired or created new brands to reach new audiences. Banana Republic, for example, is meant to be a lifestyle brand that appeals to fashion-forward audiences for their work wear or travel needs. It has successfully collaborated with shows like *Mad Men* to further enhance its lifestyle image. Old Navy attempts to provide affordable fashion for the whole family. It always has a sale and its ads feature active families. Intermix is Gap's most expensive brand and is targeted toward fashion enthusiasts. It curates an interesting assortment of established and up-and-coming designers and its stores feel fashion-forward and trendy. Athletes targets women and girls to help them achieve their fitness and health aspirations. It features models of all shapes and sizes to promote body positivity.

Each of Gap Inc.'s brands uses a different strategy to create value for its primary target market. Understanding the distinctive needs of each market allows the brands to respond and adapt to changing trends. Rather than trying to meet the needs of all consumers with one general brand, Gap Inc. offers a variety of products to meet the various needs of its unique markets.

Banana Republic sends a targeted message to its audience seeking fashion for business and travel.

Defining Target Marketing

While market segmentation provides a good first step toward reaching potential consumers, the next step is for firms to review the segments to determine which ones to target. **Targeting,** or **target marketing,** occurs when marketers evaluate each market segment and determine which segments present the most attractive opportunity to maximize sales. The segments selected are the firm's target markets. A **target market** is the group of customers toward which an organization has decided to direct its marketing efforts.

Targeting

Evaluation of market segments to determine segments that present the most opportunity to maximize sales. The segments selected are the firm's target markets.

Targeting markets helps firms design specific marketing strategies.

Market segmentation

The process of dividing a larger market into smaller groups, based on meaningfully shared characteristics.

Market segmentation helps firms identify the needs and wants of customers.

Positioning

Consumers compare products and brands based on benefits. The features that lead to benefits desired by target markets are used to develop products and brands.

Market positioning helps firms decide how to allocate their marketing resources.

The Segmentation, Targeting, and Positioning Process

Firms rely on their knowledge of a target market's needs and wants to determine a target marketing strategy. Firms then decide whether they should market their products and brand the same way across *all* target markets, or whether they should apply a more specialized approach. Coca-Cola, for instance, deliberately targets college students as a way to maintain the firm's future dominance. But Coca-Cola's marketers cannot target this consumer group with a single strategy. The college market comprises multiple segments, including college athletes, sports fans, gamers, and international students; in order to reach each of these segments, Coca-Cola's marketers have to identify each target market's set of needs and wants and how they differ. The next section discusses the three important criteria to consider when targeting a consumer group.

Targeting Consumer Groups

Selecting target markets is the key to most firms' success. But simply identifying segments and then selecting some to target is no guarantee that target marketing will be successful. Firms need to be sure that they are targeting the correct segments. To make sure the correct markets are targeted, firms use three important criteria during the target selection process:

1. **Growth potential.** Typically, the higher the future growth rate, the more attractive is the segment.

2. **Level of competition.** The more intense the competition within a segment, the less attractive the segment is to marketers. Competitors will fight extremely hard to prevent market share loss, and the potential for price wars can negatively impact success. Generally, more competitors means a firm has to work harder and invest more in promotion to earn business and increase market share. When considering two market segments in which other factors, such as size and growth potential, are constant, the one with a less competitive environment is more attractive.

3. **Strategic fit.** Marketers should work to ensure that the selected target markets fit with what the organization is and wants to be as defined in its mission statement. The SWOT analysis, which analyzes a firm's strengths (S), weaknesses (W), opportunities (O), and threats (T), provides an excellent framework for determining whether a firm will be successful at targeting a specific segment.

Growth Potential

In terms of **growth potential,** AT&T's existing residential landline business is still a multibillion-dollar market. However, revenue and profits in the business have declined every year for the past decade as more and more consumers use their cell phones exclusively and get rid of their home phones. Although a marketer looks at the residential phone business and sees large sales potential, the growth potential is very limited, making the segment far less attractive.

Level of Competition

In terms of **level of competition,** a new pizza shop might want to be aware of the risks of existing in a market where the number of restaurants offering all-you-can-eat pizza buffets and delivery service is high. The increased level of competition will mean the pizza shop will need to work harder and invest more in promotions to reach consumers in that community and to profit.

Strategic Fit

In terms of **strategic fit,** McDonald's introduced an Angus burger to its menu in 2009 as a premium choice at its restaurants. The Angus burger was one of the priciest menu items and did not fit well with the company's typical dollar-menu consumers. McDonald's target market did not want to pay four or five times more for a hamburger, even though the product was superior. McDonald's eventually removed the Angus burger from its menu in 2013.

Target Marketing Strategies

The three basic strategies for targeting markets include:

- undifferentiated targeting,
- differentiated targeting, and
- niche marketing.

Each of these is described in the sections that follow, but for a visual depiction of each of these targeting strategies, see the figures below.

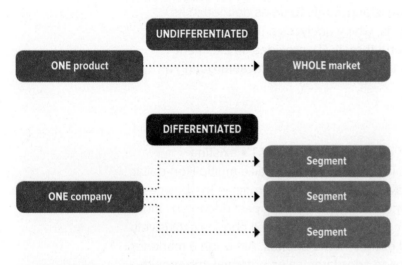

Undifferentiated versus Differentiated Targeting

Undifferentiated Targeting

An **undifferentiated targeting** strategy approaches the marketplace as one large segment. Because the firm doesn't segment the market further, it can approach all consumers with the same product offering and marketing mix. Undifferentiated targeting works best with uniform products, such as salt or bananas, for which the firm can develop a single marketing mix that satisfies the needs and wants of most customers. Only a limited number of products fall into this category; the majority of products satisfy very different needs and wants for different consumers.

Differentiated Targeting

Typically, firms can provide increased levels of satisfaction and generate more sales using a differentiated targeting strategy rather than an undifferentiated targeting strategy. **Differentiated targeting** occurs when an organization simultaneously pursues several different market segments, usually with a different marketing mix strategy for each. General Motors (GM) provides a classic example of successful differentiated targeting. Nearly a century ago, GM segmented consumers by the price they could afford and the quality they desired, then customized its products, messages, and promotions to the unique needs of each group. This practice was the beginning of the GM family—from Chevrolet to Buick to Cadillac.

Firms that market the same product to multiple regions with different preferences often use differentiated target marketing. In such cases, the product may be tweaked to ensure that it meets the unique needs of each segment. For instance, some regions may prefer food flavors that won't sell in other areas. Frito-Lay offers more than 10 regional flavors, including Wavy Au Gratin in the Midwest and a ketchup-flavored chip that has gained a following in Buffalo, New York, but was originally targeted to the Canadian market. Although the core product remains the same, select flavors may be distributed to certain regions based on local preferences.

L'Oréal is another example of a firm using a differentiated strategy. L'Oréal has a portfolio of brands that range from Maybelline (targeting younger customers just starting to use makeup) to Yves Saint Laurent makeup (targeting older customers willing to purchase cosmetics at a higher price point). The wide portfolio of brands at L'Oréal highlights its commitment to targeting a variety of segments with different products, different price points, and different distribution strategies.

A firm marketing to a region outside the United States may also need to modify products according to local government regulations and cultural preferences. For example, Walmart initially offended Chinese consumers by selling fish and meat wrapped in Styrofoam and cellophane—the same way it sells these items in the United States. But because Chinese consumers value fresh food, they viewed the prewrapped products as old merchandise. Marketers responded quickly to this problem and differentiated their Chinese offerings by installing fish tanks and selling unwrapped meat products. These moves, in combination with other differentiating strategies like elaborate cosmetics counters just inside the front door, have helped Walmart resonate with its target market in China.

Niche Marketing

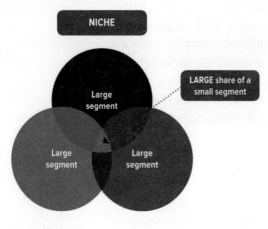

Niche Targeting

Consumers of niche marketing products typically have very specialized needs and will pay higher prices to meet those needs. **Niche marketing** involves targeting a large share of a small market segment. Niche product firms possess a unique offering or specialization that is desirable to their targeted customers. **Ties.com** is a successful Internet-based niche retail company. The business, which shares its name with its website, has been in operation since 1998. It focuses exclusively on men's neckties and related products. Fashion retailer Kathy Marrou founded the company, replacing her general clothing retail operation with one focused only on ties. The idea has been successful and the company has now added scarves to its lineup to target a niche market of female consumers who are passionate about neckwear. Whatever market niche a firm targets must still be a segment that is substantial enough to be successful. If **Ties.com** sold neckwear to only 10 people who liked exotic ties, it would be virtually impossible for the company to turn a profit and sustain itself as a business. The market segment for neckwear is substantial enough that **Ties.com** was able to generate $2.4 million in annual revenue in 2012, a 578 percent increase over just three years before.

There can be multiple niche markets within the same product category. Pizza sales in the United States reached $32 billion annually in recent years, led by firms such as Pizza Hut, Domino's, and Papa John's. However, a number of pizza providers have found success targeting specific niche segments of the pizza market. Pizza Fusion targets consumers looking for a healthier and environmentally friendly alternative to the large pizza chains. Papa Murphy's leads the take-and-bake pizza niche, and Little Caesars's $5 Hot-N-Ready Pizza targets customers looking for value and time savings.

New small firms often turn to niche marketing. The benefit of niche marketing is that often a smaller firm can identify a target market that is big enough to sustain the firm, yet small enough to be ignored by larger competitors. Initially, Gap Inc. may have considered a niche marketing strategy. It started in one location and focused on a small segment that valued good service and a broad assortment of jeans. However, as business grew, Gap Inc. moved from its original niche strategy to the more effective differentiated targeting strategy.

Once marketers have chosen one or more target markets and determined a strategic approach to address those markets, they can then start positioning their products and brand. This next step is known as market positioning.

Marketing Analytics: Using Data Analysis for Target Marketing

Targeting a Responsive Market

Targeting occurs when the firm has fully evaluated all potential segments and then has chosen the segments around which it will adjust its marketing mix. To ensure that marketers are targeting the segment that will respond to their marketing efforts, they evaluate demographic, psychographic, and behavioral data, which give marketers a full picture of who their target consumers are, as well as what they need and want based on *why* and *how* they buy. Marketers then reconfigure their marketing mix elements to best deliver value to this target market. For example, sports fans are an interesting segment for marketers. Fans will display their allegiance to a team by purchasing team-related items such as jerseys, sweatshirts, posters, and so on. For years the marketers of team regalia did not produce these products for women believing that women were not in the audience for sports-related products. Data in fact proved this assumption to be wrong for all sports but especially for football. In fact, 50 percent of the audience for football is women. Further analysis revealed that women were actively involved with their preferred team and were highly interested in products and services related to their favorite teams. The NFL finally realized how crucial this segment of the market was to them and began producing a women's line of jerseys, sweatshirts, caps, and so forth. The NFL also began licensing team logos to jewelry companies that began producing such products as bracelets, rings, and earrings. The NFL partnership with breast cancer awareness organizations has also boosted its relationship with its female fans.

Evaluating Internal Resources to Determine Target Market

In addition to studying the segment the firm wants to target, it must also do a complete analysis of its resources. It may be that some segments are just too costly to serve and do not have sufficient growth potential to merit inclusion. For example, if a segment requires intensive sales support but the firm has a very low budget for a sales force, then this segment will not be appropriate. The needs of the segment must be matched with the resources of the firm. Also, the market itself must be studied to ensure that sufficient information is available to forecast future demand. These data can come from both the firm's internal data as well as from external data provided by sources that measure total market size and market growth. Market share and the cost of growing market share are also considerations.

Determining Targeting Strategy

A firm must decide how many segments to target. *Undifferentiated targeting* is the easiest strategy because it does not require the firm to track different target markets. This can be risky, however. A firm often pursues multiple target markets simultaneously, which requires that it track each segment. The more segments it tracks the more data it gathers. The trap some firms fall into is thinking that two groups that *appear* different are truly different in their consumer behavior. Demographic differences do not always translate to psychographic differences or behavioral differences. Segments have to be continuously monitored to ensure that they are truly unique and require separate marketing programs. The other truism is just because there is a very large segment does not mean it will be more profitable than a smaller, more defined segment. For example, certain demographic characteristics mandate adaptations to the marketing program, thus creating a smaller niche market for the product—but this smaller niche market can be highly profitable. Rogaine® is an over-the-counter treatment for hair loss. For men, the problem of thinning hair is one that is often the subject of jokes and is not viewed as a serious health problem. In fact, many view it as a rite of passage, just something that happens as men grow older. For women, thinning hair is not as common nor is it ever the subject of jokes among women. Thinning hair is a serious issue for women and often is accompanied by deep emotional turmoil. Rogaine® understood that to market to women it had to dramatically change its marketing strategy. For women it had to demonstrate a sensitivity to the issue of thinning hair that it did not with men. By understanding that different strategies for men and women were needed, Rogaine® was able to meet the needs and wants of both target markets.

Responsible Marketing

What Do You Do When the Target Market Is Children?

This lesson discusses how firms should decide which market segments to target. One of the main criteria for a successful target market is if that market is substantial, which is related to the size of the market. According to the 2020 U.S. Census (census.gov), there are approximately 48 million children under the age of 12 in the United States currently. Clearly, this is a substantial market. However, firms have to consider how responsible marketing to children is.

In the late 1980s and early 1990s, Camel cigarettes used a cartoon Camel as its logo. Eventually, the company got in trouble when people realized that the Camel cartoon was almost as recognizable as Mickey Mouse to children. Obviously, instances of marketing harmful products like tobacco and alcohol, to children, are pretty cut and dry in terms of ethical violations.

But what about products like sugary cereals, unhealthy kids meals, or toys? For many products, a child is the intended user, so should firms market to them? What about when a product is potentially unhealthy, like a high-sugar cereal? Would it be more appropriate to market those products to parents?

Reflection Questions

1. If you were a marketing executive for a children's apparel brand, what would you marketing strategy be and why?

2. What are some other potential ethical issues related to targeting children?

Lesson 5-3
Positioning

Gap Inc.: Positioning Its Brands to Speak to Its Markets

Positioning relates to how consumers see a company's product or service in relation to competitors in the marketplace and is important to all firms. To signal to customers the benefits of a company's product versus the competitors', firms may focus their positioning strategies on certain product attributes, features, price points, value perceptions, or other brand dimensions. For example, Gap Inc. positions Banana Republic as a more fashion-forward retailer where consumers might shop for fashionable professional clothing. Conversely, the company positions Old Navy as a store where consumers can access affordable fashion for the entire family. Positioning allows a firm to speak to its target market directly and to highlight how the firm is different (and hopefully better) from its competitors.

Old Navy positions itself as affordable family fashion. This positioning strategy resonates with its intended target audience.

This lesson focuses on positioning and repositioning, which occur after the segmentation and targeting processes.

By the end of this lesson you will be able to

- Define market positioning and its importance in helping firms allocate resources.
- Outline the three steps to effective market positioning.
- Explain the reason for positioning and repositioning products.

- Explain how the marketing mix relates to repositioning.
- Explain how the competition relates to repositioning.
- Explain how marketers use data to define positioning strategies.

Marketing Analytics Implications

- Once a firm has segmented its market and chosen its target market, the firm then uses data to develop a positioning strategy to attract its target consumers to its product, service, or brand.

- Firms have scarce resources, so it is imperative that resources are allocated in the most effective and efficient means possible. Data must be used to develop a full understanding of the cost of the chosen positioning strategy. Remember that positioning may require an adjustment to any of the elements of the marketing mix—this is known as *repositioning,* and repositioning each target market renders a cost.

- Positioning is not stagnant, and a firm must respond to its competitors' positioning in the market. Market scanning and consumer tracking can assist the firm in maintaining a relevant positioning strategy.

Defining Market Positioning

After a firm's marketers have segmented its market and selected its target consumer groups, the final step is to position the firm's products and brand. **Positioning** is a company's efforts to influence the customer's perception about its products, services, and ideas. These efforts assume that consumers compare the benefits of competing products and brands against each other. Similarly, Sprint recently began using the former spokesperson from Verizon (the "Can you hear me now?" guy) to position itself as having the same coverage as Verizon, but at lower prices. When the marketplace starts to change, firms need to reestablish a product's position through *repositioning,* which is discussed in greater detail later in this lesson.

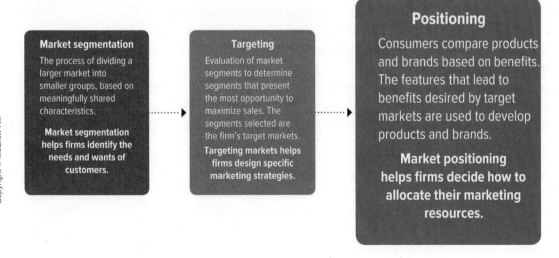

Market segmentation

The process of dividing a larger market into smaller groups, based on meaningfully shared characteristics.

Market segmentation helps firms identify the needs and wants of customers.

Targeting

Evaluation of market segments to determine segments that present the most opportunity to maximize sales. The segments selected are the firm's target markets.

Targeting markets helps firms design specific marketing strategies.

Positioning

Consumers compare products and brands based on benefits. The features that lead to benefits desired by target markets are used to develop products and brands.

Market positioning helps firms decide how to allocate their marketing resources.

The Segmentation, Targeting, and Positioning Process

Positioning helps firms decide how to allocate their marketing resources based on their positioning goals. Firms determine how to implement their strategy to avoid inefficiencies and increase profits. They often serve more than one market, each of which is different in some aspect. Positioning based on the expected potential of a target market allows firms to determine where marketing budgets should be spent. For example, Gap Inc. has been able to funnel promotional and capital resources toward establishing locations in other profitable markets. This positioning allows Gap to use its marketing resources to place its products and brand in areas where more frequent users reside, which leads to better performance in the marketplace.

The Key Steps to Effective Market Positioning

Success within the target market depends, to some degree, on how the firm positions its products. Positioning often takes into account the identity of the organization and where it fits relative to the competition. Successful positioning involves all of the marketing mix elements (product, price, place, and promotion). Marketers should follow three key steps to decide how to best position their product:

- Step 1: Analyze competitors' positions.
- Step 2: Highlight competitive advantage.
- Step 3: Evaluate consumer feedback and understand its importance.

Step 1: Analyze Competitors' Positions

As a first step, firms must understand the position other competitors have taken in the marketplace. Positioning does not occur in isolation, and it is important for marketers to have a realistic view of how customers perceive competitive offerings. Competitive analysis becomes even more important when competitors all appear to offer a similar good or service. Financial institutions such as banks and credit unions face this challenge; services such as free checking and online banking are pretty much the same regardless of the type of bank. Bank of America has tried to overcome this obstacle by using advertising to emphasize its thousands of ATMs around the world and the fact that customers can scan and deposit a check with their phones via an app. In contrast, credit unions promote customer-friendly service—including lower fees and interest rates—and highlight that they are not designed to make a profit. By explaining to potential members that a credit union is not designed to extract profit from them, credit unions may be able to position themselves as more customer focused and convince individuals to switch from for-profit banks.

A **perceptual map** provides a valuable tool for understanding competitors' positions in the marketplace. It creates a visual picture of where products are located in consumers' minds. Marketers can develop a perceptual map based on marketing research or from their knowledge about a specific market. Following is a hypothetical perceptual map illustrating how Five Guys might segment and position itself against competitors within the quick-service restaurant industry.

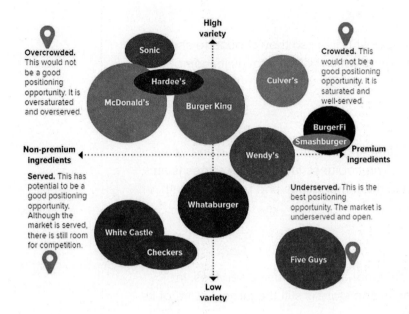

Perceptual maps provide guidance on potential market positions that might be unserved or underserved. As the perceptual map on the quick-service restaurant industry reveals, there is an underserved market segment where Five Guys may want to consider positioning itself (lower-right quadrant). Before deciding whether it should position itself to attract this segment, Five Guys would need to consider several questions, including, "How would increasing variety affect product quality, current target customers, and new competitors?" Recall that Five Guys takes a cautious approach through its limited menu. The perceptual map indicates that it may want to consider ways to attract this underserved segment, but if it chooses to try, it must ensure that its plans do not damage its existing business.

Step 2: Highlight Competitive Advantage

Great marketers understand competitive advantage and why consumers buy their firms' goods and services. There are a number of positioning strategies available to highlight a firm's competitive advantage, including the following:

1. **Price/quality relationship.** Walmart is a great example of a low-cost retailer that has successfully positioned itself using a price/quality relationship strategy. Because of its thousands of stores globally, it can negotiate bulk discounts from wholesalers and keep its distribution costs low. These cost savings translate into low selling prices for quality goods with which rivals cannot compete.

2. **Attributes.** Often a product will have multiple attributes that create a unique position in the market. Marketers should evaluate those attributes that put their product in a special category of value to the customer. Successful attributes might include leadership, heritage, product manufacturing process, or the coolness factor.

3. **Application.** Apple has had success with its iPad product; this is in part because Apple has the competitive advantage relative to applications and product use. As new versions of the iPad are introduced, Apple continues to offer such exclusive services as FaceTime and iCloud to further differentiate the iPad from other competing tablet brands; this, in turn, furthers the product's success.

Walmart's Positioning Strategy

Shoppers know that Walmart offers the lowest prices, so they choose to shop there. This in turn enhances Walmart's brand, which only improves its competitive advantage.

Nike's Positioning Strategy

Nike continued to have sales success marketing the Air Jordan shoe line years after Michael Jordan retired from sports. The coolness factor of the shoes is an attribute that creates a unique position in the marketplace and resonates with consumers.

Apple's Positioning Strategy

With each new version of the iPad, Apple emphasizes exclusive services. In addition, Apple reminds customers that its app store is still the largest store of its type with the highest security standards.

Firms can choose to promote one or multiple competitive advantages, as long as they can clearly articulate those advantages to their target market.

Step 3: Evaluate Consumer Feedback and Understand Its Importance

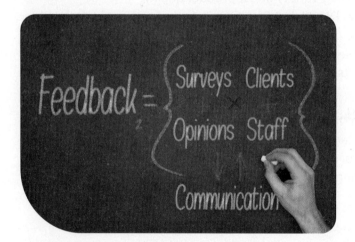

The third step in market positioning is constantly evaluating consumer feedback. Just as fashion styles change, consumer tastes for almost everything, including music, food, and even educational learning formats, change. For example, as more students began to work while attending college, universities began offering a larger number of night classes than they had before. In recent years, to accommodate fluctuating student work schedules, high gas prices, and a weak economy, universities began offering more online courses. Universities continue to position themselves as providers of quality higher education programs even as they have shifted their product to accommodate the additional features and conveniences their target market requires.

Disney provides another example. Disney positioned itself as a partner to parents concerned about childhood health, after survey feedback revealed that this was an issue on the forefront of consumers' minds. The company had announced that it would require food and beverage products advertised on its networks to meet specific nutritional standards. In 2012, it began using a "Mickey Check" seal of approval to indicate that foods sold in its stores, theme parks, and resorts met its nutritional standards. The stated objective of these programs is "to combat childhood obesity." Listening to customer feedback on this sensitive issue has allowed Disney to position itself as a partner for parents in ensuring healthy lives for children.

Repositioning

In this digital age, marketers have to change and adapt their strategies if they want to continue to reach their target markets. **Repositioning** involves reestablishing a product's position to respond to changes in the marketplace. Dr Pepper Snapple Group, the Sunkist brand's licensee for soda in the United States, initiated a repositioning strategy aimed at trend-savvy teens and young adults. The soda brand now utilizes YouTube and Facebook platforms to promote its products.

In marketing, one constant is that customers are always changing. Needs and wants shift over time and consumers' perceptions of market offerings will also shift, especially as new products and competitors come onto the scene and lifestyles change. In extreme cases, firms may decide not to reposition, but instead abandon an existing product to develop a new one. In most cases, however, as customer segments change, firms will reposition themselves after considering their own marketing mix and their competitors' positions.

Positioning and Repositioning Products

Let's review again the previous hypothetical perceptual map (Perceptual Map A) and compare it to this new one (Perceptual Map B), which reflects the hypothetical quick-service restaurant market one year later.

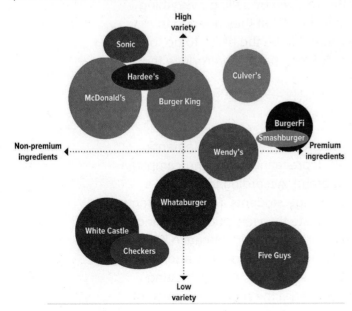

Perceptual Map A

Five Guys has found an underserved segment of the market and is positioning itself as the restaurant that can offer this market a low variety of premium products made from premium ingredients (lower-right quadrant).

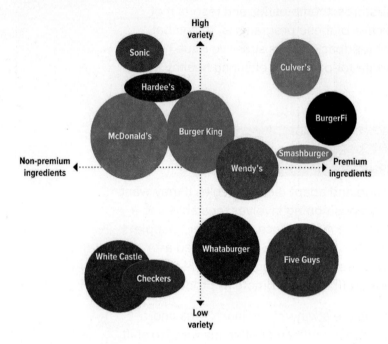

Perceptual Map B

One year later, we see how all the restaurants, except Five Guys, have slightly repositioned themselves. They all seem to be moving toward wanting to offer more premium products (lower-right quadrant). Five Guys finds itself contending with a close competitor, Whataburger. What should Five Guys do at this point?

Positioning is not just a one-time event. In fact, most firms reposition their products on a regular basis. Repositioning is necessary because customer needs and wants are always changing. Competition also changes. Consider Hardee's and Carl's Jr. At one point, these brands tried to compete with everyone by selling everything from fried chicken to roast beef. Eventually, they realized that they needed to change to differentiate themselves. They repositioned to be the place to go for a premier hamburger and were one of the first fast-food chains to introduce the Angus Burger.

Repositioning and the Marketing Mix

Repositioning typically involves changing one or more marketing mix elements, often product or promotion. Domino's Pizza's self-deprecating marketing campaign serves as a great example of how to reposition a brand. According to Domino's internal research, its biggest customer segment, college students, reject the brand as they get older and care more about quality and taste than low prices. Domino's realized this attrition offered a sizable opportunity for growth and, because it has been transparent in its repositioning campaign, its story has caught the attention of lots of people. One key reason for the campaign's success is the advertising footage highlighting consumers' negative feelings toward the quality and taste of Domino's pizza. By acknowledging this part of the story and understanding the attitudes and opinions of its target audience, Domino's message resonated with consumers. Current and past customers bought into the first part of the story in which the company admitted its problems; because of this, Domino's was given permission to tell the rest of the story about how it has changed and improved.

Repositioning and the Competition

Sometimes marketers choose to reposition the competition rather than change their own position. For example, Apple had success repositioning its PC competitors such as Dell and Hewlett-Packard using the popular "I'm a Mac, I'm a PC" advertisements that compared the capabilities and attributes of Macs and PCs. The promotion succeeded in characterizing the PC as formal and stuffy, while portraying Apple's Mac as the relaxed, easy-to-use alternative. This type of repositioning strategy is also common in political marketing as strategists look to define their opponent negatively without changing their own candidate's positions. Check out any major election cycle in your state or region and you are likely to find plenty of examples of this type of repositioning strategy.

Tying It All Together: Segmentation, Targeting, and Positioning

To conclude this lesson on positioning and repositioning, it is important to remember the overall process of segmentation first, targeting second, and positioning third. As simple as that sounds, most marketing occurs in markets that change rapidly and that are filled with savvy competitors motivated to gain market share. Thus, segmenting customers and positioning to target markets often unfolds in a different order. However, as a firm grows and matures, the goal should be to develop a regular process for evaluating these elements and to adjust as needed. Those firms don't often fail. At the time of this writing, Sears, a century-old company, is failing because it failed to keep up with its markets. There is a lesson in its failure, which is "to stay in business, keep an eye on your segments, targets, and positions!"

Marketing Analytics: Using Data to Define Positioning Strategies

Identifying Positioning Strategy

Because positioning is determined by the needs and wants of the chosen target market, a firm needs relevant and timely data from the target market to determine the most effective positioning strategy. A firm will often incentivize target market consumers to join loyalty programs or download apps for the firm's products or services to better track target market consumer behavior. A firm will also survey target market consumers to get immediate feedback. All of this information taken together will show the firm what its target consumers need or want. It will also inform the firm of how receptive the target market is to its marketing campaigns. This feedback from the target market is crucial to establishing and maintaining a competitive position in the market.

Using Resources Effectively

Because the firm is adapting the marketing mix to meet the needs of the target market, it is imperative that the firm understands how target market consumers view the firm's offerings vis-à-vis those of the competitors. Perceptual maps are drawn from marketing research data. Surveys of the target market consumers are conducted asking about the firm's products and services as well as those of the competitors. Perceptual map studies also look at the relevant dimensions or attributes upon which consumers make their purchase decisions. For example, for a car these could be price, style, comfort, and safety. Each of the competitors is ranked based on these attributes and then the map is drawn. The X and Y axes often represent the two most important attributes to consumers, but the marketer does have control over what attributes are used by choosing which data to model. For example, price is often one of the two most important attributes in consumer choice, but if the marketer does not want to model price then he or she can simply not include that variable in the analysis.

Understanding That Positioning Is Not Stable

Because positioning exists in the minds of the consumers, it is not stable. In other words, as competitors change their offerings and marketing programs, the consumers' view of the firm's products will also shift. Thus, it is imperative that through the firm's market scanning process, competitors' repositioning efforts are evaluated for their impact on the firm's positioning strategy—the data a firm gathers will help it decide how to respond. Occasionally, a firm may respond to demographic changes, or other environmental changes, by adjusting its positioning strategy (*reposition*) for certain products or services. Customer feedback through surveys as well as other data collection methods is essential to these efforts. For example, beer brands are often positioned to appeal to a certain demographic. But tastes and preferences change as people age out of certain behaviors. If you have positioned your product to the hardworking man who enjoys a beer after work and leave that positioning strategy for 10 or more years, then you will have an issue such as Budweiser did. The target market simply aged out, work life changed, and the positioning strategy no longer worked. Budweiser marketers understood that to survive they would have to reposition the product to appeal to a younger audience. The Budweiser Frog campaign was the result of these efforts. This campaign was designed to reposition Budweiser as a fun beer for a younger, cooler segment. After the launch of the campaign, sales of Budweiser improved among its younger audience.

Responsible Marketing

Will This Product Actually Change Your Life?

Positioning requires firms to consider how consumers view their product or service in relation to competitors in the marketplace. For example, some brands might position their products to be higher-quality than their competitors, while others might position their products to be lower-priced.

In a noisy competitive landscape, brands have to use language and images to help them stand out from their competitors. But sometimes these images and words can go too far, and the positioning becomes extreme.

Consider the supplement industry. According to some reports, dietary supplements are worth over $50 billion in annual sales in the United States alone.

- Some weight loss supplements show before and after pictures, with the after pictures showing extreme weight loss and muscle gains. This is usually accompanied by a tan as well. The implications are that the weight loss supplement caused these results.

- Some hair products show models with voluminous hair, promising similar results to the consumer.

- Some supplements position themselves as helping with depression and use testimonials of people who became "happy" after using them.

While the FTC (Federal Trade Commission) does mandate that these products put a small disclaimer, something along the lines of "results may vary," the images and language still position the products as improving consumers physical appearance or mental health. Unfortunately, with little FDA (U.S. Food & Drug Administration) oversight, these supplements are operating in a grey area.

Reflection Questions

1. If you were a marketing executive for a weight loss supplement, how would you position your product? How do you balance ethics with sales goals?

2. How much more oversight should there be in regulating brands' positioning strategies?

Segmentation, Targeting, and Positioning: Test

1. Red Wing Shoes focuses on designing sturdy shoes intended for industries such as mining, logging, and farming. Red Wing's focus on customers within these specific industries is known as market segmentation.

   TRUE

2. Five Guys understands the importance of market segmentation as evidenced by its marketing strategies, which are designed specifically to appeal to its market segment. It develops promotional campaigns and advertisements that focus on fresh, high-quality products rather than on lowering prices or offering quicker service.

   TRUE

3. The local women's fitness center is now offering new spin classes to post-partum women looking to ease back into an exercise routine. These classes exemplify the center's use of demographic segmentation to draw in new customers.

   FALSE

4. When marketers segment based on _____, they divide the market into groups according to the reason consumers purchase products.

   A. demographics

   B. psychographics

   C. customer convenience

   D. market size

   E. behavior

5. When Merrill Lynch decided to tailor its global investment advice and professional money management services to best meet the needs of mass affluents, it was focusing on which demographic variable?

   A. lifestyle

   B. family size

   C. age

   D. marital status

   E. income

6. Which of the following strategies exemplifies a geographic segmentation strategy based on customer convenience?

   A. Starbucks providing customers exactly what they want before they walk into the store.

   B. Netflix segmenting its market based on users and nonusers.

   C. Mercedes salespeople that hightlight specific features of their product based on customer motivations.

   D. Kraft foods's providing consumers with busy schedules a food assistant app to make meal planning easier.

   E. Cracker Barrel Old Country Store® locating its stores along interstate highways.

7. Which of the following is a variable of behavioral segmentation?

   A. family size

   B. marital status

   C. lifestyle

   D. loyalty

   E. customer convenience

8. Segments must be _____ in order to be considered as potential target markets because such segments hold promise of greater rewards (sales, profits, market share, etc.).

   A. measurable

   B. accessible

   C. differentiable

   D. actionable

   E. substantial

9. If a market segment exists, then it is therefore accessible to the firm that would like to serve it.

   FALSE

10. The allocation of promotional resources should be considered in the context of how _____ a segment is.

    A. measurable

    B. substantial

    C. actionable

    D. accessible

    E. differentiable

11. Sam, the marketing manager at Indigo Inc., is in the process of market positioning. He has finished analyzing competitors' positions. What should he do next?

    A. He should evaluate the feedback.

    B. He should clearly define the firm's competitive advantage.

    C. He should determine if the business will be profitable.

    D. He should introduce the product to a new limited market.

    E. He should conceive new product concepts.

12. Marketers should follow three major steps to decide how to best position their product. The first step is to

    A. conceive new product concepts.

    B. determine if the business will be profitable.

    C. introduce the product to a new limited market.

    D. analyze competitors' positions.

    E. clearly define the firm's competitive advantage.

13. Microshine is a brand of toothpaste. In its earlier campaign, Microshine was advertised as an effective whitening toothpaste. However, due to declining sales, its parent company has decided to rethink its marketing strategy. In its new campaign, Microshine is still portrayed as whitening toothpaste, but it also features a research study where other familiar brands in the market fail to make the grade. In this campaign, Microshine has tried to reposition

    A. the target market.

    B. its application.

    C. the competition.

    D. its price.

    E. its attributes.

14. Positioning is a company's efforts to gain competitive advantage via strategic target marketing.

    FALSE

15. Which of the following statements accurately depicts how marketers use perceptual maps?

    A. Perceptual maps provide marketers a visual picture of where products are located in consumers' minds.

    B. Perceptual maps help marketers psychographically segment their markets.

    C. Perceptual maps provide marketers with a visual picture of all the businesses located in a specific area.

    D. Perceptual maps help marketers determine where to conduct further marketing research.

    E. Perceptual maps help marketers organize different segmentation variables.

16. Which of the following best defines a target market?

    A. A market segment in which there are many competitors offering the same products

    B. A market segment for which the future growth rate is very low

    C. The group of customers that has consistently stayed unresponsive to a firm's marketing efforts over several years

    D. The group of customers toward which an organization has decided to direct its marketing efforts

    E. The demographic segment that a firm is unable to access

17. Verve, a successful soap manufacturer, sells luxury soaps exclusively for women under the brands Verve Opulent, Verve Imperial, and Verve Divine. It has now added Verve Regal to target a specialized segment of urban men who are particular about their grooming practices. The introduction of Verve Regal represents

    A. undifferentiated targeting.

    B. differentiated targeting.

    C. niche marketing.

    D. behavioral targeting.

    E. mass marketing.

18. A new smoothie cart has set up shop near the campus of the local community college. The area comprises a diverse population: college students, faculty, commuters rushing to the train or bus station, and local business people. For a marketer, each of these groups represents a potential target market.

    TRUE

    Difficulty: 1 Easy

    Learning Objective: Define target marketing and its importance in establishing a firm's target marketing strategies.

    Topic: Defining Target Marketing

Feedback: A target market is the group of customers toward which an organization has decided to direct its marketing efforts.

19. One of the criteria for targeting consumer groups is to ensure that the potential target market aligns with the organization's overall values and mission. This criterion is known as

    A. targeting strategy.

    B. strategic fit.

    C. market potential.

    D. competitive fit.

    E. growth potential.

20. If you were asked to market a product like sugar or salt, it would be best to engage in a(n) _____ targeting strategy.

    A. mixed

    B. strategic

    C. undifferentiated

    D. differentiated

    E. niche

6 Creating Value

What To Expect

What separates one product from another? Why do you choose one brand over another? You probably have specific reasons why you choose one product over another, and that reason is your perceived value that product gives. In this chapter we will look at how that value is created.

Chapter Topics

- **6-1** Product, Branding, and Packaging
- **6-2** New Product Development
- **6-3** Marketing Services

Product, Branding, and Packaging

This lesson focuses on products, branding, and packaging, and how firms use these concepts to build sustainable marketing plans. In particular, we discuss the types of decisions that marketers make regarding product, branding, and packaging.

By the end of the lesson you will be able to

- Describe the components of a product.
- Distinguish between the major classifications of products and services (specialty, convenience, unsought, and shopping).
- Define product item, product line, product mix, product mix depth, and product mix breadth and how they are connected.
- Define branding.
- Describe packaging function and design.
- Explain the strategic decisions companies make in building and managing their brands.
- Describe how packaging relates to marketing strategies.
- Explain how firms use data to define their product, branding, and packaging strategies.

Marketing Analytics Implications

- To create profitable products that bring value to consumers, firms must use data to understand their consumers' needs and wants, especially those in their target markets.
- Firms conduct extensive research to determine whether consumers will accept new products they introduce to an existing line. These data also help firms determine which product names will most resonate with their target markets.
- Brands can increase or decrease in value very quickly. Brands also exist in the *minds* of consumers. Therefore brand managers must constantly scan the market and survey consumers about their brands in order to evaluate and confirm the firm's brand strategy.
- Measuring brand equity is a complex process; currently no universally agreed-upon formula for calculating brand equity exists.
- Packaging must be carefully researched for both physical attributes and psychological impact. A package cannot be merely attractive; it must also be functional.

A New Product and a Brand Extension at Lululemon

In 2020, working out from home became more important than ever for many consumers. However, demand for home workout equipment started growing before that. In 2016, Brynn Putnam was expecting her first child and longing for the convenience of better work-from-home fitness equipment, thus leading her to develop Mirror—an interactive home fitness device where customers can stream thousands of on-demand exercise classes from ballet to boxing and everything in between. The equipment is part LCD and part mirror and allows participants to see their own reflections to check out their form and progress. While not in use, Mirror doubles as—you guessed it—a sleek mirror.

After several years of funding and development, Putnam launched Mirror in 2018. By 2020, she sold Mirror to lululemon for $500 million. Lululemon, recognizing the trend in at-home fitness, saw the acquisition of Mirror as the perfect way to reach new customers while also continuing to build its brand and provide value to loyal customers. In 2019, lululemon detailed its growth strategy on its website, saying that it aimed to "realize the full potential of the brand" and aspired to "ignite a community of people living the sweatlife." The acquisition of Mirror helps lululemon reach its aspirational goals through innovative products and brand extensions.

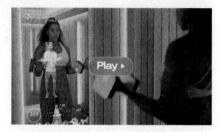

Lululemon, historically known for its yoga apparel, purchased Mirror in order to expand its brand reach with a new product. As seen in this image, Mirror is an in-home, interactive fitness device that consumers can use to elevate their workout. When not in use, Mirror doubles as, you guessed it, a mirror.

Product Components

A **product** is the specific combination of goods, services, or ideas that a firm offers to its target market. Products enrich the lives of consumers by providing designs, features, and functions that people need and want. *Goods* are tangible products, like a mountain bike, with physical dimensions. *Services* are intangible products, such as maintenance on the bike; they cannot be touched, weighed, or measured. *Ideas* are also intangible and represent formulated thoughts or opinions. The perfect vacation that the state of Michigan promotes through its "Pure Michigan" marketing campaign is an example of an idea. New goods, services, and ideas are critical to a company's survival in today's marketing environment.

Thus, a marketer's definition of a product is quite broad as it includes:

- physical goods (e.g., iPhone, Amazon Kindle, and food items),
- services (e.g., iTunes, iCloud, cell phone service providers, GrubHub, and restaurants), and
- ideas (e.g., "Got Milk," Nike's "Just Do It," political platforms, and sustainability).

Regardless of how a product is defined, products have the following three components:

- **Core product**—this component relates to the basic benefit obtained by customers of the product.
- **Actual product**—this component involves the form of the product itself such as the brand, features, quality, style, size, color, and packaging.
- **Augmented product**—this component contains the services, experiences, warranties, and financing that enhance the product value.

Marketers should understand these three components of their products so that they know why customers purchase them and also to determine how they should design the products to maximize customer value.

Classifications of Products

All products can be categorized based on characteristics of how the consumer shops, pays, or uses the product. The categories may differ between consumers, but in general a product will fall into one of four main categories, which are convenience, shopping, specialty, and unsought products. Each is defined and explained as follows:

- **Convenience products** are products that are purchased with low consumer involvement in the purchase process. Accordingly, convenience products are often purchased frequently and are rather inexpensive to buy. As a result, the consumer does not spend lots of time to make a purchase. Examples of this large category of products include a pack of gum, a bottle of water, and most items in a grocery store. If a product is considered a convenience product, marketers must pay careful attention to the distribution strategies of the product to make sure it is convenient for the customers to purchase.

- **Shopping products** require moderate consumer involvement in the purchase process. Hence, consumers typically allocate considerable time to compare brands and product features before making a purchase decision. Examples of this type of product include cars, technology, insurance, and home furnishings/appliances. For products that are considered shopping products, it is important for marketers to communicate the value proposition of the product in order to make it easy for the consumer to compare.

- **Specialty products** require high consumer involvement in the purchase process. These purchases tend to be expensive, purchased infrequently, and specific. Specialty products are usually one of a kind or rare. Because of the unique nature of these products, consumers are less likely to settle for a similar product, if one even exists. Examples include Rolex watches, first edition books, Maserati cars, and artwork. For specialty products, the product needs to be in pristine condition and worth the increased information search and perceived risk in order for the customer to justify the purchase.

- **Unsought products** require substantial marketing and sales efforts because consumers are unlikely to purchase these products independently. Unsought products are often thought of as products that you don't need, until you need them. Often unsought products become important after some unexpected change or accident. Examples include the service of a tow truck, plumber, or defense attorney. Unsought products can also be a good, such as a prosthetic leg, septic tank, or roofing shingles. For unsought products, marketers must have thoughtful promotional strategies to communicate the value and necessity of the product or service to potential customers.

Convenience products

- Purchased with low consumer involvement in the purchase process
- Often purchased frequently and are rather inexpensive to buy

Shopping products

- Require moderate consumer involvement in the purchase process
- Consumers typically allocate considerable time to compare brands and product features before making a purchase decision

Specialty products

- Require high consumer involvement in the purchase process
- Purchases tend to be expensive, purchased infrequently, specific, and usually one of a kind or rare.

Unsought products

- Require substantial marketing and sales efforts because consumers are unlikely to purchase these products independently
- Often thought of as products that you don't need, until you need them
- Often become important after some unexpected change or accident

Product Line and Product Mix Decisions

All companies sell some combination of products. A **product item** is a single, specific product such as an iPhone, whereas a **product line** consists of different versions of a product within a product category. In other words, product lines are composed of multiple product items in a product category. For example, Apple offers a variety of products. One such product is the MacBook Air. This would be one product in the product line that would include all of Apple's computers: the MacBook Air, the MacBook Pro, the iMac, the iMac Pro, and so forth. These individual products comprise the product line. Another good example is Procter & Gamble's (P&G) Febreze Small Spaces air fresheners. The Febreze Small Spaces air freshener is a product item. The product line of Febreze products includes Febreze fabric fresheners, Febreze air freshener sprays, Febreze car fresheners, and so forth. These individual product items comprise an overall Febreze product line.

Conversely, the **product mix** includes all of a company's products, which can extend beyond product lines. Product mix is typically an aggregate of all of a company's product lines. P&G, for instance, also offers product lines for Tide detergents, Charmin bathroom tissue, Crest toothpaste products, and Pampers diapers, as well as a host of other household brands. All of these product lines comprise Procter & Gamble's product mix.

The **product mix breadth** refers to the number of different product lines that a company offers whereas the **product line depth** refers to the number of products within each product line. In the case of P&G, the product mix breadth is extensive: Always, Bold, Bounty, Charmin, Crest, Dawn, Downy, Gain, Gillette, Oral-B, Pampers, Tide, and Vicks, to name several. However, the product line depth varies by product line.

Manufacturers such as P&G and Apple invest heavily in research and development (R&D) to develop successful products. However, once a successful product has been introduced in the marketplace, marketers grow revenue by expanding their portfolio of products through the introduction of new versions of the product or new products that extend beyond the original product category. Marketers assess the performance of each product individually and the product mix collectively to determine which products they should add to or remove from the product mix. Developing a product mix strategy not only helps grow revenues, it also can lead to a sustainable competitive advantage because competitors are less likely to compete directly against a manufacturer's entire product mix. Here are some examples:

Mac Product Line

- MacBook Air
- MacBook Pro
- iMac
- iMac Pro

- Mac mini
- Mac Pro
- Mac Studio

iPad Product Line

- iPad Pro
- iPad Air
- iPad
- iPad mini

iPhone Product Line

- iPhone 13 Pro, iPhone 13 Pro Max, iPhone 13, iPhone 13 mini
- iPhone 12, iPhone 12 mini, iPhone 12 Pro, iPhone 12 Pro Max
- iPhone 11, iPhone 11 Pro, iPhone 11 Pro Max
- iPhone SE
- iPhone X, iPhone XR, iPhone XS, iPhone XS Max
- iPhone 8, iPhone 8 Plus

Apple Watch Product Line

- Apple Watch Series 7
- Apple Watch SE
- Apple Watch Nike
- Apple Watch Hermes
- Apple Watch Studio

Branding

One of the most important jobs that marketers perform is developing and maintaining a brand. For current, former, and potential customers, a brand represents everything that a product (good, service, or idea) means to them. The differentiating characteristics of the brands that matter to you might be tangible and related to the product (such as the towing capacity of a Ford F-150 truck), or they might be emotional and focused on a special memory (such as your memories of Disney World). Specifically, a **brand** is the name, term, symbol, design, or any combination of these that identifies and differentiates a firm's products.

1. Deliver a Quality Product

A strong brand provides continued value and quality to customers over time. Southwest Airlines has accomplished this by offering low fares and refusing to charge baggage fees, even as other airlines have begun doing so.

2. Create a Consistent Brand Image

All of the firm's marketing decisions, promotions, and employees should reinforce the brand by providing a consistent experience for consumers. Mountain Dew's efforts to be seen as a youthful, energetic, and extreme brand would be compromised if it began sponsoring senior golf tournaments and advertising in business trade magazines.

3. Create Consistent Brand Messaging

Brand messaging should be consistent and concise. It should be easy to remember. Too many messages can confuse customers as to why they should purchase a brand. Auto insurance company GEICO has succeeded in providing one consistent brand message through a variety of ad concepts: a promise to save consumers money on car insurance.

4. Capture Feedback

Because the power of a brand exists in the minds of consumers, marketers must capture and analyze customer feedback. For example, Chick-fil-A offers random customers a free chicken sandwich if they fill out an online survey about their experience with the restaurant. Chick-fil-A then uses the data to identify problems and improve the dining experience.

The outcome of successful **branding** is the creation of brand equity. **Brand equity** is the value the firm derives from customers' positive perception of its products. The topic of brand equity is the focus of the next section.

Components of Brand Equity

Large U.S. firms typically spend millions of dollars each year developing and promoting their brand in an effort to increase brand equity. Brand equity increases the likelihood that the consumer will purchase the firm's brand rather than a competing brand. Firms with high brand equity stand apart from competitors, are relevant to a large segment of consumers, and are well known and positively thought of by the majority of their target markets. Organizations with high brand equity enjoy significant advantages over other firms. Brand equity consists of four main components, which are *brand recognition, associations, perceived quality,* and *loyalty.*

Brand recognition is the degree to which customers can identify the brand under a variety of circumstances. Firms such as Nike and McDonald's employ **brand marks,** which are the elements of a brand not expressed in words that a consumer instantly recognizes, such as a symbol, color, or design. The Nike swoosh and McDonald's golden arches are brand marks that have become powerful marketing tools for those companies. Brand recognition indicates awareness of a brand, which is important to marketers. A benefit of recognition is that it increases the likelihood of more consumers purchasing a brand's product.

Brand associations are consumer thoughts connected to the consumer's memory about the brand. Consumers form associations based on several factors, including engagement with the firm and its employees, advertising messages, price, product usage, product displays, publicity, comparison to competition, celebrity endorsement, consumer reviews, and any other connection that they have with the brand. Associations affect brand equity, as strong associations induce brand purchases and generate positive reviews. Conversely, negative associations can harm brand equity. For example, Toyota's brand equity was damaged by a series of product safety recalls from 2005 to 2010 that made consumers question their faith in the brand. Successful management of these negative associations slowly rebuilt Toyota's brand equity with consumers. (More about how Toyota attempted to revitalize its brand is discussed in the section on brand revitalization later in this lesson.)

Perceived quality is a consumer's perception of the overall quality of a brand. In assessing quality, the consumer takes into consideration brand performance as it relates to personal parameters and expectations. Higher consumer quality perceptions provide benefits for firms, such as higher sales prices, customer satisfaction, and loyalty. For example, companies that have better quality products can sell those products at a premium price. Lululemon products are often perceived to be of higher quality than its competitors, which results in lululemon being able to charge higher prices for its products. Quality is one of the main reasons for consumer preference for a brand in any product category.

Brand loyalty is a consumer's steadfast allegiance to a brand by repeatedly purchasing it. Brand loyalty typically develops because of a customer's satisfaction with an organization's products. A benefit of brand loyalty is that brand-loyal customers typically exhibit less sensitivity to price, making them an important contributor to a firm's long-term success and profitability. For example, lululemon customers tend to display fierce brand loyalty, to the point that they are willing to act as brand ambassadors.

Brand recognition

- The degree to which customers can identify the brand under a variety of circumstances.
- Some firms employ brand marks such as a symbol, color, or design.
- Brand recognition indicates awareness of a brand.

Brand associations

- Consumer thoughts connected to the consumer's memory about the brand.
- Consumers form associations based on many factors.
- Strong brand associations induce brand purchases and generate positive reviews, whereas negative associations can harm brand equity.

Perceived quality

- Consumer's perception of the overall quality of a brand.
- Higher consumer quality perceptions provide higher sales prices, customer satisfaction, and loyalty.

Brand loyalty

- Consumer's steadfast allegiance to a brand by repeatedly purchasing it.
- Typically develops because of a customer's satisfaction with an organization's products.
- Brand-loyal customers typically exhibit less sensitivity to price.

Brand Ownership

In recent years, as a sluggish economy forced consumers to closely monitor their spending, more and more retailers have begun pursuing their own branding strategy—private label brands. **Private label brands,** sometimes referred to as **store brands,** are products developed by a retailer and sold only by that specific retailer. Private label goods and services are available in a wide range of industries, from food to cosmetics to web hosting. They are often positioned as lower cost alternatives to well-known **manufacturer brands,** which are sometimes referred to as **national brands** and are managed and owned by the manufacturer. Private label brands such as Walgreens's aspirin can cost up to 50 percent less than Bayer aspirin, a manufacturer brand.

Over the past decade, annual sales of private label products have increased by 40 percent in U.S. supermarkets. In addition, over 40 percent of U.S. shoppers now say that at least half of the groceries they buy are private label brands. Walmart's Great Value brand, and others like it, are even more popular in Europe, where private label brands account for 35 percent of retail sales, a significantly higher market penetration than in the United States. In the first quarter of 2020, during the COVID-19 pandemic, dollar sales for private label brands increased nearly 15 percent and unit sales increased nearly 13 percent, representing a gain of roughly 1.5 billion products sold across all retail outlets.

Private label brands offer two advantages over manufacturer brands:

1. The potential for higher profits on private label brands can be more desirable than the profit returns on national brands.

2. Successful private label brands such as Kirkland Signature provide consumers with a reason to shop exclusively at Costco. Otherwise, the assortment of products available in one retail store versus another may be nearly indistinguishable. Consequently, private label brands can help retailers differentiate themselves from competitors and create a sustainable competitive advantage.

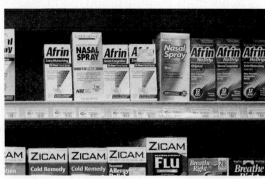

Branding Strategy

Branding strategies should align with firms' overall marketing strategies. Brands are an integral component of a firm's ability to achieve its marketing objectives. Companies have a number of choices as they select which brand strategies are best for their organization.

Brand Extension

Companies that already possess a strong brand and high brand equity may pursue a brand extension strategy. **Brand extension** is the process of broadening the use of an organization's current brand to include new products. Unlike product extensions, in which a firm expands within the same product category (e.g., Coke and Diet Coke are both soft drinks), brand extensions typically involve taking a brand name into a different product category. A brand extension enables new products to profit from the recognition and acceptance the brand already enjoys. For example, Crest extended its brand beyond toothpaste to include dental floss, mouthwash, and toothbrushes.

As a company implements a brand extension, it must remain mindful of the following two potential concerns:

1. **It must ensure that the extension lives up to the quality consumers expect from the brand.** If the quality of the extension products does not meet customer expectations, the firm jeopardizes sales, consumer trust, and brand loyalty.

2. **Brand extensions must be implemented with an eye toward avoiding cannibalization. Cannibalization** occurs when new products take sales away from the firm's existing products rather than generating additional revenues or profits through new sales.

Brand Revitalization

Brands do not die natural deaths. However, they can be destroyed through mismanagement. Some firms mismanage brands into a position from which they cannot recover, but others can be revitalized. **Brand revitalization,** or **rebranding,** is a decision to recapture lost sources of brand equity and identify and establish new sources of brand equity. Revitalization often begins with an investment in rebuilding trust with consumers. Following the largest recall in the history of the company, Toyota marketers embarked on a brand revitalization campaign with the motto "Moving Forward." Its efforts included advertisements communicating the company's desire to start fresh with consumers. By fixing defective products and promoting a brighter future, Toyota was able to begin rebuilding its brand image and was successful in increasing positive perceptions of the brand in the year following the start of the revitalization campaign.

Co-Branding

As an alternative to extending its brand through new product development, a firm can choose to increase its own brand equity by leveraging the equity of another brand using co-branding. **Co-branding** occurs when two or more companies issue a single product in an effort to capitalize on the equity of each company's brand. For example, the menu at casual dining restaurant T.G.I. Friday's has an entire section dedicated to Jack Daniel's–flavored food. The partnership started in 1997 and continues to be a customer favorite.

Co-branding has many benefits, but if one of the brands involved receives negative publicity, it could impact the co-branding partner in a negative way. For example, consider a rental car company that enters into a co-branding agreement with a hotel chain to provide additional value for business travelers. If the hotel chain receives negative publicity because of poor customer service, the co-branded rental car company is likely to be negatively impacted in the minds of consumers because of its association with the underperforming brand.

Brand Extension

The Clorox brand is known for its original bleach product. Building on the success of that product, Clorox has extended its brand to other cleaning products, such as toilet cleaning wands.

Dove has extended its brand to include men's personal care products such as shampoo and body wash.

Brand Revitalization

The Olive Garden changed its logo, restaurant design, and menu in an effort to change its brand image.

After being acquired by Nike, the Converse brand has experienced revitalization and increased sales.

Co-Branding

Many companies combine their brand equity in an effort to capitalize on a stronger product image, such as when Betty Crocker and Hershey's worked together to market brownies that include both products.

Nike and Apple recently paired up to create Apple Watch Nike.

Packaging Function and Design

Almost all physical products require some form of **packaging.** A product package can range from a simple, plain cardboard box to a high-quality, stylish iPhone box. Marketers are increasingly using packaging to not only protect their product for shipping and display, but also to set it apart from its competitors. Current packaging designs seek to fulfill four main functions, which are to contain and protect products, to advertise and promote products, to increase the convenience of the product, and to be as sustainable as possible.

Contain and Protect Products

When a manufacturer produces something, it needs to find a way to effectively get the product to market. First, a package must be durable and able to protect the product from physical damage as it travels in ships, trains, trucks, and via delivery vehicle to a retail store. Second, a package must protect a product from any changes in temperature or weather as it is transported. And finally, a package must be secure enough that theft of the product is deterred throughout the logistics process. In the past, wooden shipping crates, barrels, and metal drums would be used to protect products in their journey from the manufacturer to the consumer. Today, more sophisticated packaging performs this function, as a variety of cardboard, plastic, and recycled products are designed to efficiently enclose and secure individual products. For Apple products, durable and artistic boxes are specifically designed to protect the fragile contents that they contain and also to keep all components safely together as they travel to a retailer's shelf or a consumer's mailbox.

Advertise and Promote Products

Another function of a product's package is to promote the product's features and benefits to consumers. Consumer goods that sell in a retail setting are prime examples of the promotional impact of product packaging. In many cases, a variety of similar, competing products will be featured in close proximity on a retailer's shelf display. Because consumers compare products before purchase, packaging is one way to differentiate one product from another. Usually a package will list ingredients, contents, specifications, and features of the enclosed product. In addition, packaging can be used to promote "new and improved" configuration, changes in product volume or size, and benefits of the product.

The shape of packaging also promotes the product it contains. One way that many consumer products are promoted is based on the shape of their container. Think about a bottle of Coca-Cola; does it need a label or logo for you to identify it? For Coca-Cola, the distinctive shape of its bottle is a brand mark in itself, making it easy for consumers to identify the product. The same is true for many products; Tide's distinctive orange bottle, Heinz's ketchup bottle, and Mrs. Butterworth's maple syrup are all examples of shape-based promotional properties of packaging. In some cases, creative shapes are used to attract new customers.

Increase Convenience of the Product

Packaging is increasingly being used to enhance the convenience of a product. Packaging was once viewed as disposable but now is often viewed as an extension of a product's benefit. Perishable items, such as Kraft cheese, have added value to their products by selling them in re-sealable "zipper" packages. In recent years, the packaging of sandwich lunch meat has been revolutionized by the use of reusable food containers that become useful products in themselves after the meat product has been consumed. This type of packaging serves multiple functions as it is used to contain and protect the product, add convenience to the product, and remind consumers of the product for an extended period after the product is used up.

Sustainability

A major function of packaging is to create sustainable products that can be recycled or that degrade after use. Many manufacturers devote a great deal of research and development into eliminating any waste in their packaging. The sustainability of packaging extends beyond just its ability to be recycled or its biodegradable properties. Reductions and redesign of packaging can also increase sustainability in other areas. For example, Costco utilizes plastic milk containers that are square in shape and designed to reduce the space needed on trucks that transport them. By reducing size and increasing the amount of product that can be delivered by a truck, Costco is able to use less diesel fuel and fewer trucks to deliver milk to its stores.

The Pringles brand was able to distinguish itself from other brands because of its unique can packaging when traditional potato chips brands relied on bag packaging.

Packaging and Marketing Strategy

The two main strategic considerations for packaging—promoting brand image and reinforcing brand image—are described below.

Promoting Brand Image

Many organizations consider packaging only in the basic terms of containing, protecting, and shipping packages. This is a shortsighted view. Packaging provides marketers with an opportunity to promote the image they want others to associate with the brand. For example, packaging allows premium products to communicate class and sophistication. A classic Tiffany gift box indicates quality and suggests a lavish lifestyle, both to the person receiving the gift and the rest of the world. The Tiffany's box and shopping bag a consumer carries out of the store are seen by other consumers, particularly in large metropolitan cities, and serve as a type of mobile billboard. The shopping bag may also remain in a consumer's home for some time after the purchase, offering a constant reminder of the luxury experience, which can increase brand recall. Packaging is one of the few points of contact upscale firms have with customers that they can directly control. As a result, they cannot afford to miss the chance to extend the luxury experience beyond the store's walls.

Reinforcing Brand Image

Marketers also can use packaging to reinforce their brand image with consumers. For example, demand for environmentally friendly packaging has changed the marketing landscape for U.S. consumer goods in many categories. Research suggests that a growing number of consumers are green motivated or driven to make decisions based on concern about the environment. They make purchasing decisions based not only on environmentally friendly ingredients and manufacturing procedures but also on packaging materials. In response to this trend, Coca-Cola launched PlantBottle, a recyclable plastic bottle made partly of plants, in 2009. Regardless of the industry, packaging is considered an important indicator of brand quality. The quality of the brand, therefore, has to be communicated by good packaging and not just by promises of quality made in the text on the packaging. Effective packaging results in an engaging and persuasive marketing tool in which the product and its packaging form a coherent whole, and the consumer forms his or her image of the brand based on this consistency.

Coca-Cola's "PlantBottle" is a good example of how a product's package can serve many functions and strategies at the same time. The bottle contains and protects the product, advertises the product, is a convenient container, and is eco-friendly. The bottle also promotes the brand in its shape and reinforces the brand's message of environmental consciousness.

Marketing Analytics: Using Data to Define Product, Branding, and Packaging Strategies

Using Data to Identify and Confirm Product Value

Products create value for consumers. In order to produce profitable products firms must understand consumers' needs and wants. Both qualitative and quantitative data are gathered from consumers to gain insight into what product features and what intangible aspects of the product create value. Interviews and observational studies will yield data about how consumers view the product and what problem the product solves for the consumers—that is, what need or want the product is fulfilling. Surveys and sales data can yield data related to consumers' current purchasing behavior and unmet needs or wants.

Using Data to Inform Product Line Extensions

There is an old adage in marketing that your brand is only as good as the last product line extension. Any firm wishing to add a product to the existing product line must clearly understand how the new product will be viewed by consumers in relation to the existing products. For each new product, the firm has to carefully examine the data related to brand attributes and brand strengths. If the brand presents a line extension that is inconsistent with the brand, then consumers will reject that product, thus rendering negative impacts on the brand. Prior to any launch of a new product, concept testing should be conducted to determine how consumers view the new product vis-à-vis the existing line. These data can be used to evaluate the potential impact on brand evaluation of the new product.

Using Data to Understand and Monitor Branding Strategy

Brands exist primarily in the minds of consumers, which means that because brand evaluation is based on consumer perception, it is therefore not stable. Perceptions are impacted by a variety of factors beyond the reputation of the brand itself. For example, a scandal involving a firm or the CEO of a firm can have dire consequences for the brand. Or, a competitor that generates a lot of positive media for its competing brand can negatively impact another firm's brand evaluation. Through market scanning, firms can continuously collect data regarding any potential threats to, or opportunities for, their brand. These data are used to adjust branding strategy, which allows firms to respond to a threat or to seize an opportunity.

Using Data to Determine Brand Equity

Because so many intangible factors contribute to brand equity there is no universally agreed-upon measure for it. Interbrand, a global brand consultancy, publishes the top global brands list every year and uses a complex formula for arriving at its assigned brand values. Firms wishing to measure the value of their brands must collect data on the value of the *tangible* brand assets. They must also develop strategies for assigning value to such *intangible* brand assets as brand loyalty, brand recognition, and brand preference. There are a few metrics that companies can use to calculate brand equity: knowledge metrics, preference metrics, and financial metrics. Knowledge metrics include measures of brand awareness and brand associations. Preference metrics include measures of brand relevance, accessibility, emotional connection, and value. Finally, financial metrics includes measures of market share, pricing data, price premium, revenue and potential revenue, growth rate, and sustainability.

Using Data to Determine Packaging

Determining the correct packaging for a product requires extensive research and testing. Data gathered from such research and testing helps firms to develop packaging that is both aesthetically pleasing and functional. Packaging is functional when it protects goods in transport and ensures shelf stability. Apple's clean, sleek packaging is a good example of both elegant and functional packaging. Apple understood that "the user experience isn't solely relegated to the device itself, but begins when a consumer picks up the box itself."

Today, due to increased consumer concern for, and awareness around, the environmental impact of product packaging, many firms are now faced with the added challenge of providing sustainable packaging. Research can help firms determine how to engage in sustainable packaging that benefits them, the consumers they serve, and the environment in which they all engage.

Different forms of retailing also demand different packaging adaptations. For example, with the increase in online retailing, companies are now finding ways to differentiate themselves through their shipping packaging. According to a recent e-commerce analysis by Dotcom Distribution, "39% of online shoppers shared a product image or a video on social media." Companies like Glossier, an online beauty brand, took advantage of these data and used them to inform its distinctive pink bubble-wrap packaging, noting, "We want every detail of the unboxing experience to be exciting, surprising—and photogenic."

Since 2009, the Dieline Awards for Best Packaging have been presented. Awards are judged on "structural packaging, design, [and] branding"; a panel of consumer product experts then examine each submission according to five key criteria: creativity, marketability, innovation, execution, and on-pack branding. These awards are given to firms whose packaging is both aesthetically pleasing and functional.

Responsible Marketing

Greenwashing of Products, Packaging, and Brands

You have probably heard the term "greenwashing," which refers to a firm misleading consumers about its environmental practices in some way. Greenwashing can occur in a variety of contexts. The following are some examples of greenwashing that apply to this lesson:

- **Product:** Companies can greenwash their products in many ways. For example, companies may claim that ingredients in a product are "natural" or "organic," yet this may apply to only a fraction of the ingredients. Because greenwashing encourages consumers to believe that products are sustainable and better for the planet, companies often increase the price of products as a result.

- **Packaging:** Some brands engage in greenwashing by having environmentally sound products that are encased in wasteful, nonrecyclable packaging. Conversely, some very harmful products may be labeled "sustainable" because they are packaged using recyclable materials. In addition, some packages may claim to be recyclable when they are not.

- **Brand:** Some companies change their brand name or logo to something that is seen as representing products that are more sustainable and environmentally friendly. Examples of this include adding the term "natural" or "organic" to a brand name or slogan. Some firms also engage in greenwashing by changing their brand colors to green tones to represent the environment.

Reflection Questions

1. What are some examples you have seen of greenwashing?

2. What can consumers do to be more aware of greenwashing?

3. Should firms be held accountable for greenwashing behavior?

New Product Development

For a firm to be sustainable, and grow over time, its marketers must continually invest in developing new products and improving existing products. Most firms simply cannot survive without developing new products. Recent examples of firms that did not invest in new products, and thus failed, include Nokia, which did not change the user experience with mobile phones as the market evolved. Conversely, Apple monitored the changing mobile phone market and launched the iPhone in 2007, the first phone without a keyboard. This lesson explains the new product development process and how it relates to a firm's overall marketing efforts.

By the end of this lesson you will be able to

- Identify the reasons firms innovate their products.
- Describe different types of new products.
- Describe the consumer adoption process.
- Explain how different types of adopters affect product diffusion.
- Explain how the characteristics of a new product affect product diffusion.
- Explain the new product development process.
- Identify the major risks encountered in new product development.
- Describe the ethical issues in new product development.
- Define the product life cycle.
- Explain how marketers use data to plan a development strategy for new products.

Marketing Analytics Implications

- Marketers use data to monitor changes in the market and in consumer preferences; these changes will signal to a firm whether it needs to innovate its products.
- Marketers use data to study and track how consumers are moving through the diffusion process.
- Marketers use data to help define and guide the new product development process, as well as to identify risks.
- Marketers use data to help identify where a product is in its product life cycle.

Chicken Sandwich Wars: Revamped Ideas Creating New Products

It is strange to think but at one point, every product that is around today was new-to-the-market, even the fried chicken sandwich. Some believe that S. Truett Cathy invented the chicken sandwich in 1946 when he made an alternative to a traditional hamburger, placing a piece of boneless grilled chicken on a bun. Fast-forward 18 years to 1964 and Cathy started selling that original chicken sandwich with two pickles on a toasted butter bun at Chick-fil-A. So how did a product first sold in 1964 spark a war across fast-food restaurants 55 years later?

In 2019 Popeyes revamped the original chicken sandwich, which sparked a war across fast-food restaurants.

On August 12, 2019, Popeyes launched its now famous fried chicken sandwich, with a largely unnoticed debut. One week later, and with the help of social media, a Twitter food fight began between Chick-fil-A and Popeyes, turning the Popeyes menu item into a national icon and forcing fast-food customers to choose a side. Tensions heated and the lines for Popeyes grew longer as customers had to try this new product, until sales came to an abrupt halt when the company sold out by the month's end. In November, the war was in full swing again with Popeyes restocked and once more selling the fried chicken sandwich to hungry customers.

Now it seems that every fast-food restaurant wants a piece of the chicken sandwich action. McDonald's, which first launched a disappointing McChicken in 1980, has reentered the chicken sandwich market with the Crispy Chicken Sandwich that comes in three versions: original, spicy, and deluxe. Other fast-food restaurants, including KFC, Church's, Burger King, Wendy's, Arby's, Shake Shack, and Jack in the Box, have either entered the market with a new chicken sandwich or revamped a previously offered, yet much less successful, chicken sandwich. Even Taco Bell, whose slogan was once "Think outside the bun," has entered the war with a chicken sandwich taco.

So who is winning the chicken sandwich war, you ask? Everyone! Thanks in part to the success of Popeyes, 47.8 percent of U.S. restaurants feature chicken sandwiches on their menu, and Americans consume chicken more than any other meat. The poultry industry and restaurants have increased sales and market share—and customers have a plethora of chicken sandwich options to choose from.

Product Innovation

As discussed in previous lessons, a **product** is the specific combination of goods, services, or ideas that a firm offers to its target market. Well-designed products enrich the lives of consumers by providing designs, features, and functions that consumers need. However, the things that consumers need, want, and prefer change over time. Thus, marketers must regularly innovate their products to meet new consumer demands. For product marketers, **innovation** is the creation of a new or significantly improved product offering.

The overall reason that firms innovate is to maintain or advance their product's position in the market. The following are five specific drivers of innovation:

- **Consumer expectations:** As noted, most innovation is related to changes in what consumers need or want. Kimberly-Clark would not be competitive if it sold the same thin, scratchy toilet paper that it sold in 1890 because consumers now expect a better product.

- **Competition:** Marketers are often forced to update or add products because competitors have improved a similar product. Smartphones are a good example. Apple, Samsung, Motorola, LG, and others continually upgrade their products to maintain or advance their position in the market.

- **Globalization:** Marketers often innovate because their product is sold in an expanded global marketplace. Different cultures have different preferences, which requires product innovation to meet consumer demand. For example, when Dunkin' Donuts expanded its presence globally (3,100 stores in 30 countries), it had to update its menu to appeal to global tastes. In Korea, Dunkin' offers a Grapefruit Coolata, in Lebanon a Mango Chocolate Donut, and in Russia a Dunclair.

- **Technology:** The advancement of technology has always driven product innovation. In recent decades, the pace of technological progress has increased, thus, so has the pace of product innovation. Consider Motorola, a leader in flip-phone technology in the early 2000s. Since the advent of touch-screen technology, the flip phone has become virtually extinct. In response, Motorola has innovated its product offerings by adopting touch-screen phones.

- **Changing society:** How societies operate is constantly changing. Such change impacts the products we need. Consider how the workplace has shifted from a traditional physical location paradigm to a mobile, accessible paradigm. Instead of needing hard-wired fax machines and phones, most firms now arm their employees with scanners, tablets, and mobile phones. Similarly, book sellers, like McGraw Hill, have also had to shift from printed products to interactive, digital ones (like this one).

Finally, firms innovate because if they don't, they are likely to fail. Sears, for example, was the largest U.S. retailer until 1989, but it lost ground to Walmart and Target as business became more centered on technology and online sales. Today, Sears is closing stores and struggling to stay in business.

Consumer Expectations

The 1971 Ford pickup truck was an innovation: it was as comfortable as a car (of that era). Look at this advertisement and see how much consumer expectations of comfort, style, and safety have changed since 1971.

Competition

What SUV is this? By taking competition into account, makers of the Ford Edge have created an SUV that looks very similar to the Chevy, Mazda, Honda, and Toyota versions of the same class of vehicle.

Globalization

The Ford Mondeo is similar to the Fusion, but it is not sold in the United States. The Mondeo is available in Europe and features design characteristics that fit that market, such as manual transmission and different eco-friendly equipment.

Technology

Advances in technology often drive innovation in other products. Apple's Car Play and Android's Auto are changing how Ford builds its radio and navigation system.

Changing Society

Changes in how cars are powered, like this electric Ford Focus, can drive many product innovations. For example, how will an increased number of electric cars be "refueled"? Presumably, traditional gas stations will need to update their product offerings.

Types of New Products

A new product is one that is new to a company in any way. If a product is functionally different from existing products in the market or is not marketed in its current form by the company, it can be considered a new product. A company's core competencies and strengths influence its strategy toward developing new products. Firms that have a strong research and development (R&D) department will focus on developing new products to beat the competition, whereas firms that have a strong brand can extend or revamp their current product lines with similar but somehow differentiating attributes. Let's look at some different categories of new products.

New-to-the-Market Products

Products that have never been seen before and create a new market are considered **new-to-the-market products.** Although they make up the smallest percentage of new products, they disrupt the market because they provide innovative benefits that often render existing products obsolete. This represents tremendous upside potential for firms because getting to the market before competitors do often means increased sales, profits, and customer loyalty, as well as a leadership position in that market. However, new-to-the-market products can be time-consuming and expensive to develop, creating significant risk for a firm. In fact, the vast majority of new products fail, leaving the firm with no revenue to compensate for expenditures of financial and human resources.

The Uber app can be classified as a new-to-the-market product. Launched in 2010 in San Francisco as UberCab, the app's popularity has grown rapidly and currently operates in 58 countries. However, though first to market, Uber has faced problems from regulators and traditional taxi services, and direct competitors like Lyft, showing the risks involved in being the first one to do something.

New Category Entries

New category entries are products that are new to a company but are not new to the marketplace. New category entries help companies compete better in an already-established market or enter a new market. Entering new markets is important to businesses because it opens up a new set of customers and potentially more revenue and profits. New category entries are less risky than new-to-the-market products because they don't represent something that has never been sold before, and therefore the company can access information on sales trends, competitor products and prices, location of markets, and more.

Marriott is innovating its hotel brands to appeal to Millennials, who are rapidly increasing in buying power. These new category entries include AC Hotels, Moxy Hotels in Europe, and EDITION luxury hotels. These hotels do not have business centers, have more casual and social lobbies, and feature more technology in rooms. There is considerable risk to Marriott in developing this new product line because the company has little to no experience targeting this new demographic.

Amazon's purchase of Whole Foods represents a new category entry for Amazon. Grocery retail is not a new market, but it is a new market for Amazon.

Product Line Extensions

A **product line** is a group of related products marketed by the same firm. **Product line extensions** are products that extend and supplement a company's established product line. For example, product line extensions may add new functions, flavors, or other attributes to an existing product line. Advantages of a product line extension include that the company and brand may be easily recognized;

customers may already feel loyalty to the product line; manufacturing may be easier and more efficient because the firm already produces similar goods; and the new product can be advertised alongside existing products. Product line extensions carry risk due to uncertainty about how well the new products will be accepted by the market, but they carry far less risk than new-to-the-market products or new category entries.

Product line extensions are common, especially among companies hoping to offset reduced sales on other products due to seasonality or trends. For example, the Vaseline Intensive Care brand added SPF (sun protection factor) to its products such as lotion and lip balm. Vaseline extended its product line in part to reach new markets, and also due to slower sales of body lotions during the summer months.

Taco Bell excels at product line extensions. One line extension, the "Nacho Cheese Doritos Locos Tacos Supreme," utilizes cross branding with Doritos to extend Taco Bell's product line.

Revamped Products

A new product can sometimes take the form of a **revamped product** that has new packaging, different features, and updated designs and functions. If you have seen a label claiming a product is new and improved, it falls within this category. Legally, a company can label a product new only if, according to the Federal Trade Commission (FTC), the product has been changed in a "functionally significant or substantial respect." Also, the company can advertise a product as new and improved for only six months after it hits store shelves.

Reformulations of current products are a common type of revamped product. For example, yogurt maker Yoplait introduced a line of French yogurt called Oui in 2017 as an extension of its regular and Greek yogurts. Oui is different because it is packaged in glass jars and has a sweeter and less tart taste than other yogurts. Because firms base revamped products on existing brands, they carry less risk than new-to-the-market products or new category entries. In addition to leveraging brand recognition and customer loyalty, firms can advertise the revamped products along with existing ones and capitalize on the existing network to sell the products.

Are you aware of these products?

New-to-the-Market

Cottonelle has a new competitor in the form of new-to-the-market toilet seats that wash and dry the user after each bathroom use. These seats, new to the U.S. market, eliminate the need for toilet paper.

New Category Entries

Cottonelle Freshcare Flushable Cleansing Cloths are a new category entry. The wipes are similar to toilet paper, but are different because they were developed for older demographics that prefer a product that has more moisture.

Product Line Extensions

Cottonelle introduced a product line extension in its Gentle Care with Aloe & E, which adds healing and soothing elements to the original product.

Revamped Products

Cottonelle's CleanCare product has been revamped with "clean ripple" technology to improve the product's functionality.

The Consumer Adoption Process

When a consumer purchases and uses a product, the product has been adopted. The process by which a product is adopted and spreads across various types of adopters is called **diffusion**—how their new products are likely to be adopted, the rate at which they will be adopted, and the process through which their products will spread into markets. Marketers who understand diffusion have a better chance of successfully launching and sustaining new products.

Diffusion gives marketers a way to figure out who will likely buy their product over a period of time, plan an appropriate marketing mix, and forecast potential sales. The result of diffusion is a well-understood process of product adoption, which comprises five stages:

| STEP 1 | STEP 2 | STEP 3 | STEP 4 | STEP 5 |
| Awareness | Interest | Evaluation | Trial | Adoption |

To learn more about the specifics of this process as it applies to a real-world product, explore the following Click & Learn interactive.

Awareness

In this stage, the consumer has been exposed to the product and knows that it is available on the market.

Think of the recent fad, the fidget spinner. Do you recall when you became aware of this product?

Interest

Interest occurs when the product registers as a potential purchase in a consumer's mind and he or she begins to look for information concerning the product.

For the fidget spinner, did you become interested at all, right away, or did it take time (and seeing others with them)?

Evaluation

In the third stage, the customer thinks about the product's value and whether to try it out. The customer evaluates competing products and determines which product best satisfies his or her needs and wants.

Perhaps you wanted a fidget spinner right away, but could only find ones in colors you did not like. Or, maybe you investigated different styles online.

Trial

In the trial stage, the consumer tests the product to see if it meets requirements. Trials include test driving an automobile, borrowing a new shampoo from a friend, or signing up for 30-day access to an online streaming service.

If you bought a fidget spinner, did you try a friend's fidget spinner first? Did you play with one while waiting to pay for items at a store?

Adoption

In the final stage, the consumer buys and uses the product. Adoption happens at different times depending on the type of adopter.

Do you own a fidget spinner that you keep in your dorm or office? Do you use it for fun or stress relief?

Product Diffusion and Types of Adopters

All consumers follow a similar product adoption process, but they don't all follow it at the same time. Customers, whether individual consumers or business customers, buy and use a new good or service at different times and at different rates following product launch. When a firm launches a new product, only a few people buy and use it, but as word spreads, consumers purchase the product at an increasing rate for a period of time. Eventually, sales hit a peak and start to diminish. During this time, various types of adopters purchase the product. We can group these types of adopters into five categories: innovators, early adopters, early majority, late majority, and laggards.

Innovators

Approximately 2.5 percent of those who adopt a product do so almost immediately after the product is launched. These people are called **innovators.** If you are always one of the first people you know to try a new product, you probably fall into this category. Innovators tend to be younger and more mobile than those who adopt a product later in the diffusion process. They are often obsessed with the idea of newness and unafraid to take risks when it comes to trying new products. In addition, they tend to be very knowledgeable, have higher-than-average incomes, possess self-confidence, and choose not to follow conventional norms. While only a small number of purchasers fit into this category, firms value innovators because they share information about the product with others, which can help the product gain market acceptance.

Early Adopters

Early adopters comprise roughly the next 13.5 percent of adopters after innovators. **Early adopters** purchase and use a product soon after it has been introduced, but not as quickly as innovators. They tend to conform to group norms and values more closely than innovators and have closer ties to social groups and their communities. Though they adopt products earlier than the remaining categories, unlike innovators, they wait for product reviews and further information concerning new products before purchasing them. Early adopters are typically well respected by their peers, and marketers seek to gain their acceptance because they tend to be opinion leaders who are willing to talk to other people about their experiences with their purchases. These individuals are therefore important to the diffusion of a new product.

Early Majority

The next category of adopters comprises approximately 34 percent of the adopters of a new product. The **early majority** are careful in their approach, gathering more information and spending more time thinking about the purchasing decision than the previous two categories. Typically, by the time early majority adopters buy a product, more competitors have entered the market, so they will have some choice as to which product to buy. Members of the early majority generally are not opinion leaders themselves, but they often are associated with such leaders. If early majority adopters do not purchase the product, the good or service will likely fail to be profitable. The early majority group also serves as a bridge to the next group of adopters: the late majority.

Late Majority

The late majority category also comprises about 34 percent of adopters. Members of the **late majority** tend to be cautious about new things and ideas. They often are older than members of the previous three categories and may not act on a new product without peer pressure. They often rely on others for information, buying a good or service because their friends have already done so. Members of this group tend to be below average in income and education. When late majority adopters purchase a product, the product typically has achieved all it can from a market in terms of profitability and growth.

Laggards

The final category of adopters is the laggards. These customers make up about 16 percent of the market. **Laggards** tend to not like change; they may remain loyal to a product until it is no longer available for sale. Laggards are typically older and less educated than members of the other four categories. Many choose not to use the Internet. They tend to be tied to tradition and are not easily motivated by promotional strategies. In fact, marketers may never convince laggards to buy their good or service, making them a group on which marketers should not expend a great deal of time or effort. Individuals who have no access to the Internet are examples of laggards who will not be using Amazon as a source from which to buy products.

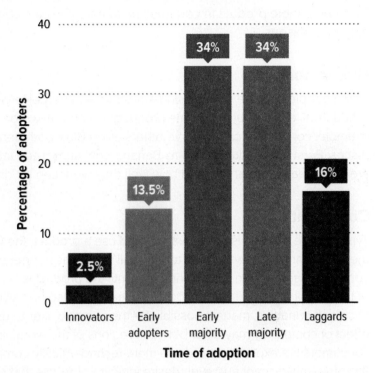

Product Diffusion and New Product Characteristics

Categories of adopters aren't the only factors that affect the diffusion of a new product. New product characteristics, including competitive advantage, compatibility, observability, complexity, and trialability, also impact the adoption rate.

Competitive Advantage

A product obtains a competitive advantage over competing products if consumers believe it has more value than other products in its category. If a product has a competitive advantage, it will be adopted quickly. For example, the Amazon Echo has a competitive advantage over similar offerings by Google and Apple by having a more responsive product to human voice commands. That competitive advantage is reflected in the fact that the Amazon Echo has an 88 percent share of the market.

"Alexa, play party music."

Compatibility

Compatibility refers to how well a new product fits a potential customer's needs, values, product knowledge, and past behaviors. For example, a new beer will not be a compatible product in countries that abstain from alcoholic beverages due to religious reasons.

Observability

When people can see others using a product and perceive value in its use, the product will diffuse quickly. Some products are naturally more visible than others. For example, consumers can observe others using Bluetooth headphones in many public places. Personal products such as Banana Boat sunscreen lotions, on the other hand, are not as easily observed, and how well they work can be difficult to confirm.

Complexity

Typically, the easier it is to understand and use a product, the faster it will diffuse. If the market finds a new product too difficult to use, as happened with the first personal computers, sales often will not increase until the product is simplified. In the case of the personal computer, the addition of easier to use operation systems, such as Microsoft Windows, made it possible for the average user to use a computer. The effect of complexity may differ between regions of the world. In industrialized societies, consumers are exposed to and use more technologically complex products and therefore may accept and even desire intricate products that offer additional features.

Trialability

Most of the time, products that consumers can try without significant expense will diffuse more quickly than others. For example, a photo app that costs 99 cents will diffuse a lot faster than one that costs $100. Similarly, consumers may adopt a product they are first exposed to through trial sizes, free-sampling programs, and in-store trials (e.g., taste testing food).

New Product Development Process

Whatever new product type a firm plans to develop, it will likely follow a formal **new product development (NPD) process.** This process, shown in the figure below, consists of seven stages: (1) new product strategy development, (2) idea generation, (3) idea screening, (4) business analysis, (5) product development, (6) test marketing, and (7) product launch. Let's take a look at each of these stages in more detail.

Stage 1: New Product Strategy Development

STAGE 1

New product strategy development
Determining the direction a company will take when developing a new product

In the first stage of the NPD process, the firm establishes a new product strategy to align the development of products with the company's overall marketing strategy. **New product strategy development** involves determining the direction a company will take when it develops a new product. A new product strategy accomplishes the following:

- Provides general guidelines for the NPD process
- Specifies how new products will fit into the company's marketing plan
- Outlines the general characteristics of the types of products the firm will develop
- Specifies the target markets to be served by new products

A new product strategy should also include an estimate of the profit the company hopes to make and when the firm can expect the product to be profitable.

The Yeti Hopper Backflip 24 is a new product that fits well with Yeti's overall strategy to market products that are durable, portable, and useful for outdoor enthusiasts.

Stage 2: Idea Generation

STAGE 2

Idea generation
Conceiving new product concepts from which possible new products can be selected

Once the firm establishes a new product strategy, it moves to the second stage of the NPD process, **idea generation.** Idea generation, which can be internal or external, involves coming up with a set of product concepts from which to identify potentially viable new products. Few of these ideas ever become marketed products. In fact, a firm must generate as many as 100 ideas to find one product that will actually make it to the marketplace.

Internal Idea Generation

Ideas for new products can come from a variety of sources. Some come from company employees. For example, an engineer at 3M came up with the idea for Post-it Notes by identifying a need in the marketplace that fit an opportunity at 3M. The engineer knew of a weak glue compound that 3M had developed but for which it had yet to find a market. Knowing this information and applying it to a situation in his church choir, where songs were marked with pieces of paper (that often fell out of the choir book), and the idea of Post-it Notes was formed. The pieces would stick well enough to hold on the book, but they could also be reused for next week's songs. This is an example of why smart companies incentivize their employees to suggest new product ideas.

External Idea Generation

Many ideas originate from external sources. Procter & Gamble, for example, gets more than half of its new product ideas from outside the company.

- A firm's customers can be excellent sources of ideas. Salespeople often help facilitate the customers' feedback on products by reporting what customers need and want.

- Competitors' products also provide an important idea source. Many companies, such as automobile manufacturers, purchase and analyze the new products their competitors offer as a basis for devising similar, but better, alternatives. For example, a number of online dating services have taken the idea developed by Tinder of "swipe right" to make a match and "swipe left" to reject a candidate. This intuitive motion for accepting and rejecting can now be found in many apps.

- Firms can also outsource their product development to independent laboratories that provide new product ideas. **Outsourcing** occurs when a firm procures goods, services, or, in this case, ideas from a third-party supplier rather than from an internal source. Other possible sources of inspiration include suppliers, universities, and independent inventors.

Companies with cultures that value all new product ideas, whatever the source, tend to develop more blockbuster products than companies that are unwilling to search for sources of innovation. Flexibility can be an asset as well. Sometimes ideas that start out going in one direction can be pivoted, which involves applying an idea in one way, then, if that does not work out, applying it in another way. Pivoting often occurs in mobile and Web sectors, where it is possible to quickly and inexpensively develop and change a software product.

Stage 3: Idea Screening

STAGE 3

Idea screening
Determining if the idea fits into the company's marketing strategy and should be developed further

Once one of the sources already described has suggested an idea, the firm evaluates the idea to determine whether it fits into the new product strategy during the **idea screening** stage. At this stage, the company often ends up rejecting most new product ideas for one reason or another. Firms may reject products on a number of different bases. Potential issues with product safety may cause a firm to reject an idea for both regulatory compliance and liability reasons.

Firms may also want to make sure that the potential product meets their return-on-investment (ROI) requirements. Companies often have a minimum ROI "hurdle" over which a new product must be expected to perform.

Can it be done by the firm?

Can the product be developed and marketed within the time and budget constraints of the company?

If a lack of resources means the company can't beat the competition, the lost sales that may result from being second in the market may render the product less attractive. Additionally, human and financial resources are finite. New ideas must be compared based on the expenditure of these resources. A perfectly good idea may need to be rejected if another new idea that requires the same resources seems more promising.

Can the company make it?

Is the proposed product within the company's ability to produce?

If a new product would require a firm to purchase new equipment, build more space, or establish different processes, the project may be rejected based on the time and uncertainty this could add to the product launch. Again, such considerations often make product line extension and revamped product ideas more attractive than other new product types.

What is the role of social media?

What will the role of social media be?

Organizations increasingly use social media to evaluate potential new products. Social media are especially valuable for small businesses and nonprofit organizations, which typically have less money to spend on the NPD process. Using social media, marketers can engage consumers in helping them screen potential ideas and ensure only those new products that will be best received in the market move ahead in the process.

Stage 4: Business Analysis

STAGE 4

Business analysis
Determining if the idea can be turned into a product that will prove to be profitable

Even if a product passes the idea screening step, the firm cannot guarantee that it will be profitable. Firms must use complex analysis before they can be reasonably sure that a new product will provide sufficient profitability to make it worthwhile to develop. Profitability, measured by subtracting costs (i.e., all the costs to produce and sell the good or service) from revenue (i.e., the price of the good or service multiplied by the number of units sold), can be difficult to determine, especially if the product is new to the market. To determine profitability, a company must estimate costs, identify pricing, and evaluate demand for the product.

Stage 5: Product Development

STAGE 5

Product development
Prototyping and testing new product ideas to determine likely consumer interest

Once a firm believes that the new product will generate a profit, it enters Stage 5 of the NPD process. In the **product development** stage, the firm determines that the product can be produced in a way that meets customer needs and generates profits. For a good, the company may create a prototype based on previous concept testing. A **concept test** is a procedure in which marketing professionals ask consumers for their reactions to verbal descriptions and rough visual models of a potential product. A **prototype** is a mockup of the good, often created individually with the materials the firm expects to use in the final product. Prototype tests ensure that the product will not be a hazard to users, that it can be produced in the company's or supplier's facilities, and that it can be manufactured at a cost low enough to generate profits.

If the firm is developing a new service rather than a good, it may use this stage to establish protocols for training employees, identify equipment needed, and determine the staffing required. The marketing department also begins developing a marketing strategy during this stage. Regardless of whether the product is a good or service, the product development stage of the NPD process can be long and costly, which is another reason only a small number of ideas make it this far in the process.

Stage 6: Test Marketing

STAGE 6

Test marketing
Introducing the product to a new, geographically limited market to see how well it sells

A product that makes it past the product development stage is ready to be tested more fully with potential customers. **Test marketing** involves introducing a new product to a geographically limited market to see how well the product sells and the reaction to it from potential users. The company selects test markets based on how well they mirror the overall target market in terms of demographics, income levels, lifestyles, and other factors. The selection of test markets is critical to ensuring that the results of the test will be representative of the sales the company can expect.

During the test marketing stage, the firm tests not only the product itself but also the marketing strategy related to it. The marketing department may try different approaches in different test markets to see which marketing mix approach works best. For example, an airline might offer more legroom in certain sections of the airplane at a higher price in one region while offering no luggage fees for first- and business-class customers in another region to see which offering generates the most seat upgrades. Although test marketing can be valuable, there are downsides. The process is expensive and it can be time-consuming. An additional risk to test marketing is that firms open themselves up to imitation from competitors, which can diminish the advantages of being first to market.

Stage 7: Product Launch

STAGE 7

Product launch
Initially producing, distributing, and promoting a new product

Once the firm feels the product is ready for the market, it enters the final stage of the NPD process. The **product launch** involves completing all the final preparations for making the fully tested product available to the market. At this stage, the firm may undertake any or all of the following activities:

- Purchasing the materials to make and package the good
- Hiring employees, such as bank tellers, to provide the service
- Manufacturing enough of a good to fill the distribution pipelines and to store as inventory for continuing distribution, or building enough capacity to provide a service for the expected level of sales of that service
- Strategically placing the good in warehouses in preparation for customer orders
- Preparing internal systems for taking service orders
- Training new employees on how best to deliver the service

Firms must carefully plan the product launch to ensure that the product hits the market according to schedule. Numerous product launches have been delayed because suppliers could not deliver on time, consumer demand was unexpectedly high, or goods couldn't be released due to quality problems. For example, in November 2017, Apple delayed the release of its HomePod until early 2018. This announcement caused the product to be unavailable for sale during the Christmas shopping season, allowing competitors such as Amazon Echo to gain more market share. Delayed product launches often cost companies a great deal of money in overtime labor and shipping charges.

In 2018, Intel stock plunged on news of product delays.

Risks with New Product Development

Firms that want to maintain or improve their competitive position in the marketplace generally must develop new products. Failure to do so carries tremendous risk, particularly for companies that develop products that quickly become obsolete. However, NPD does not come without its own risk to a company. The table below identifies some of these risks, where they fall in terms of severity, and what the consequences of the risks are to firms. Companies need to understand both the types and severity of risks and how to mitigate those risks if they are going to succeed in introducing new products to the marketplace.

New Product Development Risks and Related Outcomes

Severity of Risk	Types of Risk	Outcomes
Very high	• Product fails to meet needs and wants of customers. • Product proves to be dangerous or defective.	• Costs are not recouped; company loses money. • Company suffers legal liabilities and product recalls.
High	• Quality is not up to customer standards. • Supply of product is inadequate to meet demand. • New product is not accepted well in the marketplace. • Inadequate supply of materials delays production. • Target price is not accepted by the market.	• Customers are dissatisfied and there are excessive returns. • Company loses orders, sales, and customers. • Company loses revenue and profits and is stuck with obsolete goods. • Product launch is delayed; first-to-market advantage is lost. • Company reduces price, meaning lost revenue and profits.
Moderate	• Supplier cost savings are not achieved. • Product takes sales from existing products. • Competitors copy products and sell them at a lower price.	• Profitability is reduced. • Total company revenue and profits are less than expected. • Company loses market share and profits.

The highest level of risk occurs when products fail to generate sales or prove to be dangerous and defective. These risks can be devastating to a company if they permanently damage the firm's image or create legal liabilities, such as in the case of loss of life or injury to customers. For example, Samsung had to recall more than 1 million phones in 2016 due to phone batteries overheating and catching on fire. This led to the phones being banned on many airlines and damaging the brand's reputation. Beyond this, the firm lost sales and incurred the additional costs of removing the products from the marketplace and compensating consumers.

Listening to the customer carefully

The voice of the customer should always help guide new product development. Customers use current products and have unique insight into what will help to improve products.

Committing to the NPD process

Every stage of the NPD process must be followed for the process to work correctly. Companies trying to launch a poorly conceived or executed product just to get a jump on the market or to generate quick profits are likely to damage their brand image as well as lose money.

Understanding and anticipating market trends and changes

Gathering data requires a commitment to both marketing and technical research. A firm must stay on top of changing economic, demographic, cultural, and technological conditions so it can anticipate market demand over time.

Asking the right questions

Some questions a firm should ask include: What is the expected demand for the good or service over time? How much will it cost to produce and distribute the good or service at varying levels of demand? What level of quality will be required for the product to be competitive? Are there any potential environmental and safety issues? What is the best pricing strategy to apply at the launch of the product?

Being willing to fail on occasion

Companies that develop new products sometimes will fail. However, firms must learn from past failures. Learning should come not only from products that failed after launch but also from those that failed along the NPD process.

Ethical Issues with New Product Development

Another type of risk that a company needs to be aware of in NPD is creating an ethical problem that may hurt the company's image. Several ethical issues can arise as firms develop new products:

- **What is new?** The most common ethical issue involves the FTC's definition of the term *new*. Firms may make only minor changes to a product and then try to claim that it is changed in a "functionally significant or substantial respect." Such actions result not only in ethical issues, but also in legal issues if the FTC disagrees with the company's viewpoint.

- **Withholding innovation.** Ethical issues also arise if a company chooses not to develop important new products until an existing product has become obsolete or its patent has expired. For example, pharmaceutical companies typically put many resources into developing a drug and need many years of sales to recoup their costs. As a result, they may want to hold off marketing new drugs until they've paid for development of an existing drug. Keeping new drugs out of the marketplace protects a company's stockholders and ensures that the company has sufficient funds to invest in new R&D. At the same time, however, the company could be holding back important innovations that may improve, and possibly save, lives.

- **Planned obsolescence.** A third ethical issue related to new product development is planned obsolescence. **Planned obsolescence** occurs when companies frequently come out with new models of a product that make existing models obsolete. An example is Apple's iPhone, which has been criticized as being designed to slow down as new phone models are released into the market. For Apple, a policy of planned obsolescence has led to class action lawsuits by angry customers. Ethical issues aside, such behavior may damage a company's relationships with its customers. However, there is a difference between planned obsolescence and products that legitimately change frequently due to changes in technology. People want the latest and greatest technology, which forces companies to constantly upgrade the functionality of their products to keep up with the competition.

Apple has been accused of using planned obsolescence to make older iPhone products unusable in an effort to increase adoption of newer product models.

The Product Life Cycle

The launch of a new product begins that product's life. Just like humans, products go through various stages that mark their lifespan. This series of stages is called the **product life cycle (PLC).** The figure below illustrates the four stages of the PLC and the implications of the various stages on sales and profits during the life of a product.

Stages of the Product Life Cycle

During the new product development process, which the lesson has focused on up to this point, the product is actually costing the company money. Once the product is introduced to the market, sales increase slowly as the firm's marketing activities begin to raise awareness of the product. The growth stage brings a spike in sales and profits as consumers recognize the product's ability to satisfy their needs and wants. Sales and profits begin to drop in the maturity stage, as competition increases and customers begin to look for the next big thing, then fall off completely during the decline stage. We will discuss the introduction stage through the decline stage of the PLC in more depth in the sections that follow.

Stages of the Product Life Cycle

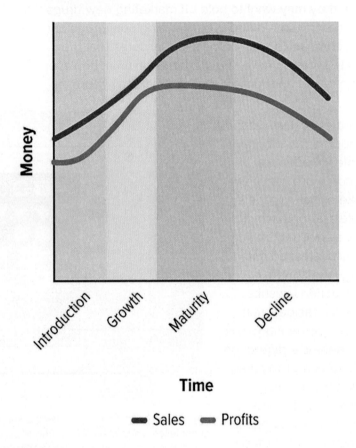

Time

● Sales ● Profits

Introduction Stage

Once the firm launches the product and innovators begin to buy it, it has entered the **introduction stage** of the PLC. This stage is characterized by few or no competitors if the product is new to the market, but sales are typically slow because customers are not yet accustomed to the product. If the firm is first to market, it may be able to capture a large percentage of the market early, giving it advantages in economies of scale and brand recognition. Also, if the product is a good, the company may be able to monopolize the capacity of available suppliers, making it more difficult for other companies to get supplies of components from which to make the product. However, due to the high cost of developing, advertising, manufacturing, and distributing a new product, profits are often low or negative at this point in the PLC.

A number of factors influence how long the introduction stage lasts, including the product's relative advantage, the amount of resources the company puts into promoting the product, and the effort required to educate the market about the product's attributes. For example, high definition televisions moved slowly through the introduction stage; smartphones, on the other hand, moved quickly.

Honest Tea represents a product that is in the *introduction* stage for Coca-Cola.

Growth Stage

Early adopters, followed by the early majority, begin to buy the product during the **growth stage.** Sales begin to rise, as do profits, as companies begin to take advantage of economies of scale in purchasing, manufacturing, and distribution. Competitors enter the market, which forces prices down. During this stage, the firm has to promote the differences between its brand and the competition and may attempt to refine aspects of the product by improving quality or adding new features. Also during this phase, a company typically tries to focus its marketing efforts on showcasing the competitive advantage of its product over others. If the product satisfies the market, repeat purchases help make the product profitable, and brand loyalty begins to develop.

Coca-Cola Zero Sugar represents a product that is in the *growth* stage for Coca-Cola.

Maturity Stage

In the **maturity stage** of the PLC, late majority and repeat buyers make up an increasing percentage of the customer base. The main objectives of the maturity stage are earning profits and maintaining the firm's market share for as long as possible. Sales level off as the market becomes saturated and competition becomes fierce. Companies not doing well drop out of the market. Marketing costs rise due to competition as each firm tries to find ways to gain market share. The firm may need to make large promotional expenditures to show the differences between the firm's product and the competition and feel pressure to reduce prices. As a result, profits typically begin to decline during this stage.

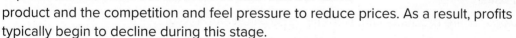

Coca-Cola's flagship cola represents a product that is in the *maturity* stage for Coca-Cola.

Decline Stage

The **decline stage** of the PLC is preceded by declining sales and profits. Depending on the product, the decline in sales may occur over a long period of time. During the decline stage, competitors drop out of the market as the product becomes unprofitable. The firm will likely cut prices to generate sales, curtail advertising, eliminate unprofitable items from the product line, and reduce

or eliminate promotion to individual consumers and resellers. Little or no effort is put into changing a good's appearance or functionality at this stage because consumers have moved on to other products. For example, travel agencies are in far less demand than they were a generation ago, before travelers could research and book trips online. In response, many travel agencies have shifted their focus from general consumer markets to niche markets, such as horseback-riding trips in South America or wine-tasting tours in Europe. Such trips generate limited demand but can still earn profits for the agency.

Diet Coke represents a product that is in the *decline* stage for Coca-Cola (due to the popularity of Coke Zero Sugar), although new packaging and flavors may reverse this trend and return the product to the maturity stage.

Variations of the Product Life Cycle

Once firms understand the stages of the PLC, they can tailor the marketing mix for their new products appropriately at each stage. Marketers should estimate the length of the product's life, taking into account marketing research and analysis of similar products. For example, Apple could have used data from its iPad to estimate the PLC of its Apple Watch because the products have many similarities and possibly many of the same customers.

Product life cycles can be of varying lengths, depending on the type of product. Technology-driven products like computers tend to have a short PLC because of rapid changes in computing power and features. Other products may remain viable in the marketplace for decades. For example, Wells Fargo has been offering banking services since 1852 and continues to be one of the largest suppliers of mortgage financing.

Marketers must also consider what type of PLC their product is likely to have. The following are four common product types, that each have a different PLC:

- *High-learning products* take longer for consumers to see the benefits of and often do not have a good infrastructure in place to support them. For example, it has taken a while for the electric car to catch on as a viable mode of transportation because consumers don't fully understand the product's usefulness and risks.

- *Low-learning products* are products with benefits customers can easily see. When a new soda flavor is launched, it is pretty easy for a consumer to determine whether they should try it or not based on familiarity with similar products.

- *Fad products* are very popular for a relatively short amount of time. As referenced earlier in this lesson, the fidget spinner was a fad that grew and faded quickly.

- *Fashion products* come in and out of favor with consumers. Most fashion designers, such as Ralph Lauren, introduce new fashions at least once per year.

It is critical for marketers to understand how long they expect the PLC to be and also what type of PLC their product should have. This information is then used to estimate product stages so the marketing mix can be adjusted accordingly.

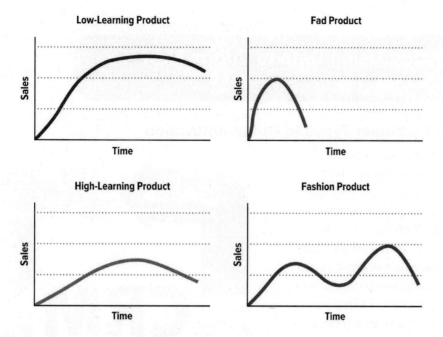

Different Types of Product Life Cycle Curves. Understanding the PLC and its implications impacts how marketing professionals manage the marketing mix for products.

The Marketing Mix and PLC Stages

Just as our interest, activities, and abilities change as we enter various stages of our life, the stage of a product's life has many implications for how a firm markets the product. The table below lists the typical strategies a firm undertakes for each of the four Ps, depending on where the product is in its life cycle.

Typical Marketing Mix Strategies during a Product's Life Cycle

Stage	Product Strategies	Place Strategies	Promotion Strategies	Price Strategies
Introduction	Offer small number of models.	Limit distribution; attract channel partners.	Promote to develop customer awareness.	Set price high.
Growth	Offer variety of models with many modifications.	Make intensive effort to expand distribution.	Promote to build awareness; increase personal selling.	Lower price as competition arrives.
Maturity	Offer full line of products.	Make intensive effort to maintain distribution.	Promote to point out brand attributes and differentiating features.	Set price equal to or below competitors'.
Decline	Reduce number of products based on profitability.	Phase out unprofitable outlets.	Reduce promotion to a minimum or eliminate.	Set price to maintain small profit; can increase if product appeals to niche market.

Marketing Analytics: Using Data to Inform New Product Development

Using Data to Track Customer Types to Guide Innovation

Current customers often voice changes that they would like to see implemented in a firm's products or services. To monitor these changes (as well as sales), a firm will employ a customer relationship management (CRM) system. The information collected through CRM can then be shared with the product development team and help a firm identify who among these consumer groups might be lead or innovative users. Often, these consumers are the ones interested in collaborating with the firm in new product development. Data from this subset of consumers are key to firm innovation. Although a firm relies on feedback from its current customers to help drive innovation, it also relies on feedback from former customers and noncustomers. Monitoring feedback from these customers gives a firm insight into why former customers no longer use its products, or why noncustomers have yet to adopt them. These additional data can confirm whether product innovation is needed. Although customer feedback is integral to determining product innovation needs, so are data about industry trends and new products entering the market. New technologies are a type of industry trend that can often displace existing products, so a firm must also track data from technology companies in order to forecast how these companies might impact its decision to innovate or not.

Using Data to Track Consumers through the Diffusion Process

Data are necessary to determine when the marketing mix needs to adjust to meet changes in the diffusion process over time. Each product moves through the diffusion process at a different rate, and it is imperative that data are collected not only on sales but also on changes in consumers' profiles over time. The type of consumer in the marketplace drives expectations regarding price and distribution, which in turn mandates changes in strategy. For example, when a product is first launched, and innovators in the market are evident, the firm will focus on meeting their needs because they are considered high-knowledge and high-risk tolerant consumers. Innovators are not concerned about price because they often purchase products based on other criteria. A recent example are drones. When drones were first sold they appealed only to a small segment of buyers. Drones were very expensive and required technical expertise to fly them. However, over time, the price of drones declined and their availability increased. Drones are now sold in toy stores as well as technology and specialty shops. Drones today appeal to a segment of buyers very different than the buyers who were initially in the market for them.

Using Data to Define New Product Development and to Identify Risks

In order to ensure the success of any new product, the firm must rely on data to identify *whom* the new product is for and *why* it is needed. New products often fit with existing products and customers. Or, new products can be designed to solve a new problem for the existing customer base. *Why* the firm is introducing a new product will determine its development process. The new product development process includes a defined set of benchmarks that must be met in order to reduce the risks of a failed launch. These benchmarks are often derived from firm and industry data about market potential and profitability or other objectives. For example, test marketing is an expensive process; often products are put in test marketing for a limited period of time to see whether they can achieve a sales target. This target is defined prior to the test market, and data from the test market will be evaluated against the benchmark data. It is important to remember that not all new products are introduced for profitability; sometimes a firm will introduce a new product simply to have a product in the market against a competitor or to confuse consumers about the value or use of a competing product. For example, years ago Pepsi introduced a product called Crystal Pepsi, which was a clear cola. Coke then introduced a product called Tab Clear, which was a clear diet cola. The only reason Coke launched this product was to eliminate Crystal Pepsi. In this case, Coke relied on data to identify Crystal Pepsi's declining sales. Once data confirmed the decline, Coca-Cola withdrew the Tab Clear product, even though its launch was successful.

Using Data to Determine Where a Product Is in the Product Life Cycle

The product life cycle is a backward-looking measure. In other words, it is impossible to know when you have changed from one stage to the next without looking backward through the data. Because the model is based on sales over time, sales figures must be measured periodically to track how the firm is doing against prior sales and against overall industry sales. The shape of the product life cycle for the firm should closely reflect the shape of the product life cycle for the industry. Because firms often do not grow at the same rate over time, the actual product life cycle for the firm may not be the smooth curve that is often shown. Sales tend to rise and fall from month to month, so firms analyze the overall trend in sales. For example, in the crucial turn from growth to maturity, it may be difficult at first to know when that turn took place because sales will often gradually level off for both the firm and the industry. The period of late growth or early maturity is difficult to define but is important to track because the marketing mix strategy must change to accommodate a mature market versus a growth market. For example, when Apple launches a new iPhone, it concentrates all of its marketing efforts on the new version. It creates new cases, earbuds, and other related accessories to accommodate it. Although earlier versions remain on the market, additional production of accessories for these older phones ceases. Eventually, these older versions of the iPhone are eliminated from the market. The timing of the launch of a new iPhone often corresponds with declining or leveling sales of the existing iPhone. When sales for the existing phone hit maturity, the new phone is launched, and the existing phone begins its decline phase. The new phone brings new users into the market.

Responsible Marketing

Controversial New Products

There are numerous examples of new product failures, from New Coke to Clairol "Touch of Yogurt" Shampoo to flavored bottled water for pets. Those are examples of products that failed because of poor wording or a lack of marketing research. However, some products fail because they are hurtful or inappropriate.

However, some products fail because they are hurtful or inappropriate and should never have been made. One example of such an egregious product was when, in 2014, Urban Outfitters released a "vintage Kent State" sweatshirt, featuring what looked like blood spatter. Kent State is a college that was the site of a shooting where four people were killed in 1970. This was highly offensive to many consumers.

But some new products lie in the ethical gray area. Another example from 2014 is when U2 announced its new album, *Songs of Innocence*. Instead of going through normal streaming options, Apple made the album available to all of its iTunes subscribers at no cost, causing almost half a billion people to wake up one morning to U2 in their music collections. Companies are always looking for clever and innovative ways to launch new products, including new music, but to many, this felt like an extreme invasion of privacy. According to *Wired* magazine, the giveaway felt "worse than spam" to customers. Within days, Apple issued instructions for how to remove the album from consumers' playlists.

Reflection Questions

1. What are some other ethical or responsible marketing considerations firms should review when launching a new product?

2. What are cultural considerations firms should review before launching a new product?

Marketing Services

As the brief case on Planet Fitness illustrates, the marketing of services poses unique challenges. In the case of Planet Fitness, how do you provide a welcoming environment for a target market seeking exercise in an atmosphere free from judgment and intimidation?

By the end of this lesson you will be able to

- Explain the role and value of services to the economy.
- Compare and contrast services marketing and product marketing.
- Describe service quality.
- Describe the Service Gaps Model.
- Explain why service failure occurs.
- Describe service recovery.

Planet Fitness: Providing Service in a Judgment-Free Zone

When we think of a service, we tend to think about those people, businesses, and organizations who provide a service to us—a server in a restaurant, the stylist who cuts your hair, or the ride-share driver who gets you from point A to point B. But the service industry is much bigger. Many businesses and organizations provide a service where we, the consumer, take an active role in the experience.

Consider our own personal health: Everyone has their own fitness goals, whether it's losing a few pounds, training for a triathlon, getting in shape, or simply being comfortable in your own skin. To reach these goals, many of us go to a gym or health club, and we definitely play an active role in the process. When exploring which health club to join, we quickly realize that while these businesses all provide a similar basic service (they give you a place to exercise), not all gyms are the same. They aren't all designed to meet the needs of an Olympic athlete, nor are they all built for the needs of the stay-at-home parent who wants to get in some quick exercise while the kids are at day care.

Planet Fitness isn't like every other gym out there either. The company designs its clubs to "provide a comfortable, safe and energetic environment where everyone feels accepted and respected."[1] With more than 2,100 locations in more than 1,400 cities in all 50 states, Planet Fitness has created a "judgment-free zone" where members can work out to the best of their own abilities without feeling intimidated or unwelcome. The bright colors and friendly staff create a unique environment in clean and spacious facilities with cardio and strength equipment to suit members' needs. Many Planet Fitness locations offer free fitness training and are open 24 hours a day for flexibility and convenience. The company's mobile app makes it easy for members to sign up for a "Design Your Own Program" session, track their fitness progress, and even use the Crowd Meter to check out how crowded the club is before they walk inside.

For those who want even more from their health club, Planet Fitness offers the PF Black Card, with exclusive benefits that include free guest passes, worldwide club access, tanning, massage chairs, and discounts with fitness-related merchants. According to Steve Spinelli, chair of the board for Planet Fitness, "Planet Fitness has grown because of a dogged focus on supporting the member's personal definition of exercise and well-being. We don't preach. We support."

Planet Fitness provides higher levels of service to those who want to get the most out of their approach to fitness and health.

Why Marketers Care about Services

As consumers, we purchase and use services daily. Think about all the things you used today that weren't, well, *things.* If you listened to streaming music, asked Alexa what the weather was going to be, checked your bank account online, sent a text message, rode in an Uber, ordered a coffee using the Starbucks app, read an Amazon review, or even just turned on a faucet or flicked on a light switch, you've been the consumer of a service.

Lots of services are essentially "invisible." Usually we think about, or even remember, service experiences only when they're either very good (like when you order a Tall coffee but get a Venti) or very bad (like when the power goes out). But just because we can't "see" services like we can physical goods, doesn't mean that they aren't strategic, calculated, revenue-generating firm activities.

In addition to consuming services, you may one day be delivering them (if you haven't already). If you intend to own, manage, or work for a business—ever—there's a very high likelihood that you will be part of the service economy. Service industries in the United States represent close to 80 percent of the gross domestic product (GDP).

In the next section we will explore what makes services unique and how service marketing differs from product marketing.

Why Services Are Unique

It is fairly easy to consider how a brand manager would develop a new product: put it into production in a factory, fill a warehouse with inventory, and then ship those goods to retailers. We're quite used to thinking about business and marketing with respect to manufactured goods. But when considering a business that sells services, the same model doesn't fit as well. A brand manager is not able to make

500,000 carwashes at once, thus benefiting from economies of scale in manufacturing. She would also be unable to inventory 10,000 sales calls at a central warehouse, ready to go whenever she needed more customers. And, certainly, Amazon can't keep haircuts in stock that you could order on Prime whenever you needed one.

At the most basic level, the difference between goods and services appears to be whether something is physically tangible or not. But this is just the first of four defining characteristics of services. Services are also different almost every time they are delivered or received (*heterogeneous*). Unlike products that can be mass produced and are the same (*homogeneous*), services differ every time they are created. Furthermore, services are produced and consumed at the same time (*inseparable*), and because it's impossible to manufacture services in bulk, they do not last (*perishable*).

The list below provides a quick overview of the four defining characteristics of a service. In the following sections, we will explore these characteristics more closely. We will also learn about the technology and trade-offs involved.

Intangibility

- Not physical
- Cannot be touched
- Cannot be stored
- Cannot be possessed

Heterogeneity

- Experience changes each time
- Cannot be mass produced

Inseparability

- Production and consumption occur at the same time

Perishability

- Cannot be stored indefinitely
- Cannot be reused

Intangibility

As we mentioned, one factor that makes services unique is their **intangibility**—services have no physical substance, and so cannot be touched, stored, or possessed like goods. Marketers put goods and services on opposite ends of a spectrum, with pure goods (for example, the gasoline in your car or the salt in your shaker) on one end and pure services (for example, a football game or your marketing class) on the other end. The reality is that, as you can see in this figure, most market offerings exist somewhere along a spectrum, not at the extremes.

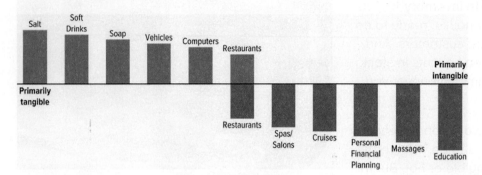

Tangibility Spectrum

A digital textbook is a perfect example of something halfway along the spectrum between goods and services: a regular textbook is almost a pure good, but the ability to read it online and interact with the content is very much a service. Fast food like McDonald's is also approximately halfway between goods and services: Your hamburger and fries are tangible goods, but the process of taking your order, cooking it, and providing it to you in a specific environment is a service.

What this means for marketers is that no matter what companies end up selling, there is almost always going to be an element of service involved.

Fast-food restaurants like McDonald's provide both goods and services. In this image, a drive-through customer is reaching for her McDonald's order, the *good,* and receiving it from the salesperson at the drive-through window, the *service.* The drive-through itself is also a service that McDonald's and many other fast-food restaurants offer.

Heterogeneity

The second unique characteristic of services is that they are heterogeneous. **Heterogeneity** means that services can be different every time you experience them. Have you ever had a bad haircut from your regular stylist that you've been to dozens of times? Or perhaps you've had a really wonderful meal at a restaurant where you've previously had a not-so-great one?

With the manufacturing of goods, production can be standardized and replicated perfectly for each unit of production, whereas with services, perfect replication is impossible. Although standardization is possible, it is not necessarily guaranteed. There are two main reasons why services cannot be perfectly standardized.

First, services usually depend on some sort of interaction between a customer and a service provider. Anytime human beings are involved in production or consumption of a market offering, there is always the chance for variability.

Second, services are difficult to measure. One of the benefits of mass production is quality control, where finished goods can be inspected prior to sale to make sure they came off the assembly line as intended. Because there is no standardized production of services, it can be difficult to know whether the service that was delivered exactly matches the service that was supposed to be delivered.

Inseparability

The third unique characteristic of services is that production and consumption are **inseparable.** Consider a can of Coca-Cola: The consumer does not need to be present at the Coke factory during production and bottling, and Coke employees do not need to be present when we drink a bottle. For most physical goods, production and consumption are entirely separate activities.

For services, however, the two occur simultaneously. It is impossible to think about getting your hair cut without having your hair stylist standing behind you with scissors, or taking a flight without an airplane and a cabin crew.

What's important to consider is how each customer and service provider, and even other customers or employees present, influence the service encounter and each other. Service production and consumption is a dynamic activity that occurs in real time between the organization and the customer.

Perishability

The final unique characteristic of services is that they are **perishable.** Just like flowers or fresh fruit, you can't store them indefinitely. Just like fresh fruit or flowers, services don't keep. Once an airplane takes off, empty seats can never be sold again. And if you have a bad flight, you are unable to return the flight and get another one. If there are no Uber drivers available when you need a ride home from the airport, you either have to wait or find another means of transportation. Or if the Planet Fitness class is only half full, those other spots can't be saved for the next day. More so, you can't buy a flight, an Uber ride, or a specific fitness class and keep it to use again the next day.

Once a flight takes off with empty seats, that airline can no longer sell seats for that flight. In this way, the service is *perishable.*

Considering that production and consumption occur simultaneously, it follows that it can be difficult to synchronize supply and demand for service offerings. A restaurant buys all of its food supplies anticipating a certain number of customers, but if nobody actually walks in, the food will go to waste. On the contrary, of course, if more customers show up than anticipated, there won't be enough food. Grocery stores and florists think in much the same way as service managers in that they try to buy just enough supply to meet demand without having too much excess.

Service Quality

Considering that service offerings are intangible and that they must be produced and consumed at the same time, and thus they are variable and cannot be mass produced or inventoried, assessing the quality of services is slightly more complicated than it is with goods. If you can't hold up a service and inspect it, how are you supposed to know whether it was high quality or not? How would you measure its quality? And just as important for marketers, how can they plan for quality so they know ahead of time that their customers will be happy?

Service quality is important for a simple but very important reason: without it, companies won't have any customers. There is a direct causal relationship between service quality and customer satisfaction. So if companies want satisfied customers, they need to provide them with quality service!

It's worth keeping in mind that customer service is not necessarily the same thing as service quality. Customer service is a *business activity*—a type of service offering that companies provide to their customers. **Service quality** is a *measurement activity* of how good a service experience is or was.

Low-quality service will lead to dissatisfied customers that will probably take their business elsewhere, causing lost revenues. Furthermore, unhappy customers are likely to tell people about the bad service, whether via word of mouth to friends or online via reviews, which can turn away prospective new customers. From restaurants to cell phone providers to grocery stores, think about the businesses you go back to repeatedly versus the ones you tried once and never went back to again. The reason is almost always service quality.

There are three ways to think about service quality, depending on which perspective we want to take:

- Quality based on objective measure of goodness or excellence of service
- Quality based on the technical and functional qualities of the service
- Quality based on evaluative perspective

Objective Measure of Goodness or Excellence

One way to think about quality is as an *objective measure* of goodness or excellence. How good something or someone is can be measured against some type of universally agreed-upon standard. Most people would probably agree that LeBron James is a very good basketball player, Chase Bank is a safe place to keep your money, and Google Maps is handy for navigation. When there is a high degree of general consensus about the goodness (or lack thereof) of a service, we can all more or less agree about its quality, but it doesn't give us much specific detail either as marketers or consumers. This perspective is more or less like looking at a product's Amazon review score, but without also reading the reviews to see why or how it got that score.

Technical and Functional Qualities of a Service

Another way to think about service quality is by considering the "what" and the "how" of the service differently. First, we can assess the quality of the *output* of the service. Because a service is an activity of doing something to something or to someone, we can always evaluate the *result* of a service, called **technical quality.** Is your hairstyle better than it was before? Do you know more about marketing than you did last semester? Was your triple-grande-in-a-venti-cup-nonfat-no-foam-with-whip-three-pump-hazelnut-one-pump-vanilla-203-degree-caramel-macchiato made correctly? We decide, retrospectively, whether the service was of good technical quality once it's completed.

But there's also consideration of "how" that service result came to be, where we can assess the quality of the service *process* called functional quality. **Functional quality** is an evaluation of the manner in which the service was provided. Was your stylist friendly and chatty during your haircut? Did you enjoy your marketing class and get value from the learning resources? Was the Starbucks ambient and playing nice music while you waited for your beverage, and did the barista make it quickly? We can evaluate functional quality *during* the service encounter.

Evaluative Perspective

A final way to think about service quality is the perspective from which we evaluate it. What a company considers quality and what the customer considers quality may not necessarily be the same. We will explore these two perspectives in much more depth later in the lesson.

Review these examples:

Good Technical Quality

Good technical quality would be represented by a hot and on-time pizza delivery.

Poor Technical Quality

Poor technical quality is when the service harms the product. In this case, perhaps the driver dropped the pizza after taking a slice?

Good Functional Quality

Good functional quality is found when the driver appears happy, is polite, wears a clean uniform, and has a clean car.

Poor Functional Quality

Poor functional quality would be if the delivery person shows up late, is rude, or insists on asking for a tip, despite offering poor and delayed service.

The Five Dimensions of Service Quality

There are five critical dimensions of consideration when it comes to evaluating service quality from the customer's perspective. You can remember these five dimensions with the acronym RATER.

1. **Reliability** is the ability to perform the promised service dependably and accurately, at the promised time.

2. **Assurance** is the knowledge and courtesy of employees and their ability to convey trust and confidence.

3. **Tangibles** are the appearance of physical facilities, equipment, personnel, and communication materials.

4. **Empathy** is the provision of caring, individualized attention to customers.

5. **Responsiveness** is the willingness to help customers and provide prompt service.

Decades and decades of service have shown that these five dimensions are the most critical factors to consider when assessing service quality from a customer's perspective. Again you will notice that these are not simply about customer service but about service quality more broadly. In the subsequent sections of this lesson, we will work through different ways to deliver great service, and you will find that each of these revolves around maximizing performance in one of the five RATER dimensions.

The Service Gaps Model

There are four gaps that companies need to consider when putting on their customer hat to evaluate service quality. These four gaps are the:

- Knowledge gap
- Design and standards gap
- Delivery gap
- Communication gap

These gaps make up what's called the **Service Gaps Model.** This model provides companies with a framework for understanding the differences between customers' perceptions and the customers' initial expectations of a service outcome. The framework presents four key criteria that firms must consider in order to ensure they are delivering optimal service to their customers.

The figure below depicts the Service Gaps Model and the relational impact among the four service criteria when one of those criteria is not fully satisfied. The goal of developing marketing strategy around these four criteria is to manage the gap between the customer's expectations and perceptions, and thus the perceived quality of service. We'll discuss these four service criteria as they relate to the four service gaps, next.

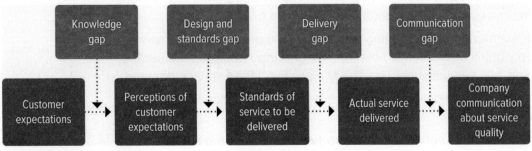

The Service Gaps Model

The Knowledge Gap

The first of four gaps in the Service Gaps Model is the **knowledge gap** (sometimes referred to as the *listening gap*). This is the difference between actual customer expectations and the company's perceptions of customer expectations. The only real way for companies to know what their customers want or expect is to ask them—and then they must listen. Companies can become better listeners by doing the following:

- Conduct more and better marketing research about the service offering and experiences. Observations, focus groups, and surveys are commonly used techniques.

- Build closer relationships with customers so they're more likely to share their thoughts and opinions. Encourage *upward flow of information*—that is, direct communication with the company—by providing customers with opportunities to share their opinions and concerns and encouraging them to do so.

- Follow up with customers after their service experiences, particularly for negative experiences. "What could we have done better?" is a powerful question to ask. Customers will be vocal about whether companies can fix or improve something in their service design.

When companies fail to listen to customers in order to really understand what they want from their businesses, this creates a knowledge gap.

The Design and Standards Gap

The second gap in the Service Gaps Model is the **design and standards gap,** or the gap between the service design and standards and the service delivery. In other words, it is the difference between the company's plan and the company's execution of that plan. Importantly, ensuring effective execution of service delivery is where, for most businesses, the proverbial torch passes from upper management to middle management and/or staff.

Service delivery is where the customer enters the picture and the service experience actually takes place. To ensure that the service has met the customer's expectations, the company must put certain standards, or criteria, in place, specifically as they relate to determining quality service, training employees, and measuring employee performance. To close the design and standards gap, companies can implement one or more of the following strategies:

- Hire and train like the business depends on it, because it does.
- Use tools like forecasting to appropriately match supply and demand; too many or too few employees can make a huge difference in a business.
- Use the price element of the marketing mix to increase or decrease demand to match supply (think airline seats, hotel rooms, or happy hour at a restaurant), keeping in mind that service supply is usually more fixed than service demand.
- Monitor and control the people and process elements of the extended services marketing mix.
- Train customers. Because services are co-produced between the service provider and the customer (for example, customers may pump their own gas at self-serve gas stations or assemble their own workbenches from Home Depot), companies should make sure that their customers know their role in the production. For simple co-production, like pumping gas, customers may

intuitively know their role. For more complex co-production, many firms, like Home Depot, invest in in-person workshops and high-quality video/interactive content to help educate customers on how to use their products.

Setting design and service standards can make the difference between a company's successful or unsuccessful delivery of a service. That success, again, is determined by customers' expectations of the quality of that service delivery.

The Delivery Gap

The third gap in the Service Gap Model is the **delivery gap.** This is the difference between what the company thinks the customer expects and how it actually delivers that service.

All companies have a plan for how they want to deliver their services, from Uber to Chase Bank to the Cheesecake Factory to the lemonade stand in the front yard. This plan comprises the business's operational strategies and tactics: basically, how the firm thinks it should run in order to ensure it is meeting its delivery standards.

If the company doesn't understand what the customer wants, or if it knows what the customer wants but doesn't deliver it exactly or completely, then the service delivery gap becomes problematic for service quality.

How can companies manage this delivery gap?

- First and foremost, they should have a plan. A service offering should be the result of strategic decisions and planning. Wise companies don't leave customer experience to chance.

- Companies should connect service design and standards (both in initial planning and later adjustment) to customer needs and company positioning.

- Companies should monitor and control their physical components in an effort to match them to their service goals (e.g., a clean restaurant supports quality perceptions of the food served).

If the company has a fairly good understanding of what the customer desires in a service, and how it should be delivered, the service encounter usually goes well.

The Communication Gap

The final service quality gap in the Service Gaps Model is the **communication gap.** This is the gap between service delivery and the company's external communications. Although at first company communications appear to be only advertising and promotions, companies speak to customers in a variety of ways. It's important that what the company says it does and what it *actually* does are very closely aligned. This is the "promise" part of the underpromise-and-overdeliver saying.

To manage the service provider communication gap, companies do one or more of the following:

- Integrate brand communications and messages across all customer touchpoints, and manage them centrally.

- Manage customer expectations, but never overpromise. Companies should be confident in their ability to add value, but if they overpromise, then they can't actually deliver the service they say they can.

- Never underpromise, either. Companies should make sure they can reach their customers and confidently show their value proposition through advertising, personal selling, physical evidence, pricing, social media, and any other communications activities. If they underdo it, customers either won't know about the value proposition of their services or customers won't know why to choose their service over the competition.

Companies can better manage customer expectations if they are clear and realistic about the services they provide.

Knowledge Gap

This is the difference between a customer's expectations and the company's perceptions of those expectations. Companies should spend less time telling customers what they *think* they want and more time listening to what they *actually* want.

Design and Standards Gap

This is the gap between the company's plan and the company's execution of that plan. To satisfy a customer's expectations, the company must put certain standards in place to determine quality service, train employees, and measure employee performance.

Communication Gap

It's important that what a company *says* it does and what it *actually* does align. For example, this image promotes all organic produce; if the company delivers nonorganic products to the customer, it will have created a communication gap.

Delivery Gap

A company needs to understand not only what and how the customer wants a service delivered, but also to make sure it delivers that service accordingly. Otherwise, that company faces a service delivery gap.

Why Service Failure Occurs

In any business, life happens. Every customer isn't satisfied 100 percent of the time, and even with the best intentions and best planning, sometimes a customer's expectations are higher than his or her perceptions of service, and the net result is a negative quality service experience. When the perceived quality of a service encounter is negative, it's referred to as a **service failure.**

Failure doesn't always mean that something broke during the delivery process; it just means that the service experience failed to go as planned. Small mismatches between customer expectations and perceptions usually aren't significant enough to count as service failures. If you ordered a glass of ice water at a restaurant and you received a glass of water without ice, that's a mismatch but it probably wouldn't lead to total dissatisfaction with the service encounter. For an expectation–perception mismatch to count as a service failure, the gap has to be notable enough to lead to customer dissatisfaction.

There are three common causes of service failure: company fault, unanticipated and unavoidable external circumstances, and other customers.

Company Fault

A company can cause the failure through poor planning, undertraining employees, overpromising in communications, or any other of the plethora of ways the firm can influence expectations or perceptions. For example, let's say you're on your way to the airport to board a flight to your vacation destination. You get to the airport, and you learn your flight is delayed by an hour due to a mechanical error. Then, when you're finally boarded, you ask for an extra packet of snack mix but the flight attendant rolls his eyes and says you're allowed to have only one. Those are examples of breakdowns in service execution that lead to a mismatch in expectations and perceptions (that the airplane should be well maintained and that the flight attendant should be more willing to respond to your simple request) and subsequently a dissatisfied customer.

Unanticipated and Unavoidable External Circumstances

Sometimes a service failure can be the result of unanticipated and unavoidable external causes that are completely out of the company's control; for example, flight delays due to bad weather or turbulence during a flight. Though most customers understand that weather is out of the control of the airline, it still can make them have a negative experience (especially if the bad weather is in a distant city). Almost all firms have some component of their service delivery that is out of their control. Have you ever had to wait a long time to check out at a store? It may be the fault of the manager's poor staffing decisions, the result of a flu outbreak that has left the retailer unavoidably understaffed, or even an unexpected increase in customers due to an accident that closed a nearby road (changing normal traffic patterns and where customers shop).

Other Customers

And, sometimes, a service failure can actually be the result of other customers in the immediate area of the service delivery. Think of that same flight, but instead of turbulence there was a crying baby on board. It's not fair to "blame" the baby for the crying, but it probably leads to dissatisfaction for most of the people sitting in adjacent rows on the plane.

The bigger the gap between service expectations and service perceptions, the more likely customers will talk about it and share their story, for better or worse. Very high-quality services lead to positive word of mouth, which is the most wonderful thing that can happen to a business. But very low-quality services lead to negative word of mouth, which is the kiss of death for almost any business.

If companies are lucky, the dissatisfied customer will complain directly to them. Why? Think back to the service provider listening gap from the previous section and how critical it is for good marketers to always listen to their customers and create opportunities for the upward flow of information. This is because not all service failures lead to complaints. A customer complaint is a *gift* for two reasons:

1. It tells the company exactly how to address the listening gap and subsequently its service design and standards to avoid future failures.

2. If a company knows who the dissatisfied customer is and why he or she is dissatisfied, it has given the company an opportunity to *recover* from the failure.

Next we'll look at some service recovery strategies.

Service Failure: Company Fault

Not having the items that people want to buy when they want to buy them is an example of a service failure caused by the company.

Service Failure: External Circumstances

Delaying a store's opening or making it hard for customers to get into the store because of a major snow storm is an example of a service failure caused by external circumstances.

Service Failure: Other Customers

High-volume retailers have many problems with other people negatively impacting customer experiences. In this case, a customer arguing with a cashier and causing a backup in the checkout lines can be interpreted by other customers as a service failure.

Meeting Customer Expectations through Service Recovery

Service recovery refers to the actions taken by a company in response to a service failure to improve the customer's level of satisfaction. The goal is to turn the customer's dissatisfaction into satisfaction, leading to a net positive quality service experience. In fact, a great company realizes that dissatisfied customers create an opportunity to overdeliver. A company that ignores dissatisfied customers usually doesn't stay in business long enough to worry about the impact on customer retention and loyalty.

For many customers, travel often provides many opportunities for service failures and service recovery. For example, if a customer arrives at a hotel to find that his or her room is not ready, even though it is time for check-in, the hotel can recover from this by quickly getting the customer's room ready and offering the customer a discount or meal voucher. However, if the hotel ignores the customer's complaint and does not remedy the situation in a timely fashion, it has failed in its service recovery efforts.

Companies follow two rules of service recovery to improve customer satisfaction:

1. Fix the customer's problem.

2. Fix the company's problem.

Rule 1: Fix the Customer's Problem

The saying "the customer is always right" can be used as a general rule to guide service recovery. In other words, a company should do whatever it realistically can to give customers the service they expected. It's important to remember that the customer is *not* always right, but the customer *is* always the customer—being right or wrong isn't nearly as important as being satisfied or dissatisfied. There are three simple rules for fixing the customer's problem:

1. *Act quickly and take ownership.* Timing and communication are critical to service recovery. Companies should always own the process of recovery, regardless of what caused the failure.

2. *Treat customers nicely, fairly, and with respect.* Remember that a business wouldn't exist without customers: people don't complain because they dislike the company; they complain because they want to feel like they're important to that company.

3. *Compensate the customer appropriately.* Compensation does not need to be financial; offers for repeat or alternative services can go a long way for a customer (and cost the company far less to provide). Compensation is an investment in loyalty and promotions, not an expense, so companies should not skimp.

Rule 2: Fix the Company's Problem

Rule 2 of service recovery is *fix the company's problem.* Whether or not the company performed the service wrong doesn't matter here. What does matter is that the customer's expectations and perceptions were so notably misaligned that there was a service failure. That's a big problem, but it's one that can be fixed by applying the Service Gaps Model. If the problem is about perceptions, that means the company needs to adjust its service design and execution to ensure that the problem does not occur again. If the problem is about expectations, that means the company needs to adjust its communications, as discussed in the service provider communication gap.

Many companies utilize an organized line system to manage customer ordering. This system eliminates uncertainty and unrealistic expectations of when a customer's order will be taken or who is next in line.

Responsible Marketing

The Human Component of Services

Uber is a beloved brand by many around the world, and completely disrupted the transportation industry. Uber now operates in over 70 countries and 10,000 cities around the world (uber.com).

Like many service-centered organizations, Uber is driven by people: its drivers and its customers. More often than not, both parties behave appropriately and drivers deliver on service expectations. However, there are some ethical concerns that can arise for Uber drivers and passengers.

- For passengers, safety is a concern. Uber drivers are alerted of pickups via their mobile device, thus often causing them to engage in distracted driving. Additionally, there are several reports of Uber drivers assaulting passengers.
- For drivers, safety is also a concern. There are also accounts of passengers assaulting Uber drivers or behaving in other inappropriate or dangerous ways.

When delivering services, firms run the risk of inappropriate human behavior, from both the service provider and the customer.

Reflection Questions

1. Besides policies and background checks, what are other ways that firms like Uber can minimize inappropriate conduct?

2. What are some other examples of irresponsible or unethical service experiences you have encountered?

Creating Value: Test

1. Haze, a beverage manufacturer, has a business strategy concentrated on developing environment-friendly products. It uses green-colored packaging to indicate the ecofriendly component of its products. This helps Haze's products to stand out and be recognized easily. This scenario exemplifies the use of:

   A. brand loyalty.

   B. brand marks.

   C. brand perception.

   D. brand consciousness.

   E. brand equity.

2. About 50 consumers are participating in a focus group moderated by marketers. They are asked to list terms and phrases that occur to them when they hear the brand name, 'Rotomatix'. The marketers of the brand use the consumers' responses to gain insight into what consumers think of the brand. This scenario is a typical example of

   A. projective association.

   B. ad hoc feedback.

   C. quantitative feedback.

   D. free association.

   E. audience optimization.

3. Which of the following is true of brand extension?

   A. It enables new products to profit from the recognition that a brand already enjoys.

   B. It must only be implemented when the firm is willing to risk cannibalization of existing products.

   C. It is a strategy in which product of two or more brands are merged into a new brand.

   D. It is not a recommended strategy for companies that possess high brand equity.

   E. It is a process in which a firm develops new products within the same product category.

4. Capitalizing on its high brand equity, ReInforay, an established retail and grocery brand, introduces a new product, ReNew, which is an improved variant of its existing hair protein serum. ReInforay soon discovers that after the successful rollout and high sales of its ReNew, the sales volume of its existing product has dropped significantly. Which of the following phenomena does this scenario exemplify?

   A. cannibalization

   B. cobranding

   C. rebranding

   D. market diversification

   E. market development

5. Healpro, a painkiller, is a low-cost product developed by S&V, a popular retail chain in the United States. It is sold exclusively in S&V stores. In this case, Healpro is an example of a:

   A. manufacturer brand.

   B. private label brand.

   C. captive brand.

   D. generic label brand.

   E. category label brand.

6. Apple iPhone's overall benefits of communication, information, and connection represent which product component?

   A. core product component

   B. augmented product component

   C. actual product component

   D. benefits product component

   E. advantages product component

7. The iPhone's smartphone is well-regarded as a stylish phone with several colors and sizes. This is an example of which product component?

   A. actual product component

   B. core product component

   C. augmented product component

   D. benefits product component

   E. advantages product component

8. Which of the following is an example of a convenience product?

   A. breath mints

   B. appliances

   C. cars

   D. gardening supplies

   E. internet service

9. Quincy enjoyed the ambiance of the theater and found its ushers to be friendly and knowledgeable. This is a matter of

   A. functional quality.

   B. evaluative goodness.

   C. technical quality.

   D. objective excellence.

   E. inseparability.

10. Technical quality of a service is based on the evaluation of the service

   A. input.

   B. objectivity.

   C. process.

   D. perspective.

   E. output.

11. Which of the following products *most likely* has an objective measure of goodness or excellence?

   A. frozen yogurt from Red Mango

   B. gym membership at Crossfit

   C. haircut from Supercuts

   D. memory subscription to Dropbox

   E. coffee from Starbucks

12. Every time Tess brings her bike in for service, the work is done well and her bike is ready to be picked up at the promised time. In regard to service quality, this is a function of

   A. empathy.

   B. assurance.

   C. tangibles.

   D. reliability.

   E. responsiveness.

13. Companies use this to understand the differences between customers' perceptions of a service outcome and the customers' initial expectations.

   A. SWOT Analysis

   B. Characteristics of Service

   C. Service Gaps Model

   D. Dimensions of Service Quality

   E. Tangibility Spectrum

14. Companies use the Service Gaps Model to

   A. inform customers of the gaps in the company's ability to provide service.

   B. bridge the gap between good service and excellent service.

   C. achieve an objective measure of goodness or excellence.

   D. manage gap between customer expectations and perceptions.

   E. train employees to provide customers with excellent service.

15. When a customer perceives the quality of a service encounter as negative, it is referred to as

    A. service quality.

    B. service gap.

    C. knowledge gap.

    D. heterogeneity.

    E. service failure.

16. Devon is considered to be tech-savvy by his friends. He is willing to take risks when it comes to trying new products and is always the first in his group to get the latest gadget. What type of product adopter is Devon?

    A. innovator

    B. early adopter

    C. laggard

    D. late majority

    E. early majority

17. Marketers seek to gain this type of adopter's acceptance because this adopter tends to be an opinion leader who is willing to talk to other people about his/her purchase experiences. This type of adopter is referred to as a(n)

    A. innovator.

    B. early adopter.

    C. laggard.

    D. late majority.

    E. early majority.

18. Keisha is in the market for a new tablet computer. She has spent quite a bit of time gathering information and thinking about the purchase. She has asked her friends for opinions about the models they own and is happy that she has a wide variety of models from which to choose. Keisha is most likely what type of adopter?

    A. innovator

    B. early adopter

    C. laggard

    D. late majority

    E. early majority

19. The marketers at Garnier included a sample of its new hair conditioner within the pages of the latest *Vogue* magazine. National sales increased dramatically within a few weeks, an example of this product's

    A. observability.

    B. compatibility.

    C. trialability.

    D. complexity.

    E. competitive advantage.

20. A company needs to simplify a product before sales will increase; it does this because

    A. an easier-to-understand product will diffuse faster.

    B. consumers can try it without significant expense.

    C. people will not buy complex products.

    D. if people see others using a product is will diffuse quickly.

    E. it will then be more compatible.

7 Capturing Value

What To Expect

Is a phone that cost $1,200 better than one that cost $400? While they may have similar functions, what you are willing to pay for them gives them a value that goes beyond their dollar amount. In this chapter we will examine how pricing can affect a product's perceived value.

Chapter Topics

- **7-1** Pricing

Lesson 7-1
Pricing

Pricing affects your life each day and is part of almost every consumer decision that you make. Thus, it is critical that you understand the strategy and tactics of pricing. This lesson will cover how marketers use pricing as a strategy to increase revenue, decrease inventory, and maximize profits.

By the end of this lesson you will be able to

- Explain the importance of pricing strategy to every organization.
- Outline the critical steps of setting a price.
- Explain why pricing objectives must extend marketing objectives.
- Explain how supply and demand sits at the heart of price setting.
- Explain the importance of accurately determining the costs of a product or service.
- Explain the importance of knowing what competitors charge for their products or services.
- Describe the most common tactical strategies for determining price.
- Explain the two common strategies for increasing prices in order to maintain price effectiveness.
- Compare the different pricing tactics that marketers can use.
- Identify the primary strategies for pricing new products.
- Explain the major legal and ethical issues associated with pricing.
- Identify the major pricing challenges facing international marketers.
- Identify the types of data firms use to make pricing decisions.

Marketing Analytics Implications

- Pricing objectives are often based on an analysis of point-of-sale data collected by a firm about its own products (and that of its competitors). Internal data are often critical sources of information for firms in setting price.
- Price elasticity of demand is determined by examining firm data regarding the impact on demand from past price increases, as well as data from pricing experiments conducted in chosen markets or modeled using artificial intelligence. When conducting a pricing experiment, the price of a chosen product is raised or lowered for a set period of time. Sales are then compared *before* and *after* this period; sometimes this set period is also compared to the same set period of the year prior.
- For break-even and other forms of pricing analysis, firms have to track and collect data regarding prices and overhead costs over time. A firm's internal database provides crucial inputs into any analysis.
- Price is the element of the marketing mix most dependent on timely data, accurate analysis, and implementation. Determining the pricing strategy for any of the firm's products or services is based on detailed analysis of firm data, market data, and competitor data.

Profitability on Demand at Hulu: Streaming a Price Point for Everyone

You probably subscribe to at least one of the three major video streaming services: Netflix, Amazon Prime, and Hulu. The second player to enter the streaming video market (after Netflix), Hulu launched its subscription-based streaming service in 2010, primarily to showcase content by founding partners NBC Universal, News Corp, and Walt Disney. The name *Hulu* comes from Mandarin to mean both "holder of precious things" and "interactive recording." Hulu's product portfolio focuses primarily on television series rather than movies. The company grew its base to 28 million active subscribers in 2019 through innovative and aggressive pricing strategies such as the following:

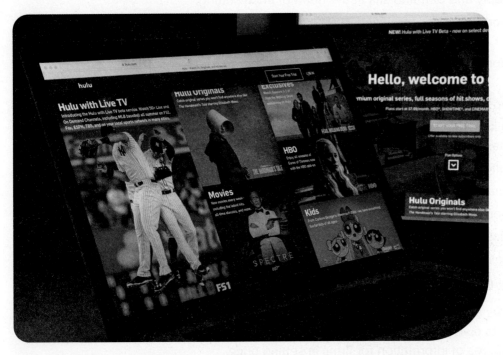

By using a number of psychological pricing techniques, together with an integrated marketing mix, Hulu was able to increase its customer base by 40 percent in just one year.

- **Tiered Pricing:** Hulu offers three tiers of pricing, designed to appeal to different customer segments with different needs and desires:
  - ▶ **Penetration Pricing:** The lowest plan Hulu offers starts at $5.99/month—a lower price than market competitors Netflix and Amazon (both $8.99/month). Hulu subsidizes this low price by streaming advertisements during its programs.
  - ▶ **Goldilocks Pricing:** The goal of a three-tiered approach is to drive customers to the middle option, the "just right" Hulu (No Ads) offering for $11.99/month. Often users will start with the lowest option, then trade up to Hulu (No Ads), which offers the same programming as the lowest plan but without ads.
  - ▶ **Premium Pricing:** In late 2018, Hulu added Live TV to its on-demand programming. This offering is at a premium price of $64.99/month with ads or $70.99/month without ads.

- **Psychological Pricing:** Every Hulu price point ends in .99—a great strategy for emotional purchases because it psychologically anchors consumers to the lower initial number.

- **Captive Pricing:** Hulu produces exclusive content like *The Mindy Project,* Marvel's *Runaways,* and *The Handmaid's Tale* that are available *only* to Hulu subscribers.

- **Promotional Pricing:** Hulu and Hulu (No Ads) offer the first month free, and Hulu + Live TV offers the first 7 days free. The goal, like any trial promotion, is to get viewers "hooked" on the content and service.

- **Add-On and Bundle Pricing:** Hulu "up-sells" its subscribers with à la carte access to premium channels like HBO, Cinemax, Showtime, and Starz for additional monthly prices ranging from $8.99 to $14.99.

This lesson looks at pricing as a complex and multifaceted strategic tool in the marketers' toolbox. We will look at how companies like Hulu use pricing strategies to attract and retain customers.

The Importance of Pricing Strategy

Price is the amount of something—money, time, or effort—that a buyer exchanges with a seller to obtain a product.

Pricing is one of the most important strategic decisions a firm faces because it reflects the value the product delivers to consumers as well as the value it captures for the firm. When used correctly, pricing strategies can maximize profits and help the firm take a commanding market position. When used incorrectly, pricing strategies can limit revenue, profits, and brand perceptions. Price has always been important, but in today's technology-enabled environment, consumers are more able to compare prices between multiple competitors, which makes having the correct price the difference between success and failure.

Pricing is the essential element for capturing revenue and profits. **Revenue** is the result of the price charged to customers multiplied by the number of units sold. **Profits** are simply revenue minus total costs.

Southwest Airlines is a perfect example of how the strategic use of price (low-price strategy) has produced sustained profitability. Until very recently, almost every U.S.-based airline (e.g., American, Delta, United, etc.) struggled to remain out of bankruptcy, much less profitable. Yet, for well over 40 years Southwest has remained a profitable and growing airline. There are many reasons for Southwest's success, but one major contributor is a consistently fair pricing structure that keeps planes flying to popular destinations filled with satisfied passengers.

The Price-Setting Process

Many factors influence how a firm sets prices. A firm's various stakeholders may voice a preference for higher or lower prices depending on their point of view. Marketing executives in search of substantial profits typically want high prices across the products they sell, whereas customers and salespeople often want low prices to increase the perceived customer value and ultimately the number of units sold.

Regardless of these factors, there are six steps involved in the price-setting process, shown in the figure below:

Step 1
Define the pricing objectives

Step 2
Evaluate demand

Step 3
Determine the costs

Step 4
Analyze the competitive price environment

Step 5
Choose a price

Step 6
Monitor and evaluate the effectiveness of the price

The Price-Setting Process

We will look at each of these steps in detail in the sections that follow.

Step 1: Define the Pricing Objectives

Step 1
Define the
pricing
objectives

Let's look at each of the steps in the price-setting process. The first step in setting a price is to clearly define the pricing objectives. Pricing objectives should be an extension of the firm's marketing objectives. They should describe what a firm hopes to achieve through pricing and, similar to the firm's marketing objectives, they should be specific, measurable, and reflect the market realities the firm faces.

Additionally, pricing is often a signal to customers about the quality of the product or service. Firms should carefully consider the quality perception that they want to develop in the marketplace when setting their prices.

Common pricing objectives include the following:

- **Profit maximization** or **price skimming:** This pricing strategy involves setting a relatively high price for a period of time after the product launches. An example of profit maximization is Apple. Every time Apple introduces a new iPhone, the price is initially very high and then decreases as the product progresses through the life cycle.
- **Volume maximization** or **penetration pricing:** This pricing strategy is designed to maximize volume and revenue for a firm, as well as encourage a greater volume of purchases.
- **Survival pricing:** This pricing strategy involves lowering prices to the point at which revenue just covers costs, allowing the firm to endure during a difficult time. For example, during the 2007 recession, General Motors reduced prices in an effort to avoid bankruptcy and sustain the firm.

Profit Maximization (Price Skimming)

The Goal

Designed to maximize profits on each unit sold.

How It's Done

Prices are set high for a period of time after the product launches and then decreased over time.

What It Needs to Succeed

Profit maximization assumes that customers value a product's differentiating attributes and are willing to pay a higher price to take advantage of those attributes, especially early in a product's life cycle.

Volume Maximization (Penetration Pricing)

The Goal

Designed to maximize volume and revenue for a firm; encourage a greater volume of purchases.

How It's Done

Products and services are offered at a lower price encouraging customers to purchase more.

What It Needs to Succeed

For this type of strategy to work over the long term, the firm must have a significant cost or resource advantage over competitors. Firms must be able to drive inventory turnover and cover profit loss.

Survival Pricing

The Goal

Designed to maximize cash flow over the short term and is typically implemented by a struggling firm.

How It's Done

Prices are lowered to the point at which revenue just covers costs, allowing the firm to endure during a difficult time.

What It Needs to Succeed

It should not be a permanent pricing objective, but it can be useful as a temporary means of staying in business.

Step 2: Evaluate Demand

The concept of supply and demand sits at the heart of setting prices.

Step 2
Evaluate demand

According to traditional economic theory, setting prices is as simple as finding the point at which marginal revenues equal marginal costs. **Marginal revenue** is the change in total revenue that results from selling one additional unit of product, while **marginal cost** is the change in total cost that results from producing one additional unit of product.

In this initial step, marketers must ensure that marginal revenue exceeds marginal cost. The question is, what is the optimal price that maximizes marginal revenue while minimizing marginal cost? The answer lies in the demand for the product.

Unfortunately, pricing in today's market is complicated. To start, environmental forces can significantly impact demand for products. Thus, before setting a price, marketers must understand the price elasticity of demand for their product, and the demand curve associated with their product.

Price Elasticity of Demand

To set price, marketers must determine the specific demand at various price points. This requires an understanding of consumers' **price sensitivity,** which is the degree to which the price of a product affects consumers' purchasing behavior. The degree of price sensitivity varies from product to product and from consumer to consumer.

Once marketers understand the price sensitivity exhibited by members of their target market, they can use this measure to calculate how changes in price will impact demand for a product and thus the amount of product the firm can sell at various price levels.

Price elasticity of demand is a measure of price sensitivity that gives the percentage change in quantity demanded in response to a percentage change in price (holding constant all the other determinants of demand, such as income). It is one of the most important concepts in marketing and should be considered when pricing any product.

Elasticity of demand is usually described in reference to products being either elastic or inelastic. **Elastic demand** is where demand changes significantly due to a small change in price. For elastic products, demand is greatly affected by changes in price, either increasing when the price drops or decreasing when the price is raised. Products with elastic demand include things that aren't critical to daily life. An example would be movies that you see in a theater. If the price of movies decreases ($5 Tuesday, 2 for 1 Thursday, etc.), the number of consumers who see movies increases. Conversely, if the price of movie tickets increases (which seems the trend), less people see movies at the theater and demand decreases.

However, products with inelastic demand are much less sensitive to changes in price. **Inelastic demand** refers to a situation in which a specific change in price causes only a small change in the amount purchased. Gasoline is a good example. If the price of gas increases by 10 percent, do you drive less? If the price of gas decreases by 10 percent, do you drive more? The answer is likely "a little more or less," but in most cases driving habits are not greatly changed by price changes to the product.

Almost no products are perfectly elastic (any amount of price change will affect demand) or inelastic (no amount of price change will affect demand), but products that are considered luxuries (boats, vacations, movies, etc.) or nonessential items are usually more elastic, while essential products (food, electricity, water, etc.) are more inelastic.

Size of expenditure

Customers are less sensitive to the prices of small expenditures which, in the case of households, are defined relative to income.

Shared costs

Customers are less price sensitive when some or all of the purchase price is paid by others.

Switching costs

Customers are less sensitive to the price of a product if there is added cost (both monetary and nonmonetary) associated with switching to a competitor.

Perceived risk

Customers are less price sensitive when it is difficult to compare competing products and the cost of not getting the expected benefits of a purchase is high.

Importance of end-benefit

Customers are less price sensitive when the product is a small part of the cost of a benefit with high economic or psychological importance.

Price–quality perceptions

Customers are less sensitive to a product's price to the extent that price is a proxy for the likely quality of the purchase.

Reference prices

Customers are more price sensitive the higher the product's price relative to the customers' price expectation.

Perceived fairness

Customers are more sensitive to a product's price when it is outside the range that they perceive as "fair" or "reasonable."

Price framing

Customers are more price sensitive when they perceive the price as a "loss" rather than as a forgone "gain." They are more price sensitive when the price is paid separately rather than as part of a bundle.

Source: Thomas Nagle, John Hogan, and Joseph Zale, *The Strategy and Tactics of Pricing*, 5th ed. (Upper Saddle River, NJ: Pearson, 2011), pp. 132–133.

Demand Curves

Directly related to a product's price elasticity is the product's demand curve. The price of all products relates to consumer demand for the product. In general, as price increases, demand decreases. Conversely, as price decreases, demand increases. The question is, how much does a product's demand change based on a price change?

For perfectly inelastic products (where price changes don't impact demand), the demand curve would be a straight, vertical line. For

perfectly elastic products (where any price change ends demand), the demand curve would be a straight, horizontal line.

For most products, the demand curve is less distinct and is illustrated by the curved line in the figure below. For somewhat inelastic products, such as gasoline and utilities, price changes don't have a large effect on demand. Most consumers need these items and simply keep buying them at a relatively constant rate regardless of price increases or decreases. Alternatively, for more highly elastic products, such as recreational sports equipment or high-end vehicles, price changes have a large impact. Most consumers don't need these items immediately, so as prices increase, demand quickly diminishes, but when the price decreases, consumers quickly start purchasing again. The curved portion of the figure illustrates the three types of demand curves that apply to most, but not all products.

The figure shows two straight lines that represent perfectly inelastic and perfectly elastic demand curves. For most products, demand and price move in a relationship represented by the curved line.

One exception to the typical demand curve is the prestige demand curve. For very high-end, prestige products, demand actually increases as prices rise and decreases as prices fall.

The prestige demand curve is shown in the figure below to illustrate the "sweet spot" of prestige pricing, which is neither too high nor too low. This phenomenon is a result of consumers who specifically equate quality, status, and luxury with higher, exclusive prices. Thus, if a product price remains high, the product is more acceptable.

For example, the Mercedes Maybach S600 costs just over $200,000. If the price were lowered to $150,000, sales would actually decline. Customers are willing to pay the higher price for the exclusivity associated with ownership. However, at some point, a very low price would increase demand for a prestige product. If the Mercedes Maybach were to be offered for $10,000, a whole new market would explode and demand would be very high, but not from the same customer base.

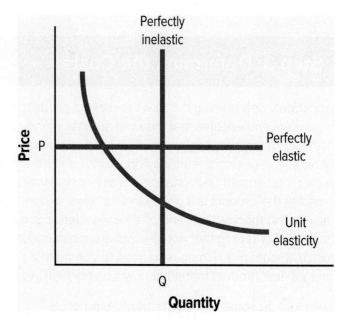

Inelastic and Perfectly Elastic Demand Curves.

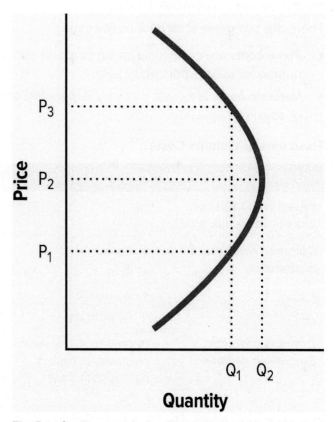

The Prestige Demand Curve. For prestige products, demand increases as price increases to a point (P_2, Q_2), with lower demand at the highest price and lowest price.

Step 3: Determine the Costs

Accurately determining the costs to make a product sets a lower price limit for marketers and ensures that they will not lose money by pricing their products too low. Although a firm may temporarily sell products below cost to generate sales as part of a survival pricing strategy, a firm cannot endure for very long employing this tactic. A marketer should understand all of the costs associated with its product offering, whether the product is a good, service, idea, or some combination of these. Costs include the money, time, and effort expended to produce and market a product. It is important to realize that not all costs are measured in dollars, as many real costs like the reallocation of human resources to complete a project or the time required to bring a new product to market may be very high yet hard to quantify in dollars.

Step 3
Determine
the costs

Let's look at some factors in determining costs.

Fixed versus Variable Costs

There are two general categories of costs:

- **Fixed costs** are costs that remain constant and do not vary based on the number of units produced or sold.
- **Variable costs** are costs that vary depending on the number of units produced or sold.

Fixed versus Variable Costs

	Fixed Costs	**Variable Costs**
Based on number of units produced or sold?	No	Yes
Common costs for businesses	• Salary • Rent • Insurance • Advertising	• Materials • Sales Commissions • Utilities • Delivery Costs
Example costs for Southwest Airlines	Annual liability insurance premium (in case of passenger injury)	Number of complimentary sodas given to passengers

Break-Even Analysis

Once a company estimates fixed and variable costs, it can incorporate them into a break-even analysis to determine how much it would need to sell to make the product profitable.

Break-even analysis is the process of calculating the break-even point, which equals the sales volume needed to achieve a profit of zero. Specifically, the **break-even point** is the point at which the costs of producing a product equal the revenue made from selling the product.

Copyright © McGraw Hill

286 Chapter 7 • Capturing Value

Once the firm has established the break-even point, it has a starting point for estimating how much revenue it must generate to earn a profit.

To calculate the break-even point, we divide total fixed costs by the unit contribution margin, which is determined by subtracting the variable cost per unit from the selling price per unit.

Consider this simple example of break-even analysis: If variable costs for a hotel room night at a Holiday Inn Express are $50, and fixed costs for a day at the hotel are $1,000, we can calculate the break-even point. Let's try this for a room rate of $100 and $250.

$$\frac{\$1{,}000}{\text{fixed costs}} \div \left(\frac{\$100 \text{ selling}}{\text{price per unit}} - \frac{\$50 \text{ variable}}{\text{cost per unit}} \right) = \frac{20}{\text{rooms}}$$

$$\frac{\$1{,}000}{\text{fixed costs}} \div \left(\frac{\$250 \text{ selling}}{\text{price per unit}} - \frac{\$50 \text{ variable}}{\text{cost per unit}} \right) = \frac{5}{\text{rooms}}$$

Note that break-even analysis analyzes only the costs of the sales; it does not reflect how demand may be affected at different price levels. In other words, it doesn't measure price sensitivity. Just because Holiday Inn Express will break even renting 5 rooms per night at a price of $250, it doesn't mean that Holiday Inn Express customers are willing to pay this much for a room.

Although marketers must understand a firm's costs to set prices effectively, costs should never dictate price. Strategic pricing requires firms to integrate costs into other aspects of the marketing mix, including what value the customer places on the product and the price strategies of other competitors in the industry.

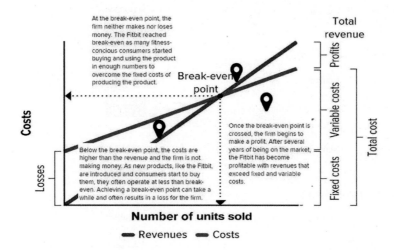

At the break-even point, the firm neither makes nor loses money. The Fitbit reached break-even as many fitness-concious consumers started buying and using the product in enough numbers to overcome the fixed costs of producing the product.

Break-even point

Below the break-even point, the costs are higher than the revenue and the firm is not making money. As new products, like the Fitbit, are introduced and consumers start to buy them, they often operate at less than break-even. Achieving a break-even point can take a while and often results in a loss for the firm.

Once the break-even point is crossed, the firm begins to make a profit. After several years of being on the market, the Fitbit has become profitable with revenues that exceed fixed and variable costs.

Costs / **Losses** / **Number of units sold** / **Total revenue** / **Profits** / **Variable costs** / **Fixed costs** / **Total cost**

━ Revenues ━ Costs

Step 4: Analyze the Competitive Price Environment

Step 4
Analyze the competitive price environment

Pricing does not occur in a vacuum. Marketers must consider what competitors charge for their products. Setting prices to compete against other firms is challenging and complex. Marketers can choose to match competitor prices, price lower than competitors thus offering customers greater value, or price higher because the firm offers a superior product.

Marketers must also consider how competitors might respond to their pricing. How a firm reacts to a change in a competitor's prices depends on whether the competitor is a stronger or weaker rival and whether the price reduction is cost justified—that is, if the lower price will allow the firm to remain profitable. Industry structure and the number of competitors are also very important.

Consider the fast-food industry. Most major fast-food chains have a $1.00 or $0.99 menu to compete with each other. Once one offers a "value meal" the competition will quickly introduce a similar "value meal" offering.

Competition is especially important because consumers are increasingly empowered by technology that allows them to comparison shop within stores and even without visiting stores. Because of this, companies like Walmart and Target have instituted pricing policies that allow customers to match prices and get the best price without having to visit a competitor. However, larger companies, like Walmart, Best Buy, Target, Home Depot, and Lowe's, also have realized that offering unique products not offered by competitors creates a situation where prices cannot be equally compared. Have you ever shopped on Black Friday, the shopping day after Thanksgiving? If so, you may have noticed that it usually isn't easy to compare sale products between stores because manufacturers have created special products or product bundles for each store that are not easily compared and/or price matched. This change in how products are marketed for different stores is a direct result of pricing strategy designed to anticipate consumer actions based on competitor offers.

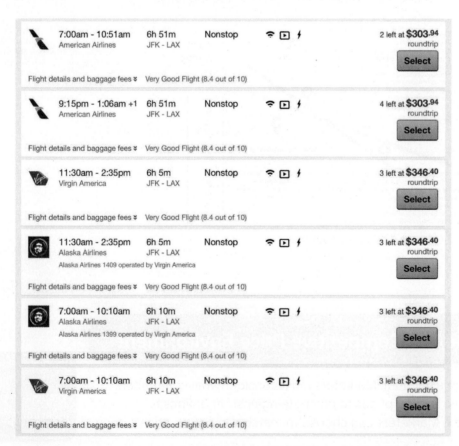

Step 5: Choose a Price

Choosing a price is a complicated process, and marketers rarely do it perfectly. The pricing decision should be made with the goal of maximizing long-term, sustainable profits.

Step 5
Choose a
price

Reference prices are the prices that consumers consider reasonable and fair for a product. Reference prices matter to marketers because consumers are typically more price sensitive the higher a product's price is relative to expectations. Marketers can capitalize on this tendency by identifying the reference prices of their targeted consumers when setting their own prices.

The pricing of Coca-Cola is a good illustration of how reference pricing works. Think of purchasing a 20-ounce bottle of Coca-Cola. What should it cost? Fifty cents, $1, $2, $3—could all be correct? The answer is yes, because reference prices are context specific. If you were purchasing a 20-ounce bottle of Coca-Cola as part of a six pack at a grocery store, an individual bottle cost of 50 cents would be close to accurate. However, the other prices might be correct too. What if you bought a bottle from a friend ($1), a gas station ($2), or at a sporting event ($3 to $5). Thus, a reference price represents a price that consumers consider reasonable and fair *in a specific context*. Prices higher than the reference seem overpriced and prices below the reference seem like a good deal.

Instead of just seeking to identify reference prices and price accordingly, marketers can also seek to *establish* reference prices for consumers. Apple, for example, launches a variety of products at various price points so that buyers can compare the products and begin to associate features with dollars.

Underpricing, or charging customers less than they are willing to pay, is also a common mistake in pricing. Customers place different values on the goods and services they buy. As the common adage goes, "You shouldn't charge less for something if a customer is willing to pay more for it." Because revenue is simply the number of units sold multiplied by the price per item, marketers too often make the mistake of setting prices too low in an effort to increase the units sold, without considering all of the other factors that contribute to the value a customer places on a product.

Airlines have considered the ramifications of underpricing for years in their pricing strategies. If you purchased an airline ticket today for a flight three months from now, you might pay $500. However, if you wait until the day before that flight to buy the same ticket, you may pay something in excess of $1,000 for the same trip. The customer buying three months in advance has plenty of time to comparison shop. However, the person who needs to fly tomorrow may have an urgent business meeting or a new baby in the family, either of which increases the value of the airline ticket.

Could the airline charge the last-minute customer the same price as the recreational traveler? Yes, but should the airline charge the same for both tickets if the value the customer places on the ticket differs? The reality is that value assessments change based on how important the customer needs the product at a particular time. Given these changes, marketers would be foolish to not adjust their prices based on consumer willingness to pay a higher price.

However, for most products, there is a line when it comes to price adjustments. It is considered unethical to overcharge, or **price gouge**, a customer in times of great need, such as during recovery from a natural disaster. For example, after a hurricane, the demand for items such as building supplies (plywood, roof shingles, etc.) may be high, but in an effort to protect consumers, ethics and laws limit the prices that firms can charge.

The difference between correctly pricing a product based on customer value perceptions and price gouging has to do with choice. The person who needs a last-minute flight has a choice—he or she could cancel the trip or drive to the destination rather than paying the current, last-minute price. However, the person whose home needs a new roof due to a hurricane has no other options. When no viable options exist, often laws or ethics step in to limit profit-taking by firms trying to capitalize on the peril of others.

Stores place their private label brands (Preferred Plus, in this case) directly next to competitors to indicate low price in reference to a similar brand name product.

Step 6: Monitor and Evaluate the Effectiveness of the Price

Choosing a price is not a one-time decision. Marketers must monitor and evaluate the decision to determine how effectively the strategy meets the pricing objectives. In addition, pricing strategy evolves and should be reevaluated throughout the product life cycle.

One of the most challenging aspects of pricing is initiating price increases. It is hard to imagine a customer being excited about paying more for the same product. However, in an effort to recover increasing costs or improve profits, firms often face situations that require price increases.

Unbundling provides value for customers who are focused on a specific price point rather than the complete product offering. Unbundling involves separating out the individual goods, services, or ideas that make up a product and pricing each one individually. Such a strategy allows marketers to maintain a similar price on the core product but recover costs in other ways on related goods and services. For example, a restaurant might unbundle a meal so that the hamburger sells for the same price, but the customer now must pay extra for French fries.

Step 6
Monitor and evaluate the effectiveness of the price

With the exception of Southwest Airlines, most airlines have pursued an unbundling strategy over the course of the last decade. They now charge separate fees for baggage rather than bundling the service of checked bags into the cost of the ticket. For many, unbundling is appreciated because it allows customers who aren't concerned about certain services to avoid paying for them. Conversely, many consumers look at the practice as a way for companies to "nickel and dime" them with many unexpected fees for services that they expected to be included in the initial purchase.

An **escalator clause** is an agreement that provides for price increases depending on certain conditions. The escalator clause ensures that providers of goods and services do not encounter unreasonable financial hardship as a result of uncontrollable increases in the costs of or decreases in the availability of something required to deliver products to customers.

An escalator clause in logistics can take the form of a fuel surcharge that allows carriers, such as trucking companies, to adjust prices based on the current price of fuel, which tends to fluctuate daily.

Many rental agreements provide a good example of how an escalator clause can be enacted. Landlords often include an escalator clause within the body of a rental contract that makes it possible to increase the monthly rent if taxes on the property go up.

Another good example of an escalator clause is a mobile data contract. AT&T offers its customers "unlimited data" but uses an escalator clause that reduces the speed of the data after a certain usage level (e.g., anything in excess of 20 GB in a month). Companies like Sprint, T-Mobile, and Verizon offer both unlimited and limited plans. Limited plans are less expensive per month but have an escalator clause that charges additional fees once a mobile data usage threshold is exceeded. This escalator clause is becoming less used as more companies move toward true unlimited data plans, but the concept applies to many settings that operate on a monthly subscription rate that has usage limits.

Firms that decide to use escalator clauses should make them as transparent as possible and specify whether price adjustments will be made at fixed intervals (e.g., quarterly, semiannually, or annually) or only at the expiration of the contract.

Step 1: Define the pricing objectives

Southwest establishes pricing objectives.

Step 2: Evaluate demand

Southwest knows there are many segments of travelers and that there is a segment that is always looking for a low-cost carrier. It also collects data that show how many seats are sold at each price so it can forecast demand.

Step 3: Determine the costs

In general, airlines have very high fixed costs. These costs don't change whether the plane has five passengers or is full. They include such things as gas, food, employee wages, and various airport charges. Because Southwest Airlines's fixed costs are so high, it is important that it fills its seats.

Step 4: Analyze the competitive price environment

Many low-cost carriers—Sprint, JetBlue, Virgin, and more—compete with Southwest, so it must constantly watching competitors' prices.

Step 5: Choose a price

Because many travelers purchase their tickets from online sites such as Expedia and Travelocity, travelers often know prices of flights when they look at Southwest prices. These reference prices influence travelers as they book their flights.

Step 6: Monitor and evaluate the effectiveness of the price

Airline prices are constantly changing due to the time of year, the price offered by competitors, changes in the popularity of travel destinations, and the economy in general. Southwest must always be ready to adjust its prices to remain competitive in the market.

Pricing Tactics

Once marketers have completed their analysis of demand, costs, and the competitive environment, they can use a number of different tactics to choose a final price. Which tactic to use depends on the value customers perceive the product to have, their ability to pay for it, and how they intend to use the product.

Pricing, like all aspects of marketing, requires firms to really understand their customers' needs and wants. Pricing requires a delicate balance of psychology as well as accounting to achieve the right strategy of maintaining profitability, creating a positive brand image, attracting customers, and reaching inventory and sales goals.

Let's look at a few pricing tactics in detail.

Markup Pricing

Markup pricing, or **cost-plus pricing,** is a pricing method in which a certain amount is added to the cost of the product to set the final price. Many marketers use a specific percentage as a markup for the products that they sell.

For example, traditionally jewelers, furniture stores, and gift shops will mark up the products they sell by 50 percent. Using this pricing method, a ring that costs a jeweler $500 would be sold for $750 [$500 cost + $250 markup ($500 × 50%) = $750 selling price]. **Profit margin** is the amount a product sells for above the total cost of the product itself.

The markup method is commonly used because it is easy to implement and it provides a good estimate of potential profit for planning purposes. However, many firms opt to use more precise methods of pricing because a standard markup policy does not take into account changing levels of customer willingness to pay. In some cases, this type of markup may be overpricing products, but it also may cause situations where products are underpriced.

Odd Pricing

Odd pricing is a pricing tactic in which a firm prices its products a few cents below the next dollar amount. The use of odd pricing is found in many consumer settings. Everything from hamburgers to cars are sold at prices that end in 3 ($2.93 for dish soap at Walmart), 5 ($4.95 for a drink at Starbucks), 7 ($12.97 for sandals at Target), or 9 ($14.99 for a pizza deal at Papa John's).

The pricing tactic of ending prices in odd numbers draws upon a basic human processing error, which is based on mental calculations that tend to see the price of $1.97 as much lower than $2. It is obvious that the difference is only three cents, but our brains tend to discount the extra change and focus on the whole number ($1 in this case). The same is true with larger numbers. A car dealer may price a vehicle at $4,995 instead of $5,000 because the odd price is viewed as closer to $4,000 than $5,000—even though the difference is only $5. Thus, marketers use this obvious yet effective practice to improve price perceptions of their products.

Prestige Pricing

Prestige pricing is a pricing strategy that involves pricing a product higher than competitors to signal that it is of higher quality. Firms that want to promote an image of superior quality and exclusivity to customers may pursue a strategy of prestige pricing. Luxury brands, such as Louis Vuitton, Cartier, and Mercedes-Benz, are perfect examples of this strategy. Such companies use high prices to suggest their products are of high quality and stylish. Simply improving the look, packaging, delivery, or promise of a product can justify a higher price and support a prestige pricing strategy.

Starbucks pursues a prestige pricing strategy by setting its prices high to convey increased value compared to other coffee purveyors like Dunkin'.

Seasonal Discounts

Seasonal discounts are price reductions given to customers purchasing goods or services out of season. Disney World pursues this strategy by offering its best rates in months like February, when demand is at its lowest due to cold weather and the fact that children are in school. Disney World promotes this time of year as its value season, which typically includes January, February, and the first three weeks of December. Seasonal discounts allow Disney World to maintain a steady stream of visitors to its parks year-round.

The strategy also exposes new customers to the brand. Young families (those with children younger than school age) can afford to try Disney World during its value season, and many go on to become loyal customers, purchasing Disney vacations during the peak summer seasons once their children start school.

Price Bundling

Price bundling is a strategy in which two or more products are packaged together and sold at a single price. Marketers often use bundling as a tool because they can charge higher prices for the bundle than they could for the elements individually. Assume you are buying a new Ford Escape SUV. Would you prefer to purchase the base model and then handpick options, such as a moon roof or satellite radio, or are you better off buying the vehicle as one all-inclusive bundle? Conventional wisdom says that à la carte pricing benefits the customer and bundled pricing benefits the firm. Undoubtedly, price bundling simplifies things for marketers. The company can sell the same bundle to everyone, leading to reduced advertising and selling costs. Think of the success bundled software packages such as Microsoft Office enjoy, despite the fact that many of the software's users need only a fraction of the available functionality.

Such success aside, price bundling has come up against increased customer resistance in some industries in recent decades. Cable television providers have always relied on a bundling strategy in which customers have to buy a package of channels rather than paying individually for the channels they want. In recent years, as the price of those channel bundles keeps rising, the practice has come under fire. In many cases, consumers are dropping cable because of bundled packages in favor of streaming services that are more flexible.

New Product Pricing

New products are sometimes priced using a different strategy than a firm's established products. In many cases, firms introduce new products and simply use the same pricing strategy that they follow for all of their products (as discussed previously in this lesson). However, two strategies specific to new products are often used by firms to stimulate demand or maximize profits of new products, as mentioned in an earlier section.

The first unique new product price strategy is *profit maximization,* or *price skimming,* which, as we defined earlier, is a pricing strategy that involves setting a relatively high price for a period of time after the product launches. This strategy may sound complex, but it is one that you are more than likely familiar with. All major smartphone marketers use some form of this strategy when they introduce a new version of their technology. For example, Apple consistently introduces new iPhones at their highest price and for an extended period to maximize profit from dedicated users. The same is true of other mobile manufacturers, such as Samsung and LG, which use a similar strategy when they launch a new product.

The second unique new product price strategy is *volume maximization,* or *penetration pricing,* which, as we discussed, is a pricing strategy that is designed to maximize volume and revenue for a firm, as well as encourage a greater volume of purchases. This new product pricing strategy should also be familiar to you. Penetration pricing is common for consumer products, such as food and services. For example, Papa John's may introduce a new pizza or type of side dish at an artificially low price to encourage trial and adoption of the product by consumers. Service providers of haircuts, nail care, or oil changes may also use penetration pricing in a competitive market to drive awareness of a new location or for expanded services.

Dollar Shave Club delivers products directly to the consumer, reducing some of its inventory and overhead costs. Because of this, Dollar Shave Club was able to offer penetration pricing to attract new customers to its models. By offering a subscription service, Dollar Shave Club is able to lock in customers and establish long-term relationships!

Legal and Ethical Issues in Pricing

Many legal and ethical issues impact pricing decisions. Pricing is one of the most watched and regulated marketing activities because it directly impacts the financial viability of both organizations and individuals. In the sections that follow, we'll discuss some of the ethical issues marketers may face as they seek to set prices for their products, including price discrimination, price fixing, predatory pricing, and deceptive pricing. We'll then look into some of the regulations the American government has put into place to combat such practices.

Price Discrimination

You may be surprised to hear that you have likely benefited from discriminatory pricing in various ways. If you've paid student prices at a movie theater or an introductory price to switch cell phone or cable providers, you've taken advantage of price discrimination. **Price discrimination** is the practice of charging different customers different prices for the same product. Price discrimination sounds negative, but it is illegal only if it injures competition. Organizations can charge customers different amounts for legitimate reasons. This is especially common in business-to-business (B2B) settings in which different customers might be charged different rates due to the quantities they buy, the strategic value of the company, or simply because one firm did a better job negotiating the contract.

Quick Facts:

Definition

Charging different customers different prices for the same product.

Example

Student discounts at movie theaters.

Illegal?

It is illegal only if it harms competition or if firms don't have legitimate reasons for the price differences.

Price Fixing

When two or more companies collude to set a product's price, they are engaging in **price fixing.** Price fixing is illegal under the Sherman Antitrust Act of 1890 and the Federal Trade Commission Act.

An example of price fixing occurred when British Airways and its rival Virgin Atlantic agreed to simultaneously increase their fuel surcharges. Over the next two years, fuel surcharges increased from an average of five British pounds a ticket to over 60 pounds (approximately $90 USD). When the price-fixing scheme was reported by Virgin, British Airways was punished with record fines. The British Office of Fair Trading fined the airline £121.5 million and the American Department of Justice levied an additional $300 million fine.

Quick Facts:

Definition

When two or more companies collude to set a product's price.

Example

British Airways and its rival Virgin Atlantic agreed to simultaneously increase their fuel surcharges to customers.

Illegal?

It is illegal because it can hurt customers and competition.

Predatory Pricing

Consider a situation in which a chain supermarket opens across the street from a locally owned grocery store. Theoretically, the prices at both stores should be similar because the costs and customer demand will be similar. However, because the chain supermarket can rely on corporate backing for support, it makes the decision to radically lower prices,

attracting more customers to its facility and eventually driving the competition out of business.

This example illustrates a strategy called predatory pricing. **Predatory pricing** is the practice of first setting prices low with the intention of pushing competitors out of the market or keeping new competitors from entering the market, and then raising prices to normal levels. This type of long-term aggressive pricing strategy could be considered an attempt to create a monopoly and is therefore illegal under U.S. law. Predatory pricing is an extreme form of volume maximization or penetration pricing (described earlier) that is used solely to eliminate competition.

Quick Facts:

Definition

Setting prices low with the intent of pushing competitors out of the market or keeping ones from entering, and then raising prices.

Example

If a chain store opens across the street from a locally owned store and offers the same product at a lower price, it will knock the locally owned store out of business.

Illegal?

It is illegal because it could be considered an attempt to create a monopoly.

Deceptive Pricing

Deceptive pricing practices can lead to price confusion, where consumers have difficulty discerning what they are actually paying. **Deceptive pricing** is an illegal practice that involves intentionally misleading customers with price promotions. The most common examples of deceptive pricing involve firms that falsely advertise wholesale pricing or promise a significant price reduction on an artificially high retail price. These types of deceptive pricing practices have come under fire in recent years in industries ranging from credit cards to home loans, where important information was often buried deep within little-noticed and hard-to-read disclaimers and information.

Quick Facts:

Definition

Intentionally misleading customers with price promotions.

Example

Falsely advertising a sale on winter coats as 50 percent off the regular price when this is not accurate or true.

Illegal?

It is illegal because customers do not have the correct information.

Global Pricing

Pricing is a critical component of a successful global marketing strategy. Historically, companies have set prices for products sold internationally higher than the same products sold domestically. However, technological advancements and growing Internet access throughout the world have made global pricing more transparent and, in many cases, more competitive. In addition, challenging economic conditions over the past decade have impacted pricing in a global context.

Let's examine a few issues related to global pricing.

Gray Market

You have probably heard of the black market, which refers to the illegal buying and selling of products outside of sanctioned channels. A lesser-known relative of the black market is the gray market. The **gray market** consists of branded products sold through legal but unauthorized distribution channels. This form of buying and selling often occurs when the price of an item is significantly higher in one country than another. Individuals or groups buy new or used products for a lower price in a foreign country and import them legally back into the domestic market. Gray market goods can be a boon for consumers, allowing them to obtain legally produced items for less than they could normally. However, gray market goods cut into a firm's revenue and profits, leaving marketers looking for ways to control and repress such activity. The increasingly interconnected nature of world economies makes gray market exchanges easier than ever, and firms find it difficult, if not impossible, to track exactly how much of their products sell in this manner.

Tariffs

Many nations place tariffs on a variety of products, especially fruits and vegetables, which have tariffs in some countries of over 25 percent. **Tariffs** are taxes on imports and exports between countries. Tariffs may raise the price that foreign customers must pay for goods produced in the United States, negatively impacting a U.S. firm's ability to be price competitive in those markets. The international pricing strategy of any U.S. firm must take into account the potential tariffs foreign countries will place on its goods.

Marketers typically prefer targeting international markets with low tariffs or with which the United States has an international agreement, such as the United States-Mexico-Canada Agreement (USMCA), to lower tariffs. The absence of tariffs among the United States, Mexico, and Canada allows for easier transactions between companies and customers in those countries.

Tariffs levied on products can impact the overall price of a product in other countries and thus how companies set their baseline prices.

An American-made surfboard that carries a 20 percent tariff in another country may need to be priced lower to be competitive in that market.

Dumping

In recent years, the removal of tariffs due to international agreements has caused countries to switch to nontariff barriers, such as anti-dumping laws, to protect their local industries. **Dumping** occurs when a company sells its exports to another country at a lower price than it sells the same product in its domestic market. Though the World Trade Organization hasn't classified dumping as illegal, many countries enact their own laws to curb the strategy. As they develop their international pricing strategy, companies must monitor how anti-dumping laws affect similar companies in the industry and calculate the potential impact of anti-dumping regulations on sales.

Marketing Analytics: Using Data to Make Pricing Decisions

As data-driven decision making becomes the norm in marketing, firms must collect the data necessary to make appropriate decisions—this is especially true of pricing. Managers must analyze both internal and external data to ensure they are setting prices that not only meet the needs of the firm, but that consumers also deem acceptable. Technology allows firms to create computer models that accurately simulate the resulting outcomes and market response to pricing decisions.

Setting Price Objectives Based on Point-of-Sale Data

Every product now contains a stock-keeping unit (SKU), which is the barcode that appears on every product and is assigned to every service. The SKU scanners at checkout register each product or service and record how much the consumer has paid. These point-of-sale data are incredibly valuable to firms in setting price objectives and can be used to demonstrate how the market is responding to a firm's price point. Each product or service is assigned a specific objective, so the point-of-sale data can inform the firm of whether or not that objective is being met.

Determining Price Elasticity of Demand Using Artificial Intelligence

Price elasticity of demand measures the market's response to a change in price. Remember that an increase in price may not result in increased revenue. If the number of consumers purchasing the product drops too much, then the price increase may result in a decrease in revenue. Thus, understanding what that market response will be is crucial for firms. In the past, firms relied on test marketing to analyze the impact of a price increase or decrease. One or more markets would be selected for this price change, and the point-of-sale data collected before the price increase would be compared to the data collected after the price increase. These studies took time and were costly to conduct. Today, many firms are using artificial intelligence (AI)—the heart of which is machine learning—to model these price changes. The process uses data to identify data patterns, and when new information is introduced, it readjusts accordingly. AI programs are taught about the market, the consumers, the competitors, and so forth. The programs then demonstrate what the market response will be when the price increases or decreases. Such predictive analytics has allowed firms to avoid costly pricing mistakes.

Conducting Break-Even Analysis Using Internal and External Data

Prior to launching a new product or service, firms calculate the time it will take for that product to earn back the costs required to create and launch it. In order to complete this analysis, firms need to have a full accounting of costs associated with that product. There also needs to be an understanding of both the demand for, and the price of, that product. This analysis requires both internal firm data to determine costs as well as external data necessary to determine price and market demand.

Determining Pricing Strategy Using Firm, Market, and Competitor Data

Pricing strategy is just one part of a broader strategy; all elements of the marketing mix must support one another. Thus, the pricing strategy must be aligned with the product strategy, distribution strategy, and the promotion strategy. For example, a firm specializing in luxury items would not sell those items at low cost because that would be inconsistent. Firms have to use all data available to them to establish a pricing strategy that is consistent with their overall marketing mix strategy. Firm data are used to ensure that the products or services are meeting objectives for profitability. Market data are used to determine market share objectives. Competitor data are used to ensure that prices are in line with competitors' pricing.

Responsible Marketing

Pricing Ethically and Fairly

Pricing is an extremely important issue for a company because the prices it charges for its products directly affects whether it can succeed in the marketplace. Unfortunately, there are many examples of unethical pricing strategies that take advantage of consumers' vulnerability or hyper-willingness to pay.

One example of an irresponsible pricing strategy is price gouging. Price gouging occurs when a firm raises prices of items that are temporarily in high demand. You may see examples of this, called surge pricing, when the cost of an Uber or Lyft increases significantly during peak hours.

While the price of a ride-share may genuinely reflect both supply and demand principles in a fair way, consider other examples of such tactics that are more opportunistic. For example, some brands dramatically increased the prices of necessities at the beginning of the COVID-19 pandemic, making essential items more accessible to consumers with greater financial resources. For many consumers who were priced out of the ability to buy common necessities during the panic shopping that occurred, such pricing tactics heightened the inequities they experienced during the pandemic.

Another example of potentially unethical pricing strategies is price discrimination, which involves charging different customers different prices for the same products. You may notice that some websites ask for your zip code before telling you the price of an item. This is an example of price discrimination that isn't necessarily unethical, as it's related to the cost of shipping and delivery in the area. However, what if that product were a pharmaceutical product? In that instance, is charging two different consumers two different prices ethical?

Reflection Questions

1. What are other examples you have seen of unethical or irresponsible pricing strategies?

2. Should brands be held accountable for unfair pricing behaviors? Why or why not? What should the consequences be?

3. When does supply and demand pricing seem fair to you and when does it not?

Capturing Value: Test

1. Which of the following is the first step in the price-setting process?

    A. analyze the competitive price environment

    B. determine the costs

    C. define the pricing objectives

    D. choose a price

    E. evaluate demand

2. Carla, the marketing manager at Regal Inc., is in the process of setting prices. She has finished analyzing the competitive price environment. According to the price-setting process, what should be Carla's immediate next step?

    A. evaluate demand

    B. define the pricing objectives

    C. monitor and evaluate the effectiveness of the price

    D. choose a price

    E. determine the costs

3. The prices that consumers consider reasonable and fair for a product are termed

    A. sensitive prices.

    B. survival prices.

    C. break-even prices.

    D. markup prices.

    E. reference prices.

4. Isabella's landlord has included a clause in the rental contract that makes it possible for him to increase Isabella's monthly rent if taxes on the property go up. Which clause was included in the contract?

    A. elevator clause

    B. escalator clause

    C. variable costs clause

    D. tariffs clause

    E. elastic demand clause

5. Frozen Yummy is an ice-cream manufacturer. Its marketing strategy from the month of November to February is "buy one get one free." Which of the following best describes the strategy applied by Frozen Yummy?

    A. markup pricing

    B. odd discounts

    C. generous pricing

    D. prestige pricing

    E. seasonal discounts

6. The marketers of Santé, a health drink, have decided to sell a combination package of three flavors—vanilla, chocolate, and strawberry. This package is priced marginally higher than the individual flavors. Given this information, we can come to the conclusion that the marketers of Santé have applied the _____ strategy.

    A. dynamic pricing

    B. price fixing

    C. odd pricing

    D. price bundling

    E. prestige pricing

7. Which of the following is true of gray markets?

    A. Gray market exchanges have become more difficult than ever.

    B. Firms find it easy to track exactly how much of their products sell in the gray market.

    C. Gray markets raise the prices that customers must pay for goods.

    D. Gray market goods can be a boon for consumers.

    E. Gray markets have helped to shift the balance of power from customers to companies.

8. High Tea, a retail store, claims to sell a pack of 20 tea bags at a discounted rate of $20 from an advertised original price of $40. The original price was verified by regulators to be $25. The pricing strategy applied by High Tea is an example of

A. fair pricing.

B. price discrimination.

C. price fixing.

D. predatory pricing.

E. deceptive pricing.

9. Pricing is one of the most important strategic decisions a firm faces because it reflects the value the product delivers to consumers

A. as well as the value it captures for the firm.

B. as well as the value of the competition generated.

C. as well as the value of the target markets generated.

D. as well as the benefits that retailers provide.

E. as well as the advantages that consumers provide.

10. At which point in the price-setting process does a firm decide whether it might engage in profit skimming?

A. Define the pricing objectives.

B. Evaluate demand.

C. Determine the costs.

D. Analyze the competitive environment.

E. Choose a price.

11. Which of the following is a product with elastic demand?

A. airline tickets

B. gasoline

C. bottled water

D. bananas

E. dishwashing liquid

12. All of the following are examples of common fixed costs *except*

A. sales commissions.

B. employee salaries.

C. rent for office space.

D. health insurance.

E. advertising budget.

13. Once marketers have completed their analysis of demand, costs, and the competitive environment, they can use a number of different tactics to choose a final price. Which tactic to use depends on the value customers perceive the product to have, their ability to pay for it, and how they intend to use the product.

A. TRUE

B. FALSE

14. Godiva Chocolates pursues a _____ strategy by setting its prices high to convey increased value compared to other chocolate brands like Lindt.

A. prestige pricing

B. markup pricing

C. odd pricing

D. price bundling

E. seasonal discounts

15. Marketers may use a price bundling pricing strategy for all of the following reasons *except*

A. it allows marketers to vary the price for the bundle based on product demand of each individual element.

B. it allows marketers to charge higher prices for the bundle than they could for each individual element.

C. It allows a company to sell the same bundle to everyone

D. It simplifies matters for marketers.

E. It helps to reduce advertising and selling costs.

16. Service providers of haircuts, nail care, or oil changes may also use _____ pricing in a competitive market to drive awareness of a new location or for expanded services.

    A. penetration

    B. skim

    C. markup

    D. odd

    E. prestige

17. Price discrimination is illegal only if it injures competition.

    A. TRUE

    B. FALSE

18. Consider a situation in which a coffee chain opens across the street from a locally owned cafe. Theoretically, the prices at both stores should be similar because the costs and customer demand will be similar. However, because the coffee chain can rely on corporate backing for support, it makes the decision to radically lower prices, attracting more customers to its facility and eventually driving the competition out of business. This scenario illustrates which ethical issue in pricing?

    A. predatory pricing

    B. price discrimination

    C. price fixing

    D. deceptive pricing

    E. pricing penetration

19. Gray market goods can be a boon for consumers, allowing them to obtain legally produced items for less than they could normally.

    A. TRUE

    B. FALSE

20. One reason why firms may use a penetration pricing strategy in a competitive market is to drive awareness of a new location or of expanded services.

    A. TRUE

    B. FALSE

8 Delivering Value

What to Expect

How good is a product if a customer can't get it? You can make the greatest product in the world, but if you can't get it to your customer then its value is lost. In this chapter we will look at what it takes to produce and then deliver goods.

Chapter Topics:

- **8-1** Supply Chain and Channel Management
- **8-2** Retailing and Omnichannel Marketing

Supply Chain and Channel Management

As you will read in the opening case on L'Oreal, *supply chains* are networks that allow goods to be produced from raw materials and then delivered to stores where consumers can purchase them. Terms related to supply chains include *marketing channels* and *logistics.* In this lesson, you will learn about how marketers create value through efficient supply chain management.

By the end of this lesson you will be able to

- Describe the supply chain and its value to a firm.
- Describe the different types of marketing channels.
- Explain the three key types of supply chain strategies.
- Describe supply chain management.
- Explain the role of logistics in supply chain management.
- Explain the benefits of supply chain integration.
- Describe the distribution center concept.
- Compare and contrast the advantages and disadvantages of various transportation methods.
- Explain how marketers use data to plan supply chain and channel management strategies.

Marketing Analytics Implications

- Firms use data to identify the costs and benefits associated with each choice of supply chain partner.
- Firms use data to determine how best to manage the costs within the supply chain.
- Firms require a high level of coordination and information sharing to manage their supply chain functionality; data collection assists in this effort.
- Firms can make decisions on which products to stock in their retailers and distribution centers and in what quantities using multiple data points.

Managing the Supply Chain at L'Oréal

Supply chain management consists of all the steps involved in getting a product or service from development and production to customers. As consumers' expectations have changed over the past few years, brands have had to quickly adapt to provide shorter delivery times, better service, and more robust returns policies, among other things.

Typically, when we think of companies with great supply chains, we think about big brands such as Walmart, Amazon, or Alibaba. However, independent supply chain consulting firm Gartner frequently ranks the French beauty products company L'Oréal as having one of the top supply chains in the world. Gartner evaluates brands' supply chains based on their resilience and agility, data and technology, and sustainability:

- **Resilience and Agility:** It is no secret that the COVID-19 pandemic had catastrophic effects on many supply chains, effects that will continue to linger for years to come. L'Oréal used data and innovative technology to develop more resilience in its supply chain and keep its delivery commitments. One way that L'Oréal achieved this was by turning physical stores into temporary distribution centers.

- **Data and Technology:** L'Oréal has set out to not only be a leader in beauty, but to be a leader in beauty tech as well. The company uses a variety of data sources to inform its decisions and continuously develop best practices for its business. From sales forecasting to automated distribution centers, L'Oréal is leading the way in technology innovation in beauty.

- **Sustainability:** L'Oréal is reducing carbon dioxide emissions in the production and transportation of its products. By 2030, L'Oréal aims to reduce its greenhouse gas emissions by 50 percent. Sustainability thus remains at the forefront of supply chain innovations and transformations for L'Oréal.

As L'Oréal continues to work toward its ambitious sustainability and technology goals and to meet the needs of today's ever demanding customers, it will have to continuously adapt its supply chain.

Consumers are demanding more transparency in the supply chain, from carbon footprints to human rights. This is especially true in the health and beauty industry. Brands like L'Oreal are working tirelessly to improve their environmental and social impact in their supply chain.

The Value of the Supply Chain

Have you ever passed a truck on the highway and wondered where it was headed and what it was carrying? Chances are it was carrying finished goods toward the end of a journey that originated halfway across the world. The truck is performing one of many functions necessary for companies to operate a supply chain. A **supply chain** is a set of three or more companies directly linked by one or more of the upstream and downstream flows of products, services, finances, and information from a source to a customer.

A **downstream flow** is the movement of goods (and other things, like information, promotion, etc.) toward the final customer. All along the downstream flow, supply chain members add value to the product by changing its form and location. The flow of information affects all members of a supply chain because without accurate, timely information companies cannot make good decisions.

As the product travels downstream, an upstream flow also occurs. **Upstream flow** is the movement of payment (and other things, like information, returns, etc.) from the customer toward the manufacturer and others in the supply chain.

Organizations make use of and are part of many supply chains, without which they cannot survive. For sure, supply chains provide value to firms. First, as noted, supply chains add value to products (availability, manufacturing, packaging, etc.) as they move downstream and return value (payment, feedback, product development, etc.) that moves upstream. In addition, supply chains have become a strategic function of businesses by adding value in additional ways.

The COVID-19 pandemic has fundamentally changed supply chains around the world. Producers of essentials goods are shifting from "just-in-time" supply chains to "just-in-case" contingencies. Within the United States, businesses are shortening national supply chains to be more flexible and responsive to market conditions. And firms outsourcing manufacturing to only one location are creating flexibility in their supply chain by outsourcing manufacturing to multiple facilities around the globe.

Production Costs

Supply chains reduce production costs by streamlining processes. Efficient supply chains add value by improving manufacturing efficiency by continually improving things such as shipment size, packaging, and inventory levels.

Location

Supply chains add value through better logistics (movement of goods). Value is added by taking better advantage of locations, improving access to expanded markets (more customers), and lowering distribution costs.

Time

Supply chains add value by reducing how long it takes upstream and downstream flows to occur. Value is added by having goods and services available when required along the supply chain (e.g., lower lead times) with better inventory and transportation management. In addition, upstream movement of payment and feedback adds value to firms by reducing financial costs and speeding product improvement cycles.

Control

Supply chains add value by allowing firms better oversight of upstream and downstream flows. Value is added by controlling stages along the supply chain, from production to end consumer. By supply chain partners working together, in an organized fashion, all chain members can better anticipate what will be needed and when from their firm and their partners.

The concept of a supply chain is actually a simplified version of what is, in actuality, a network of companies made up of many suppliers and customers. A company producing cosmetics may have hundreds of suppliers, some of which may be selling the same or similar raw and packaging materials to other cosmetics manufacturing companies. In addition, that same cosmetics manufacturer may sell to a retailer that buys similar cosmetics from a number of other manufacturers.

Downstream: The cloth manufacturer converts cotton to cloth.

Downstream: A truck delivers the finished T-shirt from the wholesaler to a store where a consumer like you can buy it.

Upstream and Downstream: A retailer may pass its sales forecast for T-shirts upstream to its wholesaler, who then passes the information to the T-shirt supplier. In return, the T-shirt supplier might pass information about when the T-shirts will be available for delivery downstream to the retailer.

Upstream: The retailer paid the wholesaler, who paid the T-shirt manufacturer, who paid the cloth manufacturer, and so on upstream through the chain.

Upstream: When you purchase a T-shirt, the money flows out of your pocket to the retailer.

Types of Marketing Channels

Within the supply chain are various marketing channels. With a **direct marketing channel**, products travel directly from the producer to the consumer with no intermediaries involved. With an **indirect marketing channel**, supply chain intermediaries, including wholesalers, distributors, and retailers, are a part of the product's sale and delivery. Each channel serves a specific purpose in the downstream and upstream flows of a supply chain.

Let's look at some specific participants in indirect channels.

Retailers

Retailers like Walmart represent an important marketing channel. A **retailer** is a company that purchases and resells products to consumers for their personal or family use. Retailers can take the form of large chain stores directed by corporate offices that centralize purchasing, advertising, human resources, and other functions. Walmart, as well as Target and Dick's Sporting Goods, are examples of chain stores. Independent retailers, such as a local bookstore or gift shop, may have only one or a few stores.

Target is one of the top 10 retailers and the second-largest discount retailer in the United States. Walmart is the largest.

Retailers can also be franchised from a large corporation. In such a case, the corporation sells the franchise store to an individual or individuals to operate, while maintaining control over management training, advertising, the supply of products being sold, and many other aspects of the store. Subway is one of the world's largest franchise operations. Subway has 26,291 U.S. and 18,317 non-U.S. franchise locations and no company-owned locations.

Not all retailers operate brick-and-mortar stores. Some companies perform direct retailing, selling to customers at their homes, and others use catalogs to sell their products. Finally, online e-commerce is the fastest-growing retailer category. Online retailers such as Amazon have taken advantage of the spread of electronic devices and the Internet throughout the world to sell goods anywhere there is an Internet connection.

Wholesalers

A **wholesaler** is a firm that buys large quantities of goods from various producers or vendors, warehouses them, and resells them to retailers or other businesses. To understand the function that a wholesaler performs, think about a grocery store.

A grocery store is considered a retailer, but for consumers, a grocery store brings together a wide variety of products, so consumers can stop at one place to purchase all of the items. Imagine having to go to 10 different stores to buy 10 different things (eggs, milk, paper towels, etc.)—it would be inefficient and frustrating. Wholesalers perform essentially the same function as grocery stores, but they typically sell to retailers or other firms that sell to consumers. For example, when a consumer shops at a Costco, Costco is performing the function of a traditional retailer. However, when companies shop at Costco for items needed to perform business functions, Costco is acting as a wholesaler.

Wholesalers can sell a variety of things. An example of a very successful wholesaler is Christmas Trees Worldwide. Christmas Trees Worldwide sells bulk Christmas trees for other companies to resell. For many industries, wholesalers are critical to the supply chain because they bring in supplies from various manufacturers that are resold to a large set of resellers further downstream in the supply chain.

Distributors

Distributors are similar to yet different from wholesalers. A **distributor** buys noncompeting products, warehouses them, and resells them to retailers or directly to end users. In many ways, distributors are wholesalers, but they sell products for only a single manufacturer, or for a set of companies that are not in competition with each other.

For example, the Dr Pepper Snapple Group has more than 50 different brands in its portfolio, from Dr Pepper to Snapple to Clamato to Schweppes products. For the Dr Pepper Snapple Group to reach every potential retailer would require a significant distribution infrastructure. Instead, Dr Pepper Snapple Group will partner with distributors in local areas who buy the company's products and sell them within a specific geographic range. The distributor brings together all of the manufacturer's products in a local distribution warehouse and sells them within their approved territory. The difference between a wholesaler and a distributor is that the distributor is usually only partnering with one company to resell its products and is essentially acting as an intermediary for that company.

Distributors typically also offer service for the product they sell. Distributors provide labor and cash support to the supplier or manufacturer's promotional efforts. They usually also provide a range of services (such as product information, estimates, technical support, after-sales services, credit) to their customers. For MillerCoors, a local distributor organizes promotions, advertising, and in-store merchandising.

Retailers

Retailers

Retailers are well-known to consumers. It should be no surprise that Walmart is among the top largest retailers in the United States.

Walmart operates more than 11,700 retail units under 65 banners in 28 countries and e-commerce websites in 11 countries. Walmart employs approximately 2.3 million associates around the world, with 1.5 million in the United States alone.

Wholesalers

Wholesalers

Wholesalers often operate large businesses without being well-known by consumers. For example, C&S Wholesale Grocers is the largest wholesale grocery supply company in the United States. From more than 75 high-tech facilities in 15 states, C&S supplies supermarkets and institutions with more than 170,000 different products. C&S's continuing commitment to new technologies is dramatically changing the cost curve and improving service and value.

Distributors

Distributors

Large distributors often exist without being known by consumers. For example, Reyes Beverage Group (RBG) is the largest beer distributor in the United States. It is the brewers' distributor of choice, leveraging its scale to provide unparalleled service and continually striving to grow its customers' profitability. RBG sells thousands of products and distributes more than 151 million cases of beer to more than 57,000 retail accounts each year. The company operates from 23 warehouses across the United States.

Supply Chain Strategies

Effective supply chain management and integration require a thoughtful supply chain strategy that helps the firm establish the right channel partners to meet its objectives. Supply chain objectives should be based on the firm's marketing objectives and can include:

- The types and locations of markets to be served
- The market share and customer service desired
- The speed with which new products should be developed
- Cost reduction and profitability goals

In the sections that follow, we will focus on three supply chain strategies: push, pull, and hybrid (push-pull).

Push Strategy

If a company's marketing strategy includes objectives related to cost competitiveness and customer service, it should consider a push strategy. A **push strategy** is one in which a company builds goods based on a sales forecast, puts those goods into storage, and waits for a customer to order the product through the marketing channel. The strategy is called a push strategy because the manufacturer works with channel members to "push" its products through the supply chain to the end consumer.

Big brands like Frito-Lay rely on sales forecasts to plan production well in advance. The firm makes large batches of one good at a time, such as the different flavors of Lay's and Cheetos brands shown here, to help reduce manufacturing costs.

The snack food industry (think of big brands like Frito-Lay), for example, relies on sales forecasts to plan production well in advance. Frito-Lay will use the previous year's data, as well as current trends in the environment, to forecast and predict snack trends for the year. Using this information, Frito-Lay will schedule production and regular distribution of products in advance. The main advantage of this type of strategy is that it allows firms to achieve economies of scale; because the firm makes large batches of one good at a time, it can reduce manufacturing, transportation, and other costs. The main disadvantage to using a push strategy is that sales forecasts are often inaccurate, resulting in excess inventories that are difficult to push through the marketing channel to consumers because demand is low. This is a problem because inventory carrying costs are expensive, typically 25–30 percent of the cost to produce and deliver the goods, and products can become outdated.

Pull Strategy

Firms whose marketing strategy requires agility and product customization should pursue a pull strategy. In a **pull strategy,** customer orders drive manufacturing and distribution operations. In a pure pull system, the good is not made until a customer order is received. The strategy is called a pull strategy because the manufacturer depends on channel members to "pull" its products through the supply chain based on finding customer orders.

There are several advantages of a pull strategy. The pull strategy reduces inventory carrying costs, allows the firm to customize products specifically to customer requirements, and gives firms the ability to respond rapidly to changing market conditions. For example, Boeing uses a pull system. It builds aircraft to customer specifications and only when an order is received. There are also disadvantages in using a pull strategy, such as it doesn't often allow firms to take advantage of economies of scale and it requires a production facility that can change rapidly to produce different products.

Ads for pharmaceutical products that say "Ask your doctor (or dermatologist) about..." use a pull strategy.

Push-Pull Strategy

A third strategy is a hybrid of the other two strategies. In a **push-pull strategy,** some channel partners operate on a push system, but completion of the product is based on a pull system. Thus, a company may forecast sales, build an inventory of components based on the forecast, and hold those components in inventory until a customer order is received, as it would using a push strategy. Then, as in a pull strategy, the company finalizes the product based on the order.

Dell, for example, builds component inventory based on its sales forecast but doesn't completely assemble computers until a customer order is received. A push-pull strategy combines the key advantages of the other two methods: companies can achieve economies of scale in purchasing, while engineering flexibility into manufacturing. However, the push-pull system does have drawbacks. It may not be as cost competitive as a push system because the firm cannot take advantage of manufacturing and transportation economies of scale. In addition, the firm will still incur costs to store components in inventory.

Supply Chain Management

A firm that recognizes and responds to the impact that supply chain flows have on its business possesses a supply chain orientation. A **supply chain orientation** is a management philosophy that guides the actions of company members toward the goal of actively managing the upstream and downstream flows of goods, services, finances, and information across the supply chain. This implies an outward focus on the activities and performance of other companies, rather than an inward focus on one's own company. It also implies a willingness to get involved in coordinating those activities to add value to the end customer. When management sets up technology to enable the sharing of demand information with suppliers, such as Walmart's Retail Link program, it is practicing supply chain management. For example, Walmart's Retail Link program "provides a way for [Walmart] suppliers to manage their own products by allowing them to monitor their data, including sales and inventory volume, in-stock percentage, gross margin and inventory turnover. As a result, they [Walmart suppliers] achieve low levels of inventory risk and associated costs."

Supply chain management refers to the actions the firm takes to coordinate the various flows within a supply chain. Companies that view the total system of interrelated companies that make up the supply chain as something to manage have greater control over the customer value they provide. This in turn generates higher revenue for the company.

Supply chain management involves making decisions that involve trade-offs, for example, the amount of inventory to keep in stock. Too much extra stock can end up costing the company a great deal of money over time, but too little can leave the company unprepared to meet demand.

The underlying philosophy behind this concept is that no company is an island. All companies rely on related companies whose activities add value to the final product. This reliance can take the form of sharing sales forecasts and product delivery information or enlisting another company to perform a function if the other company can do it more efficiently or effectively than the firm itself.

Supply chain management decisions often involve trade-offs. Decisions made about one area of the supply chain will affect other areas, sometimes negatively. For example, when sales personnel offer a volume discount to a customer—without knowing whether that amount of product can be delivered on time without incurring extra cost—the sale may actually end up costing the company money. Another important trade-off exists between inventory and customer service. When there is not enough inventory to satisfy customer demand, the firm may incur additional costs or lose sales or both.

The Role of Logistics in the Supply Chain

When you go to a store to purchase a product, you expect that it will be on the shelf. If you order an item from an online site, you expect it to be delivered quickly. However, you may not have given much thought to all that goes into ensuring the product's availability or shipping it to customers like you. Logistics personnel are a big part of making that happen.

Logistics is that part of supply chain management that plans, implements, and controls the flow of goods, services, and information between the point of origin and the final customer. People who work in logistics support supply chain strategies to achieve cost containment, innovativeness, flexibility, customer satisfaction, and other important marketing objectives.

Transportation

Transportation management is one of the most important logistics functions for retailers and every other member of the supply chain. Effective transportation management creates value that is essential to keeping customers satisfied. Cost-effective transportation allows companies to market goods to greater distances, helping companies build global supply chains and compete in new markets.

Copyright © McGraw Hill   Antonio Gravante/Shutterstock

Inventory

Logistics is responsible for the inventories that firms carry. Effective inventory management policies balance product availability with the costs of carrying inventory. The challenge of inventory management is to manage inventories in a way that balances supply and demand in a cost-effective manner. However, while managing inventories, stockouts should be avoided. A stockout occurs when a company does not have enough inventory available to fill an order. In such cases, a customer can either place a back order or cancel the order. A lost sale means lost revenue immediately and potentially in the future if customers do not return.

Purchasing

Logistics is also responsible for sourcing and procuring raw materials, component parts, components, finished goods, and services for companies. Purchasing is an important logistical function for a number of reasons. First, materials purchased for manufacturing typically account for 40 to 60 percent of the product costs, meaning that any savings in purchase costs can add to a company's profits. Second, purchasing is a major factor in good and service quality. Selecting suppliers that provide high-quality goods and services enables companies to compete with or surpass the quality of other companies.

Finally, purchasing can improve product design and time to market for products by involving appropriate suppliers early in the product design process.

Materials Management and Warehousing

Once purchasing managers procure raw and packaging materials or component parts, logistics personnel must manage them in a cost-efficient manner that supports manufacturing operations. Materials management involves the inbound movement and storage of materials in preparation for those materials to enter and flow through the manufacturing process. Storage of these materials is called warehousing, which includes storage, movement, and production of materials and finished goods.

Through these various functions, logistics adds value to products moving through the supply chain. A short summary of how logistics adds value can be remembered through the "7 Rs" of logistics—delivering the right product, to the right place, to the right customer, at the right time, in the right quantity, in the right condition, and at the right price—which determine how satisfied customers are.

Product

Logistics managers affect the product aspect of the marketing mix through feedback on packaging. Milk is often sold in gallon containers. These gallon containers leave empty spaces when boxed together, making boxing dozens of milk gallons inefficient. A more efficient packaging strategy would be square containers that stack neatly with no empty space in a box. However, this design would not appeal to customers. The design of the gallon milk container is intended to appeal to both consumers and logistics managers.

Place

It's easy to see how logistics operations impact the place aspect of the marketing mix.

Can you imagine going to the grocery store and needing to buy milk and the grocery store does not have any? That would be extremely confusing for customers. Also, if there is even a one-day delay in delivery, grocers have missed a valuable selling period for milk before the item perishes.

Price

Price is affected by logistics operations in several ways. Purchasing activities impact the cost of buying or making goods through volume discounts, reliability of suppliers, and the quality of the goods procured.

Part of the components of milk are often distributed via a pipeline, making the process of milk production less expensive. If milk had to be air freighted, the price would double, or maybe even triple.

Promotion

When marketers plan a promotion, they must make logistics managers aware of any increase in expected sales so that sufficient inventory, equipment, and personnel are available to handle increased volumes.

If a company runs a promotion advertising a sale on milk, it needs to ensure it has the appropriate quantities on hand to support the promotion.

Supply Chain Integration

The ultimate objective of supply chain management is to integrate related companies, thus facilitating the coordination of activities across the supply chain in a manner that improves the entire chain's performance. In essence, the supply chain becomes an extended enterprise. Individual firms integrate their activities to such a degree that they function as one organization. Integrated supply chains typically benefit from reduced costs, better customer service, efficient use of resources, and an increased ability to respond to changes in the marketplace.

For an integrated supply chain to work, companies must be willing to embrace relationship-based strategies that involve close, long-term collaboration for mutual benefit. For example, if a materials manager in a manufacturing facility establishes a close relationship with a box supplier that handles the plant's box inventory, the manufacturing plant could share production schedules with the supplier. The supplier then would be able to efficiently schedule the delivery of boxes, and the plant would not have to expend any manpower determining when they need boxes and how many they need or ordering them from the supplier. Both members of a supply chain gain from such a relationship-based strategy.

Today's global business environment makes supply chain integration even more of an imperative for firms trying to overcome the time and distance obstacles that can stand in the way of efficient supply chain operations.

Distribution Centers

Many firms create large distribution centers to internally control supply chains. **Distribution** occurs when a company ships its goods to its customers. A **distribution center (DC)** is a type of warehouse used specifically to store and ship finished goods to customers. DCs differ from wholesalers and distributors because they are owned by a company that is accumulating and selling products to end customers.

Walmart is a good example of a company that uses DCs. Walmart has large DC facilities that continually move products in from suppliers and out to its stores. In a direct-to-consumer setting, Amazon is well known for its distribution centers, which *accumulate, sort, allocate,* and *assort* products used to fulfill customer orders.

These four functions are common to DCs:

- **Product accumulation** involves receiving goods from various suppliers, storing the goods until they're ordered by a customer or other company-owned facility, and consolidating orders to achieve transportation economies of scale. Orders from customers are collected until enough goods are ordered to fill vehicles, which reduces transportation costs.

- **Product sortation** refers to gathering goods with similar characteristics in one area of the DC to facilitate proper inventory controls and effectively provide customer service. For example, a company that sells sunscreen, which is regulated by the United States Food and Drug Administration (FDA), must keep expired products off store shelves. Product sortation enables such companies to maintain control over their inventory and avoid accidentally shipping goods that are no longer fit for sale.

- **Product allocation** involves picking available goods to fill customer orders. Distribution centers can be set up to pick full pallets of goods, full cases of goods, or individual goods. The ability of a DC to allocate goods down to a single piece allows customers to buy in quantities that make sense for their business. Walgreens, for example, would not want to order a full pallet of ruby red lipstick that would take years to sell. The DC's allocation function enables Walgreens to order only the quantities it needs.

- **Product assortment** occurs when the DC mixes goods coming from multiple suppliers into outgoing orders so that each order includes a variety of goods rather than just one type of good. Customers often require small quantities of goods from a variety of suppliers. The product assortment function of a distribution center allows companies to meet this need in a cost-effective way by combining orders for a variety of products going to the same location in the same truck to save money.

The number and location of DCs are among the strategies a company can use to support its marketing objectives. Should the firm establish a few, large centralized DCs, or more DCs that are closer to the customer but hold less inventory? The answer has to be based on the company's marketing and supply chain strategies. The four main strategic factors that firms must consider when designing their distribution networks are *inventory levels, operating expenses, customer service, and transportation costs.*

Inventory Levels

The more facilities in the network, the higher the overall inventory in the system. The additional inventory can benefit the company if demand unexpectedly increases, but it adds to inventory carrying costs.

Operating Expenses

The fewer facilities in the network, the less operating expense incurred. With fewer facilities, the company will require fewer management and clerical employees and less information technology equipment to conduct distribution activities.

Customer Service

More facilities result in faster delivery to customers because facilities will be closer to markets. However, centralized facilities generally result in better product availability, leading to superior customer service with less inventory investment.

Transportation Costs

More facilities lead to higher transportation costs from production sites and suppliers to the warehouses or DCs. However, transportation costs between the facilities and customers may be lower due to their close proximity.

Transportation Methods

Transportation plays a critical role in the logistics function of supply chain management. A company can transport goods in a number of ways, but not all forms of transportation (modes) are suitable for every type of product. Highly profitable items such as fashionable clothing or perishable goods such as exotic fruit tend to be shipped by air because of its speed and expense. On the other hand, crude oil is most efficiently transported by pipeline. As with other supply chain decisions, mode selection involves trade-offs.

As you can see below, there are six types of transportation modes, each of which has advantages and disadvantages.

Rail

Advantages

- move wide variety of goods in large quantities
- low cost.

Disadvantages

- potential for damage
- inconsistent service
- distance from rail depots

Truck

Advantages

- highly accessible
- reliable
- fast

Disadvantages

- high cost per ton-mile
- limited vehicle capacity

Air

Advantages

- speed
- risk of damage is low
- good for shipping fragile

Disadvantages

- expensive
- limited Accessibility

Water

Advantages

- low shipping cost
- high capacity

Disadvantages

- slow
- weather dependent

Pipeline

Advantages

- efficient
- low cost

Disadvantages

- can only carry a few product types

Digital

Advantages

- instant deliver
- low cost

Disadvantages

- can't transport physical goods

Understanding transportation modes is essential to building an efficient and effective transportation network.

- **Railroads:** Railroads transport a variety of products over long distances. Over 40 percent of the ton-miles (one ton of goods carried one mile) moved in the United States are moved by rail. Historically, railroads have carried bulk products such as chemicals, coal, produce, and automobiles. Today, intermodal transportation makes railroads a transportation mode of choice for almost any consumer good. **Intermodal transportation** is transportation of freight (consumer goods) in a container using multiple modes of transportation (e.g., ship, rail, semi-truck).

- **Trucks:** Trucks (motor carriers) transport 10.5 billion tons of cargo in the United States every year. Trucks primarily move small- to mid-size shipments. Companies can either hire carriers or maintain their own fleet of trucks. For-hire carriers fall into two categories: *common carriers,* which sell their services to any business, and *contract carriers,* which move goods exclusively for certain customers. Due to the universal accessibility of motor carriers, for most shipments, some part of the journey will be completed by a truck.

- **Air transportation:** In 2017, air freight traffic recorded 54.3 million tons of freight carried through commercial airlines worldwide, and the revenue of cargo airlines amounted to around 50.7 billion U.S. dollars worldwide. For certain types of goods and shipments, air is the only viable option. For flowers to arrive fresh in New York City from South America they must be flown in; no other mode will do. Emergency medical shipments provide another example. When a life hangs in the balance, air freight may be the only viable option.

- **Water transportation:** Water transportation has taken on great significance as production has moved to low-cost producers in Asia and other parts of the world. Ships cross oceans on regular schedules, carrying thousands of containers of goods across thousands of miles. Containers can be loaded and unloaded from ships efficiently, making ocean transport the mode of choice for international transportation of not only traditional products such as raw materials and other bulk commodities but consumer products as well. Goods are also transported on lakes, rivers, and inland waterways. These shipments tend to be bulk commodities, such as farm products and minerals, rather than finished goods.

- **Pipelines:** Pipelines make up a very specialized mode that services only a few industries. Pipelines transport oil, oil products, and natural gas. Pipelines have a huge impact on national economies and play an important role in providing fuel for industrial, personal, and utility use.

- **Digital transportation:** A mode of transportation that you likely use frequently but may not think of as a transportation mode is digital delivery. Anything that can be digitized can be sent through the Internet and delivered to businesses and homes. Software, documents, and media (e.g., music, pictures, and film) can all be sent electronically. Retailers such as Amazon, with its wide selection of Kindle-ready digitized books, make digital transportation part of their supply chain strategy.

Firms must match the modes of transportation they choose to use with their overall marketing strategy to get the most advantage from their logistics operations.

Companies must consider many factors when choosing transportation modes, including mode *accessibility* (the ease of getting freight from one loading dock to another), *speed* (transit times), *capacity* (the ability of the mode to handle necessary volumes), cost, the ability of the mode to consistently *service* customers, and *intermodal capabilities* (many companies are working to create driverless carriers capable of transporting containers from port to rail and beyond).

Comparing the Six Modes of Transportation

Mode	Accessibility	Speed	Capacity	Cost	Service	International Capabilities
Rail	Low	Moderate	High	Moderate	Moderate	Intermodal
Truck	High	Moderate	Low	Moderate	High	Intermodal
Water	Low	Slow	Very High	Low	Moderate	Intermodal
Air	Low	Fast	Low	High	High	Intermodal
Pipeline	Low	Very Slow	Very High	Very Low	High	Limited
Digital	Very High	Very Fast	Very Limited	Very Low	Very Limited	Very High

Marketing Analytics: Using Data to Inform Supply Chain and Channel Management Strategies

Using Data to Identify Supply Chain Partners

The supply chain for each product has to be designed to best meet the needs of target market consumers. A firm will often select the initial partners through a request for proposal (RFP) process, which is described in greater detail in the business-to-business lesson. However, choice of supply chain partners is also based on cost data related to all the elements within the supply chain—from the input components to storage facilities to distribution channels. Data about input costs can help a firm devise strategies for reducing the costs of the final goods that get stored and ultimately delivered. Firms also gather data on the timeliness of delivery, quality of goods, and service responsiveness. Once supply chain partners are chosen, the firm evaluates their performance through ongoing data collection to ensure that its supply chain is running as efficiently and effectively as possible. To gather these data, a firm must coordinate across all of its supply chain partners and have in place a strong data-sharing program.

Using Data to Manage Supply Chain Costs

Once all of the supply chain partners are selected and in place, a firm must continuously collect data on cost, profitability, margins, as well as qualitative service elements such as communication effectiveness and timeliness, to inform how it can manage its supply chain costs. Over time costs within the supply chain can increase, which can result in lower profitability and higher prices. Costs can increase as a result of changes in production costs, as well as in logistics costs. One way to reduce logistics costs is to move the distribution center closer to the consumers. Amazon is currently building a number of very large distribution centers around the United States as a strategy for lowering the transportation costs for each good.

Using Data to Manage Supply Chain Functionality

A firm requires a high level of coordination and information sharing to manage its supply chain functionality. Collecting performance data on each supply chain partner as well as data from customer satisfaction studies contributes to this effort. Feedback mechanisms throughout the supply chain allow members to understand how to improve their performance to optimize the functionality of the whole supply chain. Amazon, for example, reviews all of the performance elements for each of its suppliers. If the reviews reveal a number of supplier service failures, such as delayed or missed deliveries, that supplier will be put on notice and perhaps even removed from the Amazon list of suppliers.

Using Data to Inform Inventory Decisions

Sales data must be used to determine which products to stock in each retailer and in what quantities. Forecasting future demand requires data from not only past sales but also from growth projections and consumer confidence ratings. For example, to ensure stock is in place by the holiday sales season, retailers must place their orders by early summer. They must then determine which products are going to be popular over the holidays and then estimate how much consumers might be willing to spend on these products. These estimates are based not only on the prior year's sales but

also on marketplace sales trends. These data are available from various sources, including industry associations that publish data about which products are trending up and which products are coming into market.

Industry trade shows provide marketers with a wealth of data about which new products are expected to be popular among consumers. Knowing *which* products to stock is important, but so is avoiding overstock or stockout of products. Overstocking or understocking a specific item can lead to a loss for the company. As noted earlier in this lesson, if a firm overstocks inventory, then those products must be sold at a loss to clear that inventory, which results in reduced profitability. For example, Macy's orders its winter coat stock for each of its stores based on a region's average temperatures and on the prior year's winter coat sales within that region. If there is an extremely warm winter, consumers may purchase fewer coats and leave Macy's with an overstock of inventory that it must now sell at a drastic price reduction. On the other hand, if Macy's understocks its inventory based on predictions of a mild winter, and it turns out to be an extremely cold one, then Macy's risks losing the sales it could have made had stock been more sufficient.

Responsible Marketing

Ethical Supply Chain and Channel Management

Unfortunately, when many of us are shopping, we don't give much consideration to the manufacturing and sourcing of the products we buy. Rarely do we ask ourselves, "where did this come from?" However, as consumers, we are increasingly becoming more aware of the environmental sustainability and social justice issues related to the brands and products we buy. The supply chain is an important consideration for firms as they consider what organizations they partner with and how they do due diligence to ensure that their partner firms adhere to the same ethical, environmental, and social standards.

Many brands outsource their manufacturing to other firms, meaning that they don't control the actual production of their products. Yet if a firm claims to be environmentally friendly, it must partner with firms that also uphold these claims.

Consider fashion, for example. Very few fashion brands actually make and produce their clothing. Unfortunately, fashion—especially fast fashion—is one of the worst contributors to environmental degradation and is often inefficient in its use of resources, especially water and cotton.

Additionally, firms must consider the social consequences of their actions. For example, facilities may have unsafe or exploitative working conditions and treat their employees unfairly. Not only is this behavior unethical, it could also harm a firm's brand image.

Reflection Questions

1. Think about a brand that you love. Would you still buy from the brand if you knew that it worked with other organizations that engaged in environmentally unsustainable behaviors or treated employees unfairly? Why or why not?

2. Review the brand's responsible management or sustainability report. Do you think the brand is doing enough to ensure responsibility throughout its supply chain? Why or why not?

3. How can consumers become more aware of the environmental practices of the companies from which they buy products?

Lesson 8-2

Retailing and Omnichannel Marketing

In this lesson we'll explore the world of retailing, including the roles of retailers and types of retailing. We'll see how companies are ushering in changes to the retail world. We'll also look at omnichannel retailing and wholesaling.

By the end of this lesson you will be able to

- Explain the role of retailers in the distribution channel.
- Describe the major types of retailers.
- Describe the methods of non-store retailing.
- Identify the benefits and challenges of omnichannel retailing.
- Describe the different levels of distribution intensity.
- Define wholesaling.
- Explain how marketers use data from retailers to improve products and services.

Marketing Analytics Implications

- Retailers are the link between producers and consumers and have the most direct contact with consumers. Therefore, marketers rely on data from retailers to inform them of consumer preferences and purchasing behavior.
- Technology has played an increasingly important role in how retailers design brick-and-mortar or online storefronts. Using virtual reality, marketers are able to study consumers in virtual environments and how they interact with products. Data gathered from these studies can inform store design.

Redefining Retail: How Retailers Responded to a Global Pandemic

The global COVID-19 pandemic will forever change the face of retail. Retailers learned that changes they thought they had years to prepare for and behaviors they thought would be around forever could change in an instant.

In the United States, the threat of quarantine caused many customers to "panic purchase" items such as toilet paper, nonperishable food items, and cleaning supplies at the beginning of the pandemic. As physical brick-and-mortar retailers had to shutter their doors, customers turned to e-commerce for everyday or common purchases. Many retailers that had never sold anything online had to quickly configure an infrastructure to allow them to reach customers through e-commerce.

Months after the COVID-19 vaccine became available, many customers felt safe returning to retail stores. However, this return was not "business as usual." The pandemic permanently altered interest in and expectations for in-store shopping. Customers now have higher expectations for in-store services and experiences. For example, curbside pickup and contactless checkout, which emerged in the pandemic, are steady fixtures for many retailers.

The physical store is still a critical asset for in-store shoppers and is now acting as an omni-fulfillment center to serve customers across all channels. Kroger is testing an artificial intelligence (AI) "smart" shopping cart for use in stores. This new cart and program "offers customers a seamless shopping experience where they can scan items and pay, all on the cart." Scan & Go at Sam's Club allows members to use their mobile app to scan the barcode on selected items, pay through the app, and then show the digital receipt to an associate at the door before leaving. The IKEA Planning Studio is a new concept store focusing on providing city residents solutions for small living spaces.

The role of the retailer is evolving in the post-pandemic world. The shifts in consumer behavior are here to stay. Retailers need to embrace the new ways of doing business to retain current customers and acquire new customers by updating the in-store experience, investing in technology, and creating brand differentiators.

During the COVID-19 pandemic, many retailers were unprepared to meet customer demands. Numerous customers faced empty grocery shelves for important household goods. Supply chains took months to right themselves as they responded to significantly altered consumer needs.

Retail Management

Often, when people think of retailing, they think of grocery stores (Kroger, Publix, Albertsons), department stores (Macy's, Nordstrom, Neiman Marcus), or specialty stores (Gap, Sephora, Anthropologie). However, retailing encompasses much more than that and is a critical component of distribution and, more broadly, marketing strategy. **Retailers** purchase and resell products to consumers for their personal or family use. The retail-to-consumer transaction is usually the end of the supply chain.

Retailers may sell products through brick-and-mortar stores, catalogs, the Internet, mobile apps, television home shopping channels, and other avenues. Retailers add value to consumers by providing the products to them when and where they want them. Retailers also add value to customers by providing assortment and additional services.

Consider Trader Joe's, one of America's favorite grocery stores. At Trader Joe's customers can find an assortment of products, from turkey meatballs to cocoa almond cashew milk. For the consumer, being able to go to one store to buy various items is a convenience. Without retailers, consumers would have to go directly to the manufacturers to buy individual products. Additionally, retailers provide consumers with other services. Trader Joe's, for example, also provides consumers with services such as offering recipes, nutrition information, sales, wine tasting, and other in-store or community events.

Retailers provide value to product manufacturers as well. Manufacturers can sell units in large quantities to retailers, who then distribute the goods in smaller quantities to consumers. It is much easier for a manufacturer to sell 10,000 units of soup or bottled water or socks to a retailer versus selling 10,000 individual units to consumers. Retailers allow manufacturers to focus on the development and production of goods rather than the final distribution to end consumers.

Retailers and manufacturers work together to make sure that their products are accessible to customers.

Retailers and manufacturers can also work together to align target markets as well as reach new audiences. Manufacturers want to collaborate with retailers who share similar target markets to them and want their products to be in locations where consumers would expect to find their products. Produce and frozen food purveyors, for example, would choose to distribute their products through supermarkets and grocery stores as consumers would expect to find those products there. Similarly, new products might hope to capitalize on a retailer's success and gain access to a retailer's target market by distributing through its stores. A brand-new producer of handbags would grow quickly if companies such as Nordstrom or Macy's began selling its products.

Retailers choose to work with manufacturers whose products are in demand by consumers and have a reasonable outcome of success. For retailers, having inventory that does not sell is a big problem. In this way, retailers and manufacturers depend on each other for mutual success.

Types of Retailers

There are 10 general categorizations of retailers. Each type of retailer has different advantages and disadvantages that manufacturers must consider depending on the product category and desired target market. Let's look at each of these categories in more detail.

Supermarkets

Supermarkets are large, self-service retailers concentrating on supplying a wide variety of food, beverage, and kitchen products. Many supermarkets have started to expand into home, fashion, and electronic products as well. Supermarkets tend to have significant buying power and focus on quick inventory turnover and lower profit margins. Examples of traditional supermarkets include Kroger, Publix, Stop & Shop, and Safeway. During COVID-19, many supermarkets had to act quickly and drastically to meet customer demand while also keeping customers and employees safe. During the pandemic, these frontline service workers were considered essential and allowed all of us to stay stocked with food and necessities.

Supercenters

Supercenters offer traditional grocery items as well as apparel, beauty products, home goods, toys, hardware, electronics, and various other merchandise. Supercenters have a wide variety and assortment and often are very competitive on price. Supercenters provide value to consumers with their lower prices and one-stop shopping format. Examples of supercenters are Walmart Supercenters and Target Superstores.

Warehouse Retailers

Warehouse retailers offer food and general merchandise products usually in larger quantities and at discounted prices. Because of the low prices offered, customers can expect limited service at warehouse retailers. However, even successful warehouse retailers like Costco offer free product samples every weekend.

Convenience Retailers

Convenience retailers offer a limited variety and assortment of merchandise, usually snack foods and minor essentials, at an easily accessible location. Consumers can make purchases quickly at a convenience store; however, the prices are usually higher than those of a supermarket or warehouse store. 7-Eleven, Speedway, Cumberland Farms, Sunoco, and Pilot are all examples of convenience retailers.

Department Stores

Department stores offer a wide range of products displayed as a collection of smaller "departments" within the store. Department stores carry merchandise ranging from cosmetics, to apparel, to home goods. Department stores also offer customers high levels of customer service and a uniquely curated inventory. Department stores are currently struggling as they try to compete with more online retailers and specialty stores. JCPenney, a traditional department store, attempted to attract younger audiences several years ago by changing its image and pricing policies. However, not only was JCPenney unable to attract new customers, but the change alienated its existing customers as well.

Specialty Retailers

Specialty retailers concentrate on a specific product category. Specialty stores offer limited variety, salesperson expertise, and high levels of service. Sephora is a specialty store that provides customers with a selection of cosmetics and beauty products. Additionally, Sephora sales associates have strong product knowledge and are able to demonstrate how to use different products.

Off-Price Retailers

Off-price retailers offer an inconsistent variety and assortment of branded products from an array of suppliers by reselling products that did not sell at another retailer. For example, at the end of the season, a department store might return large quantities of unsold merchandise back to the manufacturer. In order to try to make a profit off the inventory, the manufacturer may choose to sell the merchandise to an off-price retailer, or an outlet store, at a significant lower price. Because of the ambiguity, the merchandise offerings in off-price retailers varies significantly.

T.J.Maxx is an example of an off-price retailer and is one of the largest retailers in the world.

Drugstores

Drugstores primarily sell pharmaceuticals, health and wellness products, over-the-counter medicines, beauty products, as well as a limited assortment of food and beverages. Drugstores typically also compete on convenience as they are easy to access and usually have an efficient checkout process. In many metropolitan areas, competing drugstores are often located across the street from each other, each trying to capitalize on the convenience of the right-hand turn into the store. Walgreens, Rite-Aid, and CVS are examples of drugstores.

Service Retailers

Service retailers mostly sell services rather than merchandise. Examples of service retailers include salons and spas, banks, automotive repair shops, yoga studios, and so on. Many service providers require face-to-face interactions and really struggled during the COVID-19 pandemic.

Massage Envy is an example of a service retailer.

E-tailers

E-tailers allow customers to shop for and buy products online (or via a mobile device) for home delivery. E-tailers are rapidly reducing the delivery time to consumers, allowing them to compete more aggressively with traditional brick-and-mortar stores. E-tailers are highly convenient for consumers and are often able to offer lower prices because of reduced overhead costs. Even though Amazon now has brick-and-mortar stores, it is still the benchmark for e-tailers. Everlane is also an example of an e-tailer.

Supermarkets

Advantages

- Offer a wide variety of food, beverage, and kitchen products.
- Offer customers a large assortment of perishable items that other retailers do not offer

Disadvantages

- Provide limited service to customers.
- Can sometimes be in inconvenient locations.

Supercenters

Advantages

- Provide customers with a one-stop shop for food, beverage, and many home goods.
- Are often very competitive on price.

Disadvantages

- Provide limited service to customers.
- Sometimes the shopping experience can be overwhelming to customers.
- Often in inconvenient locations.

Warehouse Retailers

Advantages

- Offer customers the opportunity to buy merchandise in bulk at discount prices.

Disadvantages

- Provide limited service to customers.
- Sometimes inventory is inconsistent.

Convenience Retailers

Advantages

- Provide customers with a convenient location.

Disadvantages

- Assortment and variety can be limited.
- Customers might pay a higher price point.

Department Stores

Advantages

- Offer a broad variety and deep assortment of merchandise.
- Offer services to customers to improve the shopping process.

Disadvantages

- Can be inconvenient for customers to access.
- Many of today's customers want a more specialized shopping experience.

Specialty Retailers

Advantages

- Offer a narrow variety but deep assortment of merchandise.
- Employ knowledgeable sales staff to assist customers.

Disadvantages

- May charge higher prices.
- May not have the variety the customer is looking for.

Off-Price Retailers

Advantages

- Provide customers with a deep assortment of merchandise.
- Offer merchandise at a deeply discounted price.

Disadvantages

- Inventory is inconsistent.
- Offer little to no sales assistance.

Drugstores

Advantages

- Are in convenient locations.
- Now offer an expanded merchandise assortment.

Disadvantages

- Offer little to no sales assistance.

Service Retailers

Advantages

- Provide unique services to customers.
- Employ highly trained and knowledgeable sales staff.

Disadvantages

- Services are perishable.
- Often customers need appointments to be seen, making service retailers sometimes inconvenient.

E-tailers

Advantages

- Offer a wide variety and assortment of merchandise.
- Provide convenient delivery options.

Disadvantages

- Lack personalized service.
- Returning merchandise can often be time-consuming.

Non-Store Retailing

Non-store retailing involves the selling of goods and services outside the limitations of a retail facility. It is a growing industry as companies look for unique ways of engaging customers. In fact, according to the U.S. Bureau of Labor Statistics, more than 563,000 people were employed in the non-store retailing sector in 2017.

Historically, e-commerce has been considered a type of non-store retailer. However, as many e-tailers have begun to open brick-and-mortar stores, the boundaries of non-store retailing have begun to blur. Let's look at some traditional methods of non-store retailing.

Direct Selling

Direct selling is a non-store retail method that involves salespeople interacting with customers directly, usually at home, work, or at an organized "party." Direct salespeople provide high levels of service, including product knowledge and demonstrations. Direct selling extends beyond Avon, Mary Kay, and Amway. Traveling Vineyard is a direct selling organization that hosts wine tastings for consumers. Pawtree offers unique animal and pet-care products.

Vending Machines

Vending machines are a non-store retail method that dispense merchandise to consumers via an automated vending machine. In the past, vending machines primarily sold snacks and beverages. In the past 10 years, vending machines have grown in sophistication. In many airports, consumers can purchase cosmetics (from Sephora) and electronics (from Best Buy). In some restaurants, consumers can even purchase their food from vending machines.

Television Home Shopping

Television home shopping is a non-store retailing method that allows customers to see a product demonstrated on television and purchase the product by calling a number or purchasing online. In television home shopping, the television station acts as the primary communication with consumers. The most successful television home shopping stations are Home Shopping Network (HSN) and QVC. Television home shopping has come a long way. HSN, for example, partners with celebrities and athletes such as Danica Patrick, Wendy Williams, and Serena Williams to sell products.

Kiosks

Kiosks are a non-store retailing method that offer a temporary, inexpensive, movable format. Kiosk retailers can move their carts or portable stores to go to areas of high foot traffic. Kiosks can be found in malls, airports, sporting events, or even within traditional retail stores. Because of the small format, the inventory selection is usually small, but highly specialized.

Direct Selling

Direct selling offers a nontraditional shopping experience. Often, customers can shop in the comfort of their own homes and with their friends in a "party" atmosphere.

Vending Machines

Vending machines offer customers convenience in that they are easily accessible and the transactions are quick. For retailers, vending machines require little inventory and overhead expenditures.

Television Home Shopping

Home shopping has changed significantly from when it was first introduced. Today, home shopping partners with celebrities and athletes. Customers have access to a ton of information about a product.

Kiosks

Kiosks allow retailers to reach customers with a smaller format and lower inventory investment.

Omnichannel Retailing

In today's highly technological world, consumers, retailers, and manufacturers are placing a greater emphasis on omnichannel retailing. **Omnichannel retailing** is a multichannel retailing approach that allows the customer to have an integrated customer experience across all of a retailer's distribution platforms. In other words, with omnichannel retailing, customers can shop from their smartphones, in a brick-and-mortar store, on a desktop, or even by catalog and enjoy a seamless experience.

REI is an example of a retailer that has a successful omnichannel strategy. REI prioritizes offering a consistent experience for customers across all channels. For REI, the use of mobile technology has been integral to its omnichannel success. REI gives all of its in-store sales associates mobile devices to allow them to provide greater service to shoppers. This allows associates to walk around the store with customers, pull up inventory and product information quickly, and check out customers quickly without being tethered to a cash register. Additionally, REI provides customers with free Wi-Fi to access information while they shop.

In order for omnichannel retailers to be successful, they must invest in infrastructure, process, and people.

Infrastructure

The basic criterion for omnichannel retailing to be successful is that a firm has the infrastructure in place to complete the integrated shopping experience. This involves having successful relationships with suppliers and other vendors. This also means having the appropriate supply chain and logistics capabilities to deliver across multiple channels in an expedited fashion. Retailers have historically segregated channel operations. An effective omnichannel strategy requires retailers to run the operations of all channels seamlessly.

REI invests heavily in its infrastructure to provide a seamless experience to customers across channels. Additionally, REI builds its supply chain to be environmentally stable.

Process

As omnichannel initiatives expand, retailers have to implement processes and strategies to ensure that the experience is seamless across channels. This requires an investment in technology as tools like enterprise resource planning (ERP) systems can help expedite distribution processes to facilitate delivery. ERP systems allow companies and supply chains to integrate applications to more efficiently manage their businesses. This also requires processes on pricing. In order for consumers to trust a retailer's omnichannel strategy, they must trust that the price will be the same regardless of what channel they use. If customers research a product online, they would expect the price to be the same if they

REI's systems ensure that customers can check out quickly in-store and online and receive the same price and attention to detail.

visited a brick-and-mortar store. This continuity requires retailers to institute pricing processes to maintain pricing integrity across channels.

People

Retailers depend on the power of their employees. Success or failure of an omnichannel retailing strategy is totally dependent on the firm's employees. People are needed to effectively execute the firm's omnichannel strategy. This requires retailers to invest in training for employees across all channels. People are also necessary for maintaining the retailer's brand. Employees reflect the culture of a retail firm, and

REI's employees, especially its sales staff, are a huge part of REI's continued success.

this in turn impacts the retailer's brand. If employees are unhappy or disgruntled, this will damage the brand. It is important for all employees, from the sales floor, to the buyers, to the distribution center employees, to be on the same page when it comes to omnichannel retailing.

Distribution Intensity

Manufacturers must choose not only the right retailers to sell their products but also the best distribution strategy. This means the marketer has to decide what distribution intensity is appropriate for achieving its marketing objectives. **Distribution intensity** refers to the number of outlets a marketer chooses to sell through.

The level of distribution intensity depends on the target market, the pricing strategy, the promotion strategy, and the level of service required by the consumer during the purchase. There are three levels of distribution intensity: intensive, exclusive, and selective.

Intensive Distribution

Intensive distribution is a distribution strategy that involves placing a product in as many outlets as possible. Coca-Cola products, for example, are available at grocery stores, vending machines, convenience stores, and so on. A customer would not have to travel very far to find a Coca-Cola. An intensive distribution strategy is most appropriate for items that do not require much explanation and that benefit from more exposure.

Exclusive Distribution

Exclusive distribution is a low-volume distribution strategy in which products are sold through very limited channels. Exclusive distribution is best for high-involvement products in which the consumer would require salesperson interaction. Exclusive distribution benefits the manufacturer because it ensures that the retailer will uphold the brand image of the product and provide the customer with an optimal experience. Exclusive distribution benefits the retailer as it becomes the only outlet to sell a brand, limiting its competition. Automobile manufacturers typically use an exclusive distribution strategy for dealerships, allowing only one or two dealerships to sell their vehicles within a geographic area. Similarly, Nike might provide sneakers to Foot Locker that are exclusively sold at Foot Locker.

Selective Distribution

Selective distribution allows brands to sell through fewer channels but still capture economies of scale in production. In selective distribution, products are not sold exclusively through a certain retailer or geographic area, nor are they sold through every available retail outlet. Selective distribution allows brands to maintain strategic control of how their products are sold. Most apparel brands are sold using selective distribution.

Intensive Distribution

Coca-Cola is distributed through every logical channel, including convenience stores, grocery stores, vending machines, and a variety of other methods. Customers should not have to search far to purchase a Coca-Cola.

Selective Distribution

Consumers can purchase iPhones at Apple stores as well as other non-Apple specialty stores. Availability through multiple channels allows customers easier access to iPhones, but still relies on the expertise and knowledge of the sales associates to help answer customer questions.

Exclusive Distribution

Tesla uses a company-owned store model to reach its customers. These stores are limited to high-end, affluent areas and are usually limited to one store per geographic area.

Wholesaling

Wholesaling is the sale of goods or merchandise to retailers; industrial, commercial, institutional, or other professional business users; or other wholesalers. In general, a **wholesaler** is a firm that sells goods to anyone other than an end consumer.

Wholesalers frequently purchase a large quantity of a good, and then sell off smaller quantities of that good at a higher per-unit price. Wholesalers are usually physically closer to the markets they supply than the original source, thus adding value to the manufacturer. For example, Procter & Gamble (P&G) sells dozens of brands with thousands of items in its product portfolio. Rather than sell its products to many different retailers, it is easier for P&G to sell to a wholesaler, who then reduces the quantities and manages the distribution of products to individual retailers. In fact, P&G's website offers a wholesale finder option. Costco is an example of a wholesaler that can act as a retailer. When it sells products to small businesses, it is functioning as a wholesaler. However, when it sells products to individual consumers, it is functioning as a retailer.

Some large retailers act as both wholesaler and retailer. Costco, for example, performs retailing functions when it sells products to individual consumers. Conversely, Costco also acts as a wholesaler when it sells to small businesses.

Marketing Analytics: Using Retailing Data to Improve Products and Services

Using Retailers as a Data Source

In addition to providing traditional retail services, today's retailers have also become a key source of essential marketing data for firms. For example, being able to track consumers through a retail environment via an app allows product developers to better understand how to make their packages more appealing or how to design their products to be more attractive on the shelf. Data from retailers also show trends in purchasing behavior as well as the cost of stockouts and other issues. Retailers may also share data on pricing and other category-specific information, which can help firms plan their marketing strategies.

Using Data to Inform Store Design

Every retail space has to be designed to ensure profitability per square foot. Many retailers have carefully calculated how much per square foot is needed for their stores, as well as which locations are likely to prove the most profitable. These calculations are then used to design stores that reflect the way target market consumers interact with and within the retail space. This same design is then applied to store layouts across different stores within a chain of stores.

Increasingly, retailers are using technology to assist in store layout and design. Technology can track consumers as they explore the real and virtual store environments, which helps marketers identify how to make certain areas in the store more productive. For example, using data from studies that show consumers' growing interest in healthier living and whole food sources, many grocery stores have redesigned and relocated their produce sections, moving them to the center or front of the store, and replicating the farmer's market experience. Apple retailers have found that consumers value high levels of interactivity; as such, they have designed their stores to facilitate interactivity throughout the retail space, keeping merchandise displays against the side walls, the floor plan open to facilitate free movement, and the interactive product stations at the center.

Responsible Marketing

Amazon: A Responsible Retailer?

For many consumers, Amazon seems to anticipate their needs before they can. With features like Subscribe and Save, the Dash Button, or predictive shopping, it's almost too easy to shop on Amazon. (And, of course, we can always just "Ask Alexa.") During the COVID-19 pandemic, Amazon provided access to critical goods for many consumers who were uncomfortable shopping in stores. Amazon also provided a platform for various smaller businesses to engage as third-party providers on its site so that they could continue selling their products. Amazon's seemingly limitless assortment and variety, coupled with its operational efficiencies, helped prop up the economy during the pandemic.

Although for the most part Amazon was able to keep its costs low for customers, a price was still paid in other ways. Because of the increased demand, Amazon employees were asked to work additional hours, often in crowded warehouses. For many workers, the risks of this environment during the ongoing pandemic felt overwhelming, but they didn't want to lose their jobs. Even prior to the pandemic, there were numerous reports of "grueling working conditions" at Amazon, where productivity was prioritized over employee health. Amazon critics also point to Amazon's environmental degradation, suggesting that the company has an increasingly large carbon footprint. Meanwhile, former CEO and current executive chair Jeff Bezos is the richest person in the world while employees argue they are not being treated fairly.

Some could argue that these productivity-based working conditions and aggressive supply chain functions are what allow Amazon to lower costs and increase consumer convenience. This, they say, improves the quality of life for individuals and families globally. But is the price workers and the environment pay in other ways too steep?

Reflection Questions

1. What is the consumer's responsibility in demanding change from retailers?

2. Would you continue to support Amazon if its prices increased or delivery times took longer, but you knew working conditions improved and its impact to the environment was minimized?

Delivering Value: Test

1. A company producing cosmetics may have hundreds of suppliers, some of which may be selling the same or similar raw and packaging materials to other cosmetics manufacturing companies. In addition, that same cosmetics manufacturer may sell to a retailer that buys similar cosmetics from a number of other manufacturers. This scenario illustrates

   A. logistics.

   B. distribution management.

   C. a supply chain.

   D. distribution center.

   E. a value chain.

2. Supply chains add value to a business in all of the following ways *except*

   A. reducing production costs.

   B. providing better logistics.

   C. making sure goods are available when required.

   D. improves oversight of goods and payment flows between production and end consumer.

   E. simplifying the environmental scanning process.

3. C&S operates large businesses but is not well-known by consumers. It has more than 75 high-tech facilities in 15 states and supplies supermarkets and institutions with more than 170,000 different products. C&S is a

   A. wholesaler.

   B. distributor.

   C. retailer.

   D. franchise.

   E. brick-and-mortar store.

4. Which of the following is a distinguishing characteristic of a distributor?

   A. sells products for one manufacturer.

   B. warehouses goods.

   C. resells its products to retailers.

   D. buys products for resale.

   E. buys goods from all producers or vendors.

5. Hyun's Handicrafts is a company that provides custom linens for both businesses and individual consumers. Hyun does not make any products until she receives an order from the customer. What type of marketing strategy does this represent?

   A. push strategy

   B. supply strategy

   C. demand strategy

   D. forecast strategy

   E. pull strategy

6. Dell builds component inventory based on its sales forecast but doesn't completely assemble computers until a customer order is received. Which supply chain strategy is Dell applying?

   A. push strategy

   B. supply strategy

   C. demand strategy

   D. push-pull strategy

   E. pull strategy

7. Among the Trees is a manufacturer of deck furniture and other outdoor furnishings. Typically, customers purchase the company's products as the spring season begins and as such, do not want to wait for the product to be manufactured. Which supply chain strategy would work best for Among the Trees?

   A. forecast strategy

   B. supply strategy

   C. demand strategy

   D. push strategy

   E. pull strategy

8. Walmart's Retail Link program makes it possible for Walmart suppliers to manage their own products by allowing them to monitor such data as sales and inventory volume, in-stock percentage, gross margin, and inventory turnover. This program helps suppliers reduce low inventory risks and the costs associated with them. Walmart's Retail Link program is an example of a

   A. logistics approach.

   B. distribution orientation.

   C. marketing orientation.

   D. supply chain orientation.

   E. channel management approach.

9. A company that practices supply chain management will

   A. set up technology to enable the sharing of demand information with suppliers.

   B. have an inward focus on the activities and performance of other companies.

   C. recognize the importance of sharing supply and demand information with suppliers and customers.

   D. have an outward focus regarding the activities of its own company.

   E. not rely on other members of the supply chain to add value to the final product.

10. When a company does not have enough inventory available to fill an order, this is known as

    A. stockout.

    B. overrun.

    C. obsolete inventory.

    D. shortage.

    E. put-away.

11. The three basic functions of warehouse activities are

    A. storage, shipping, and restocking.

    B. sorting, allocating, and movement.

    C. production, assembly, and storage.

    D. receiving, storage, and movement.

    E. storage, movement, and production.

12. Which of the following statements best describes the benefits of supply chain integration?

    A. It makes each channel as efficient as possible, thus saving the company time, money, and human resources.

    B. It utilizes as few channel members as possible in the manufacture of the product so that a company can allocate its resources to addresses other needs in the supply chain.

    C. It reduces costs, improves customer service, makes efficient use of resources, and increases the ability to respond to changes in the marketplace.

    D. It allows a company to accurately predict production levels in order to minimize inventory carrying costs.

    E. It gives the company the freedom to choose intermediaries that share the company's mission.

13. Which of the following consumers visited a convenience retailer?

    A. Soraya was starving and stopped in at the 7-Eleven to purchase a bag of pretzels. Although it cost 50 cents more than she'd usually pay, she bought it anyway.

    B. Farah needed to set up a checking account, so she visited a local branch of Wells Fargo.

    C. Tony purchased all of his holiday gifts off Amazon.com this year.

    D. Because she was starting to get a headache, Ilise stopped by CVS to pick up some generic-brand ibuprofen.

    E. Josiah was unable to find a rain coat because the store only had a few and they didn't have his size.

14. Stella's Beads sells handmade beaded jewelry on the river esplanade where lots of passersby walk or stop to take pictures. In the winter, she brings her stand to the airport and sells her goods there. Stella operates a

    A. service retailer.

    B. vending machine.

    C. kiosk.

    D. convenience store.

    E. specialty store.

15. Paolo is a consultant for Pampered Chef. He hosts cooking shows in people's homes and sometimes the local mall using Pampered Chef products and then offers the products for sale at the end of his demonstrations. Paolo is demonstrating which type of nonstore retailing?

    A. e-tailing

    B. home shopping

    C. direct selling

    D. kiosk sales

    E. convenience retailing

16. Fatima is able to shop from Apple using her smartphone, desktop, or by visiting a physical location. Apple is engaging in

    A. direct selling.

    B. omnichannel retailing.

    C. service retailing.

    D. convenience retailing.

    E. specialty retailing.

17. Which of the following conditions is *not* necessary to ensure the success of omnichannel retailing?

    A. pricing processes

    B. investment in technology

    C. segregated channel operations

    D. effective relationships with suppliers

    E. appropriate supply chain logistics

18. The cost of a retailer's products in omnichannel marketing

    A. tends to be more expensive online.

    B. is not a concern.

    C. must be the same across all channels.

    D. tends to be more expensive in-store.

    E. usually differs across channels.

19. Coca-Cola is available at retail outlets worldwide; its products are available in vending machines, grocery stores, fast-food restaurants, and even online. This product availability is an example of which type of distribution strategy?

    A. administered

    B. exclusive

    C. intensive

    D. saturation

    E. selective

20. Exclusive distribution is best exemplified by which of the following?

    A. A maker of natural household cleaning detergent sells its goods only to specialty retailers.

    B. A manufacturer of an expensive energy drink distributes only in urban markets.

    C. A new brand of running shoe is marketed only to large warehouse retailers.

    D. Apple TV is only available at Apple stores and other specialty electronics stores.

    E. A maker of fine jewelry sells only through official dealers, which are limited in number.

9 Communicating Value

What To Expect

One of our most valuable abilities is our ability to communicate. For a Marketer, the ability to communicate the value of a good or service can ultimately determine its success. Let's take a look at the importance and value of good communication.

Chapter Topics

- **9-1** Integrated Marketing Communications
- **9-2** Digital Marketing
- **9-3** Advertising and Sales Promotion
- **9-4** Public Relations
- **9-5** Personal Selling
- **9-6** Customer Relationship Management

Copyright © McGraw Hill   Fizkes/Shutterstock

Lesson 9-1
Integrated Marketing Communications

In this lesson, you will learn how marketers communicate with customers. Marketers understand interpersonal communication and work hard to send clear and concise messages. Marketers also must be certain to send consistent messages across the promotional mix, which is used to send messages in a variety of different ways to fit different marketing situations.

By the end of this lesson you will be able to

- Explain the importance of integrated marketing communications.
- Outline the steps of the traditional communication process.
- Describe marketing communication methods commonly used by marketers.
- Explain direct marketing and its benefits to consumers and companies.
- Explain elements of the promotional mix.
- Explain how marketers use data to inform their integrated marketing communications strategies.

Marketing Analytics Implications

- The communication process begins with the sender encoding messages. Data from consumers are used to define how messages should be structured in order for the target market receiver to be able to decode them correctly.
- Data are used to determine what the best way is for senders to break through the noise in the market and reach consumers with the sender's message.
- Each communication method and element of the promotional mix are able to deliver messages in different ways; as a result, they have different strengths. Which communication method is used is based on data to determine the most effective IMC strategy.
- Each element of the promotional mix has its own measurement methods and standards. Data are used to evaluate the effectiveness of the IMC campaign.

Red Bull Gives You Wiiings: An Unconventional and Original Approach to Marketing

Red Bull's marketing strategy has evolved since its introduction in Austria in 1987, but one thing remains the same: The brand is unconventional and original. Red Bull focuses its strategy around five key pillars:

1. **Content Creation:** Red Bull creates world-famous content for its over 10 million subscribers on YouTube, over 14 million followers on Instagram, almost 48 million likes on Facebook, and 2 million followers on Twitter. The company

even created the Red Bull Content Pool offering photos, videos, and news about Red Bull events and athletes for editorial and news purposes.

2. **Publicity Stunts:** The trademark slogan "Red Bull gives you wiiings" shows the target audience that nothing is impossible. To demonstrate pushing the limits, Felix Baumgartner—as part of Red Bull's skydiving project Red Bull Stratos—broke a 50-year-old record when he fell 128,000 feet in four minutes and twenty seconds from a helium balloon in the stratosphere. This stunt alone generated more than $500 million in sales for the company.

3. **Extreme Sports:** Through extreme sport sponsorships and hosting its own events, Red Bull connects with adrenaline lovers, raises brand awareness, and generates hype. From motorsport, surfing, cliff-diving, and more, Red Bull's events are packed with action and excitement that raise awareness, and photos and videos create social media content its audience craves.

4. **User-Generated Content:** Not only does Red Bull create content for its audience, it encourages users to create and share content too by holding contests. These contests make followers feel acknowledged for contributing and provide Red Bull with a wealth of content.

5. **Influencer Marketing:** Through influencer marketing, Red Bull builds a strong brand identity that resonates with its target audience. Even in the beginning, Red Bull used brand evangelists handing out free samples on college campuses well before social media became mainstream.

Red Bull has demonstrated success using integrated marketing communications (IMC) with its unconventional marketing tactics for over 30 years. The brand brings together promotional elements to deliver a clear and consistent message to its customers by integrating each of these five pillars.

Defining Integrated Marketing Communications

A firm's promotional mix includes different ways to communicate its marketing message, such as advertising (newspaper, streaming, outdoor, etc.), sales promotions (coupons, samples, rebates, etc.), personal selling, direct marketing, or

Red Bull follows an unconventional and original approach to marketing as seen through its sponsorships of extreme sports, publicity stunts, and content creation for over 10 million YouTube followers.

public relations. An **integrated marketing communications (IMC) strategy** is used to coordinate the various promotional mix elements to provide customers with a clear and consistent message about a firm's products, which in turn allows firms to more effectively create and develop relationships with customers.

Communicating a clear marketing message across the promotional mix is important. If the look and sound of the message are inconsistent in different places, it is destined to be less effective. For example, imagine if Red Bull used Illume adventure and action sports imagery contest on social media, but then used performing arts such as ballet and theater in its marketing communications in magazines and on television. Both may be effective approaches, but by using a different look and theme, Red Bull is making its advertising less recognizable across settings. An IMC approach adapts the communication for each element of the promotional mix, but keeps the messaging consistent across all aspects of the promotional mix. By following an IMC strategy, marketers make it easy for customers to recognize their messages, regardless of where they see or hear them.

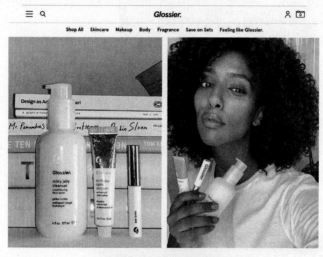

Glossier uses its website to sell and promote its products.

Glossier uses its YouTube channel to show customers how to use its products.

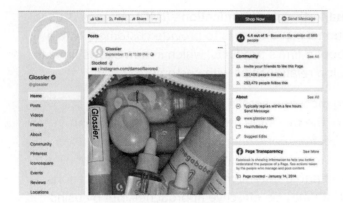

Glossier uses Facebook to connect with and inform its customers about products, events, and promotions.

Glossier uses pop-up shops that are open for a limited time to reach potential new customers.

The Communication Process

The ability to effectively communicate a marketing message is the difference between success and failure. If customers don't know about a product or service, they won't be able to buy it! Consumers are inundated with thousands of promotional messages a day. In order for a firm's message to stand out, marketers must understand the communication process.

The **communication process** describes how a message is transmitted from a sender to a receiver. The process of communicating a message is often difficult to accomplish. The process is familiar because you use it to communicate every day with friends, teachers, and family members. Communication requires a *sender* and a *receiver* who *encode* and *decode* messages.

To transmit a message, some *method of communication* must be used, for example, a text message, a television advertisement, a billboard, or most powerful, word of mouth.

Not all messages are received because sometimes *noise* interferes with the transmission of a message: a crying child can distract you from a television commercial, or talking to friends can divert your attention from an online advertisement—both instances are considered noise that disrupts the reception of a message.

Feedback closes the communication loop when the message is received and a response is returned to the sender.

In the sections that follow you will learn how each step of the communication process works and how this process specifically relates to marketing.

Sending and Receiving Messages

The **sender** starts the communication process by transmitting a message. Marketers are in the business of sending messages, so they are good at deciding what needs to be sent. Usually marketers send messages that relate to their product or brand.

The Communication Process. This figure depicts how a message is transmitted from a sender to a receiver in the communication process.

In addition to creating marketing messages (e.g., sales-related or informational messages), marketers must be adept at determining to whom the messages should be sent. The days of placing a single ad on network television for the whole world to see are long gone. Today, marketers take a targeted approach to identifying exactly who their target market is and how to most effectively send messages to those potential customers. For example, travel agencies specializing in adventure travel might target young adults who prefer exotic locations and unique experiences satisfying their need for adrenaline rushes. These travel agencies would be extremely thoughtful about which outlets to use to best promote their message.

The **receiver** interprets the message transmitted by the sender. For marketers, receivers are usually consumers (although they could be other things, such as stockholders, governments, or even competitors). As a consumer, if you follow a daily routine of work, family, school, and activity you are likely to encounter at least 3,000 marketing messages every day. If you don't have plans for a vacation, a message promoting exotic locations might get lost in the mix of your daily activities. However, when you are longing for a vacation, you might be especially open to receiving messages from airlines offering to transport you to tropical locations.

Of course, not every message that we see is actually received. In theory, you see every billboard along an interstate highway as you drive, but do you actually receive the information on every roadside sign? (Likely not.) Many messages are not received because their transmission is disrupted. Marketers call this disruption **noise,** which results from distractions such as competing messages, poor clarity of the messages, or an ineffective communication method. In the case of a billboard, noise can come from things such as heavy traffic, weather, or even other road signs.

Yet despite the prevalence of noise, many messages do get through to the sender, especially if they are based on IMC principles that make them easy for the receiver to identify and process.

Marketers must compete with noise when they invest in a highway billboard—traffic, in-car entertainment, other signage, and passengers all distract from the advertising message being communicated.

Encoding and Decoding Messages

When marketers decide to communicate, they need to encode their thoughts into a message that consumers can understand. **Encoding** is the transformation of the sender's ideas and information into a message that usually includes words, symbols, and/or pictures. It is very important for marketers to properly encode their message so consumers can understand it. If marketers do not properly encode their message, the money and other resources used to create and transmit it are wasted. During the 2016 Summer Olympics, Chase Bank featured an advertisement of a retired couple walking a small pig (**www.youtube.com/ watch?v=NaFlfaXJW4Y**)—most receivers of the message did not understand the message that Chase encoded. Despite the baby pig on a leash, the commercial encoded Chase Bank's ability to plan retirement investments. However, many receivers simply thought the pig was cute and did not make the connection to how Chase's services had made it possible for the couple in the ad to have a pig. Because the encoding of the commercial was confusing, it received very little response or social media attention.

Decoding a message occurs when a consumer receives a marketing message. More specifically, **decoding** is how the receiver perceives and interprets the sender's message. Decoding is one of the more important parts of the communication process because it really doesn't matter what message the marketer sends; rather, it matters how it is decoded by the consumer. Thus, marketers work hard to create messages that consumers will decode properly, and that the encoding is aligned with decoding. How an individual decodes a message is different for everyone depending on attitudes, experiences, beliefs, and so forth. If messages are not encoded properly, problems decoding the message may occur.

This was the case with Pepsi's "Come Alive" slogan, which was decoded by many as "Come out of the grave." Chevrolet experienced a decoding issue too when its Nova model (meaning "new star") was decoded in Spanish as No Va (meaning that it does not go).

Feedback

The final step of the communication process is feedback. **Feedback** is communication from the receiver to the sender. Traditionally, it has been hard for marketers to receive feedback because mass media communication methods (television, billboards, newspapers, etc.) do not allow for consumers to respond. One way that marketers gauged feedback was by measuring sales volume before and after an advertising campaign to determine changes caused by the campaign. Today, feedback is much more prevalent. Social media have revolutionized how consumers engage with companies and with each other. In many cases, consumer feedback may not be shared directly with a firm, but marketers can monitor popular blogs and rating sites to observe consumer feedback concerning their products and marketing messages.

An advertisement that goes "viral" is one form of feedback. Consumers are now able to transmit messages further than they formerly could. Social media and other easy-to-use communication methods (text messaging, e-mail, Snapchat, etc.) make it simple for consumers to indicate their interest in a marketing message. Sharing a funny or provocative commercial is a form of feedback (you "Like" or "Dislike" it, for instance). During Super Bowl LI (following the 2016 NFL season), Budweiser ran a commercial titled "Born the Hard Way" (**https://www.youtube.com/watch?v=HtBZvl7dlu4**), which was seen (received) by many of the approximately 111 million viewers who were watching specifically for the advertisements. However, the real power of this message came from additional, social media–driven distribution of

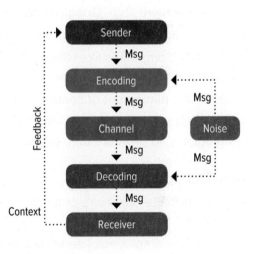

the message, as the commercial has been viewed more than 30 million times on YouTube. The viral success of the message is a feedback indicator to Budweiser that its commercial effectively communicated its message.

Marketing Communication Methods

Transmission of messages requires a **communication method**—a means for marketers to get a message to consumers. For example, consider "snail mail." The U.S. Postal Service is a reliable channel for marketers (and all of us) to deliver a message that we have prepared to a receiver at a physical address. Marketing communication methods are much like the postal service: They are used to deliver a marketer's messages efficiently to consumers. Although many marketers still use the postal service to deliver their messages, often called "junk mail" by consumers, they use many other methods as well. Communication methods take on many forms that you will recognize, and some that you may not. Marketers predominantly communicate through mass media, electronic media, interpersonal communication,

Copyright © McGraw Hill

point of purchase, physical space, and the press. Here are some examples of the ways communicate:

Mass Media

- Television commercials—Super Bowl ads are almost more interesting than the game.
- Newspaper advertising—*The New York Times, USA Today*, and so on, all feature advertising.
- Outdoor billboard advertising—Do you notice billboards?
- Wrapped buses/vehicles—Many cities sell the entire exterior of their buses to advertisers.
- Product placement—Have you noticed things like Coca-Cola in movies? They were likely a purchased product placement advertisement.

Electronic Media

- Social media advertisements—Facebook, Twitter, LinkedIn, and so on.
- Digital advertising—Google AdWords, Banner Ads, and so on.
- Mobile advertising—Hooked.com, push notifications, and so on.
- E-mail—Companies like Hilton often promote specials via e-mail to their loyalty club (Hilton Honors) members.
- YouTube placement of commercials.
- Opinion leader endorsers—Many firms work with well-known product users, or bloggers, to promote their products.

Personal Communication

- Salespeople—Verizon's employees are salespeople who help you upgrade your phone.
- Front-line employees (clerks, product specialists)—Apple's "Geniuses" are product specialists.
- Customer service representatives—American Airlines employs thousands of customer service representatives to handle customer delays and missed flights.
- Online live chat—Increasingly, customers talk to a virtual service representative, sometimes one who is artificial intelligence.

Point of Purchase

- In-store promotional signage—For example, a poster for Doritos in a convenience store.
- Brochures—These are common in hotel lobbies for local attractions and events.
- Product package design—Coca-Cola has a distinctive bottle that is easily recognizable.
- "Gas pump" TV or signage—Even gas stations have screens.

Physical Space

- Store design—For example, Starbucks has "comfortable stores."
- Website design and look—Ease of use is key.

- Permanent signage—For example, McDonald's "Golden Arches."
- Store layout and atmospherics—This includes smells, lighting, and so on.
- Employee uniforms—This also includes things like buttons, name tags, and so on.

Press

- Newspaper and TV news reports (from press releases)
- Magazine feature stories
- Blog comments
- Industry magazines and trade publications
- Online (TED Talks, instructional videos, etc.)

Direct Marketing

Direct marketing has undergone an enormous transformation in the last decade. In fact, the Direct Marketing Association, which has been in existence since 1917, recently changed its name to the Data and Marketing Association to reflect the changes and the analytical opportunities within direct marketing.

Defining Direct Marketing

What is direct marketing and what does it include? According to the American Marketing Association, **direct marketing** is the total number of activities a seller engages in to encourage the exchange of goods and services with the buyer. The seller communicates with the target audience using one or more forms of media in order to solicit a response from a prospect or customer. Direct marketing includes both offline media (such as catalogs, direct mail, telemarketing, and televised

infomercials) as well as digital media (such as e-mail, text-message marketing, and mobile advertising). Pottery Barn, for example, offers a portfolio of catalogs. A family with small children might receive a Pottery Barn Kids catalog, while college students might receive a Pottery Barn Teens catalog. This differentiation is an example of direct marketing. Amazon, however, uses a more customized approach of direct marketing by sending carefully crafted and curated e-mails to consumers based on individual preferences.

Often direct marketing is trying to elicit an immediate response, such as a purchase, donation, vote, or a call to action. Due to the needs for immediacy and action, this type of advertising is termed **direct response advertising.**

Many of you have probably seen an infomercial, which is a program-length commercial for a product. A couple of well-known ones are those sponsored by Pocket Fisherman and P90X. Infomercials want the viewer to "ACT NOW!" and often offer incentives such as "Buy One, Get One Free," but wait there's more, "Free Shipping If You Order Now!" The focus is on action.

Catalog

Over 100 million U.S. adults made a catalog purchase in 2017.

Direct Mail

Direct mail response rates are between 3 percent and 5 percent.

Email

In general, Mondays have the highest e-mail open rate of 13.3 percent, and Fridays have the lowest at 11.9 percent.

Online Direct (Digital)

38 percent of marketers plan to increase their digital marketing budgets over the next 2 years.

Mobile

123 million U.S. adults will use a mobile coupon this year.

The Benefits of Direct Marketing

How a firm develops its direct marketing campaign can lead to very different response rates from consumers.

These response rates, as well as the response *types,* tie directly to potential benefits for the firm. Response types can include:

- Registering for further information
- Requesting to be contacted by a salesperson
- Making a purchase

Quick consumer responses allow marketers to also respond quickly and to know immediately whether their messages are motivating consumers to "act now."

One of the benefits of direct marketing is the ability for firms to test different marketing message elements, such as word or visual choices, to see which is most effective with a target audience. Here we see testing of two colors on the "Add to Cart" button. A simple change of button color from green to red increased the number of clicks by 21 percent.

Direct marketing can result in a range of benefits for firms, including:

- The development of new customer relationships

- An increase in sales

- The ability for consumers to share compelling content with other consumers and users

- The ability to test the appeal of a product or service

- The ability to test various marketing approaches to see which is most effective with a target audience

- The ability to test different marketing message elements, such as word or visual choices—for example, the word *purchase* instead of *buy,* or an ad's background color

The Promotional Mix

Marketers not only need to be consistent in *how* they present their message, they also need to be smart regarding *where* they communicate their message. Marketers communicate the value of their product by using different elements of the promotional mix—which includes advertising, sales promotions, public relations, and personal selling—to engage their customers. The **promotional mix** is a combination of tools used by marketers to promote goods, services, and ideas and to accomplish their communication objectives. Typically, more than one element of the promotional mix is used, which is why it is called a *mix.*

In the sections that follow, you will learn more about how marketers choose the specific mix to market their products.

Marketing Communication and Promotion

Marketers use different elements of the promotional mix to communicate messages based on what product they are promoting. Each of the promotional mix elements can exert influence on different segments of the target market, and marketers should try to understand the impact of each element within the IMC strategy.

The basic elements of the promotional mix include advertising, sales promotion, public relations, and personal selling:

1. **Advertising:** Nonpersonal promotional communication about goods, services, or ideas that is paid for by the firm identified in the communication. Advertising is communicated through media channels, which can include any medium that connects a firm with an audience for a fee.

2. **Sales Promotion:** Used to support the other elements of the promotional mix. Sales promotions typically include some type of action-oriented marketing message. Things like coupons, rebates, samples, contests/sweepstakes, and loyalty programs are all marketing promotions used to encourage a customer to make a purchase decision.

3. **Public Relations:** Nonpersonal communication focused on promoting positive relations between a firm and its stakeholders. Public relations strategies provide information and build a firm's image with the public, including customers, employees, stockholders, and communities. Public relations differs from other promotional mix elements because it is unpaid promotion that is facilitated through media, other outlets, and community relations.

4. **Personal Selling:** The two-way flow of communication between a salesperson and a customer that is paid for by the firm and seeks to influence the customer's purchase decision. Personal selling is a highly adaptive and personal way that firms promote their products and services directly to customers.

Firms use marketing research techniques to continuously assess which promotional mix elements have the most influence on different market segments at different points in the product life cycle. Based on this research, the marketing department can allocate more of its promotional budget to the elements that are most effective—those that can increase sales, build brand equity, and improve customer relationships when designed efficiently and implemented throughout the organization.

In addition to marketing research, firms can determine which elements to use based on which communication method fits best with their marketing goals, characteristics of the product, target market characteristics, and their marketing budget.

Nature of the Method

Marketers follow the AIDA process in their communications. First a consumer must focus **a**ttention (A), to generate **i**nterest (I), and finally **d**esire (D) for the product. If they get to desire, they are more likely to take **a**ction (A) and purchase the product. Each element of the marketing mix has different strengths in regard to the AIDA model. Marketers can choose one method over another based on these strengths:

- **Advertising Strength:** Attention and Interest
- **Sales Promotion Strength:** Desire and Action
- **Personal Selling:** Interest, Desire, and Action
- **Public Relations:** Attention and Interest

Nature of the Product

The promotional mix elements can be chosen based on characteristics of the product. Compare each element and the product types they are most associated with:

- **Mass Advertising:** Simple products sold to a large audience, typically at a lower cost.
- **Sales Promotion:** Often used for newly introduced products to increase trial of the product.
- **Personal Selling:** Typically used for complex, higher-cost products.
- **Public Relations:** Used to increase awareness of an overall brand or product line.

Target Market Characteristics

Characteristics of the target market may dictate which promotional element is selected:

- **Advertising:** Best for markets that are large and diverse. Also, digital advertising may be the only option for a younger market segment.
- **Sales Promotion:** Best for markets that have highly informed buyers because a sales promotion can encourage use for customers who already are informed about the product.
- **Personal Selling:** Best suited to markets where a small set of customers for higher-priced products can easily be defined.
- **Public Relations:** Best for markets that are very large and where the firm has multiple products.

Marketing Budget

Often a firm may be constrained by its marketing budget. Many choices of which method to use are determined by the costs associated with the method. In many cases, a television advertising campaign may not be cost-effective, and creating a YouTube video may be used as a less expensive alternative. It is important to remember that marketers often make decisions based on what they can afford.

Measuring the Success of IMC

In the end, marketers have to be able to know whether their marketing communications are effective. Measuring the effectiveness of IMC can be difficult because it is often hard to know what element of the promotional mix was successful. On average, a consumer must see a marketing message at least three times before making a purchase decision. Because an IMC strategy involves multiple methods, consumers can receive the message from multiple "touchpoints" of the promotional mix. The starting point for measuring success is to set clear marketing objectives for each IMC promotional campaign. For example, Red Bull may estimate the number of social media shares or image submissions for the Red Bull Illume image contest. Different marketing metrics are used for each element of the promotional mix. Let's look at examples of metrics for each.

Measuring Advertising

Companies should measure the effectiveness of advertising before and after a campaign to understand its impact. Marketers typically conduct a *pretest* in which a sample of targeted consumers evaluates awareness of a product before the ad campaign begins. Pretests help set a baseline measure for marketers to evaluate the success or failure of the subsequent campaign. After the campaign, marketers conduct a *posttest*, which involves an evaluation of awareness of the campaign after its completion with the same targeted segment of consumers. For example, Kimberly-Clark Corporation launched a campaign of online videos designed to increase brand awareness of its Cottonelle moist wipe's product. Prior to the campaign, a pretest determined awareness levels of the new product. After the campaign, which resulted in more than 1.5 million people viewing videos about "cleaning logic," and some funny social media posts, a posttest determined that awareness of the product had improved by 6 percent. Pretests and posttests help the firm gauge the ad campaign's success.

Measuring Sales Promotion

The nature of sales promotions makes them easier to measure. Sales promotions are characterized by coupons, rebates, and product samples. Because marketing objectives can be directly tied to a coupon or rebate, metrics that indicate success (or failure) report unit sales as a result of the promotion. Even product trial activities can be measured in regard to sales revenue. A typical setting for product trial is a grocery store. For example, the marketers of Kashi energy bars may sample their products at Publix and Kroger grocery stores. After the free sample promotion is complete, Kashi can compare sales at each store versus a similar day that did not include a sales promotion. Positive increases to sales can then be attributed to the promotion to measure its success.

Measuring Public Relations

Measuring the effectiveness of public relations is often difficult. The nature of public relations activities often makes them the least direct contributor to any single product's success. For example, it may be easy to suggest that sponsorships such as cliff diving, Esports, and motocross will help build awareness for Red Bull. However, because these events are not tied to any specific product or sales promotion, it would be hard to track revenue or sales gains directly. Rather, metrics that measure public relations rely on subjective measure of things such as goodwill, positive corporate image, or brand awareness.

Measuring Personal Selling

Evaluating salespeople, although a common way to measure promotional success, presents challenges for firms. A salesperson with a large territory might sell more than a salesperson in a smaller territory with fewer customers or greater economic challenges. Does that mean that the salesperson who sells more in the large territory is better? Not necessarily. How then can organizations measure the effectiveness of their salespeople? A good example of how companies effectively (and fairly) measure salespeople is Frito-Lay, which employs many salespeople, selling in many different settings. Some sell to big-box stores, whereas others sell to convenience stores and even schools. For Frito-Lay, measuring salespeople based on a variety of sales metrics, or sales outcomes, including customer satisfaction, sales growth, and new product placements, has helped avoid the flaws of simply evaluating total dollars of product sold.

Metrics to evaluate salespeople generally fall into two categories: objective measures and subjective measures. *Objective measures* reflect statistics that can be gathered from the firm's internal data, such as sales revenue, gross profit, and total expenses. *Subjective measures* rely on personal evaluations of the salesperson's performance based on observation by managers and executives. Subjective measures might include job knowledge, sales territory, and customer relationships.

Marketing Analytics: Using Data to Inform IMC Strategies

Using Data to Determine How Senders Encode Messages

Sending a message to consumers is often not as straightforward as one may assume it to be. Marketers must use data to ensure that the messages they send communicate the messages they intend. To ensure that the meaning of the intended message does not get lost, misunderstood, or misinterpreted, researchers test consumers' responses to words, phrases, images, music, celebrities, and so forth on various target markets. This communication "test" allows a firm's marketers to verify whether the components of its message are communicating the message effectively and accurately. Recently, a clothing brand came under fire for having an offensive image in its campaign. It was extremely clear that this image should have never been used in any campaign for any product. Clearly this firm did not test this message with an audience; if it did, the message would have been quickly discarded. This example once again demonstrates how crucial it is to test the elements of your campaign with your audience.

Using Data to Identify Strategies That Break through Noise

Consumers receive more than 3,000 messages every day from multiple senders—whether that sender is a friend, a colleague, or a marketer trying to promote the firm's product or service. Breaking through all of this noise to ensure these messages are received, and received *accurately,* can be a challenge. How do firms ensure that their messages are not simply getting lost among the noise? Marketers collect data from consumers regarding their media habits in the hopes of uncovering innovative ways of delivering messages. In today's interconnected world, consumers have options for consuming media. Consumers can now watch their favorite shows on a variety of platforms—other than traditional network television—anytime they choose in or out of their homes. This has become a challenge for IMC managers who must figure out how to grab the attention of consumers in a variety of environments. For example, streaming platforms now allow consumers to order products directly through the advertisement by linking accounts to Amazon, Google, or Apple. Another example is the use of two-second advertisements that are over even before the viewer has a chance to react to them.

Using Data to Choose a Communication Method

Each communication method has strengths and weaknesses in terms of delivering messages. The type and purpose of the message must be matched with the communication method that can most effectively deliver that message. Research has been conducted that helps marketers understand how consumers process data from print, audio, and video formats. These data allow the IMC manager to understand which media would best match the message with the way consumers process it. Print media are processed more deeply because the audience must actively engage with the content through reading them. Audio material is passive engagement; the audio can be playing in the background while the consumer is focusing on other things. Video is less passive and requires more focus than audio only because of the interplay of images and sounds. For example, a very complex product that requires a lot of explanation is not well suited for a 15-second radio or TV ad, but it could be perfect for a digital media ad where the sender can provide more detail in the form of a how-to video.

Using Data to Evaluate Effectiveness of Promotional Mix Elements

The effectiveness of each promotional mix element is measured differently, but each includes some measure of audience size. Nielsen is the most established of the firms providing data on audience size, broken down by demographic and geographic segment. Nielsen has expanded beyond traditional television and now offers data on streaming TV, as well as on website use and online content viewing. Arbitron (which is also a division of Nielsen) primarily provides data related to radio listenership and publishes ratings for radio programming. Mediamark publishes data related to magazine readership including audience size as well as segmentation information. Of course individual media companies provide data as well; Google, Facebook, Twitter, and so on all publish audience data.

Because of the growth of data aggregation, IMC managers can now examine an individual consumer's media usage across all types of media, and across multiple platforms. For example, if you begin to watch a program on Netflix at home using Amazon Fire and then you resume watching that program on your tablet, only to finish viewing on your phone, the ad bundle that you see can be continuous. The IMC manager can establish a program where you do not see the same ad twice within a program even if you change platforms. Thus ratings for programs do not simply include those who watched the program when it aired originally, but also can include those who accessed the program later through other devices. This makes the audience numbers for programming more accurate, and the IMC manager can have confidence that the number of consumers who saw (or were exposed to) the firm's ad is accurate. Today's IMC managers truly do have to have a deep understanding of data and how to aggregate data in order to plan effective IMC programs.

Responsible Marketing

The Highs and Lows of Customer Reviews

A common marketing tool now for many firms is the customer review. How many of us look at Yelp or Trip Advisor before making a decision on where to travel or what to eat? For many of us, if a product or a service doesn't have at least four stars, we won't even consider it. Similarly, many consumers rely on the testimonial and reviews of their favorite celebrities or influencers and will blindly buy products based on their reviews.

However, in order for a product or service review to be truly ethical, it must represent a real experience of a real person. Many firms these days are offering incentives for customers to fill out reviews or asking them to complete a review on the spot. Even more egregious, some sellers will pay for positive reviews. Or they will pay for negative reviews of their competitors. This happens quite frequently on Amazon as third-party sellers try to move up in the search results.

In 2017, *Vice Magazine* reporter Oobah Butler got a restaurant to the top of the list on Trip Advisor as London's #1 restaurant. However, the restaurant didn't even exist and the entire hoax was predicated on fake reviews.

Reflection Questions

1. What should firms do to try to minimize fake reviews?

2. Should consumers reduce their dependence on reviews?

3. Would you still buy a product from a company if you found out they had falsified their reviews?

Lesson 9-2
Digital Marketing

In this lesson, we explore how social media and their availability almost anywhere in mobile form are changing how firms promote their products and brands to improve sales.

By the end of this lesson you will be able to

- Explain digital marketing and how it creates customer value, customer relationships, and customer experiences.
- Identify the characteristics of digital buyers and sellers that influence digital marketing.
- Explain how marketers influence online purchasing behavior.
- Explain the value of social media marketing.
- Compare and contrast social media and traditional advertising media.
- Define mobile marketing.
- Outline the steps in developing a social media marketing campaign.
- Describe unique ethical and legal issues encountered by marketers in social media marketing.
- Identify the different types of career positions available in social media marketing.
- Explain how data can help inform digital marketing strategies.

Marketing Analytics Implications

- Firms use data to determine how they need to modify their digital marketing campaigns to create value for the consumer.
- Marketers use data to understand consumers' behavior online.
- Whenever a new digital technology emerges, marketing managers must conduct research to learn how these new technological developments can be used to deliver effective marketing messages to the right target audience.
- Data are at the heart of digital marketing campaign planning. Digital platforms produce immediate analytics, therefore making it possible for marketers to make immediate adjustments to their campaign strategies.
- It is important to remember that advertising delivered in digital format is still advertising and is therefore a regulated form of marketing communications. The FTC, FCC, and individual states all play active roles in regulating these messages. Regulation of digital marketing is evolving along with the media. Thus, just like with traditional advertising, data are needed for the regulators.

#ShareaCoke: Social Media and Mobile Marketing at Coca-Cola

Coca-Cola has been on the leading edge of promotional tactics from its beginning in the 1880s. Back then, promotions for Coca-Cola included couponing to promote free samples, which was considered innovative at the time. This was soon followed by newspaper advertising and then branded promotional items featuring the company's logo. Throughout the 20th century, Coca-Cola's brand grew through the use of emerging media such as radio and television.

It's coming...

WITH OVER A THOUSAND NAMES

#ShareaCoke.

The recent "Share a Coke" campaign allows consumers to engage directly with the brand through its social media hashtag: #ShareaCoke.

More recently, the company has leveraged social media and mobile marketing in novel ways that have continued to contribute to the brand's success. In 2011, the "Share a Coke" campaign was first launched in Australia where the logo on Coca-Cola bottles was replaced by "Share a Coke with" followed by a person's name. The focus was to create a more personal relationship with consumers and to inspire shared moments of happiness.

The call to action of the "Share a Coke" campaign encouraged consumers to share their memorable moments with friends and family via social media using the hashtag #ShareaCoke. The catchy and easy to remember "Share a Coke" slogan served as a constant reminder for consumers to not only purchase a Coke, but to also share it online. In the first year alone, consumers shared more than 500,000 photos using the hashtag, shared more than 6 million "virtual" bottles, and gained Coca-Cola over 25 million new followers on Facebook.

Following the initial success, Coca-Cola updated the campaign regularly to keep the interest of consumers. In 2015, it opened an e-commerce shop for consumers to order bottles with their names on them. The company launched the "Share a Coke and a Song" campaign in 2016 in the United States with popular song lyrics printed on the bottles. Over 75 holiday destinations were printed on the labels in the UK in 2017, and shareacoke.com allowed consumers to hear a short song with their name in it. And finally, in 2018, it launched name labels that consumers could remove and place on other items such as notebooks and phones.

Because of its innovative promotions throughout the years, Coca-Cola has grown into the single most valuable nontechnology brand in the world, with an estimated value of $64.4 billion.

Understanding Digital Marketing And Its Value

Digital marketing is online marketing that can deliver content immediately to consumers through digital channels, devices, and platforms. It is an effective method for building or promoting a company's marketing message.

Forms of Digital Marketing

The primary forms of digital marketing include:

- **Inbound marketing:** This form of digital marketing utilizes such tools as blogs, webinars, or follow-up e-mails. The intent is to entice consumers to find a firm's products or services without "forcing" an interaction. Inbound marketing attempts to provide value and information to the customer at every stage of the buying journey.

- **Search engine optimization (SEO):** SEO is the process of driving traffic to a company's website from "free" or "organic" search results using search engines. Try typing in a word or phrase in a search engine and take note of which websites appear first. These companies have utilized SEO to ensure that their sites appear at the top of the list of search results. Thus, SEO is the process of increasing high-quantity and high-quality traffic to a website.

- **E-mail marketing:** This form of digital marketing is a cost-effective method of retaining, nurturing, or attracting a new customer base. The primary advantage to e-mail marketing is that firms know that most consumers have e-mail addresses.

- **Social media marketing:** This is one of the most popular forms of digital marketing and is the focus of this lesson. It encompasses marketing activities that utilize online social networks and applications as a method to communicate mass and personalized messages about brands and products.

Creating Customer Value through Digital Marketing

In previous lessons, you learned that customer value is equal to the benefits the customer receives minus costs to the customer. Digital marketing creates customer value in a number of ways. The North Face, for example, offers free three-day shipping, free returns, and customer service via phone, e-mail, or live online chat.

One way digital marketing creates value is by allowing consumers to search for products from different media, no matter where they happen to be; this, in turn, helps reduce search costs—consumers no longer need to travel from store to store to locate the products they need. Digital marketing also provides consumers with the ability to compare prices across a variety of sellers; this increases the chance of consumers finding the product they desire at a lower cost.

Additionally, digital marketing helps companies build customer relationships through exceptional customer experiences. Such experiences can be achieved through excellent website design—especially one that builds in strong customer support.

Ease in Navigation

Users must be able to find what they are looking for in as few clicks as possible.

Quick Load Time

Users expect a seamless browsing experience, so sites must load images and content quickly.

Mobile Friendly

With more people using their phones than computers now, it is important that sites are designed for mobile users.

Customer Service Tools

Companies can offer superior customer service through customer service tools, such as online chat features and the ability to try clothes on a virtual model.

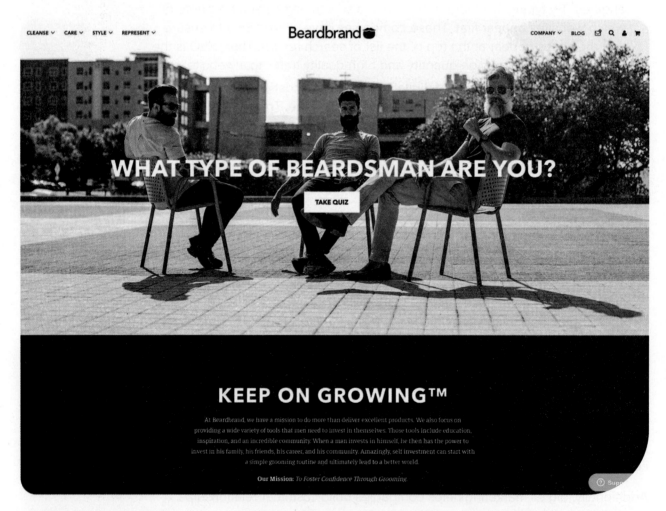

Beardbrand, a company that specializes in beard care and accessories, is an excellent example of exceptional website design built with the consumer in mind. Its drop-down menus, located at the top of the screen, make it easy for consumers to navigate and locate all the information they need—from product lines to information about how the company came to be. It also makes it easy for consumers to contact the company, locate account information, and access their shopping carts.

The Influence of Digital Buyers and Sellers

The digital marketplace is evolving quickly. Over the last 10 years, the dollar amount spent online has increased exponentially. This has only been expedited by the recent pandemic. In 2020, e-commerce increased by 44 percent over the previous year.

Digital Buyers

Just as in any retail environment, different segments of buyers exist in the digital environment. Understanding which buyers tend to turn to digital retail can assist marketers in creating compelling and effective digital marketing campaigns. Online purchasing behavior falls into five distinct types: product-focused shopping, browsing, researching, bargain hunting, and one-time shopping.

- **Product-focused shopping:** Consumers that engage in **product-focused shopping** are either replacing an existing product or purchasing a product that has been chosen for them—a textbook, for example. This type of shopping might include regularly repurchased household staples such as paper towels and shampoo.

- **Browsing:** Consumers that engage in **browsing** behavior are not really looking to make a purchase. Browsers tend to fill shopping carts that they later abandon, although some may reactivate those carts at a later date. They will typically browse their favorite sites just for fun, to kill time, or to look for new ideas.

- **Researching:** Consumers that engage in **researching** behavior are looking to buy a product for the first time. Unlike browsing, which has no expected outcome, researching is more deliberate and will result in a purchase either online or offline.

- **Bargain Hunting:** Consumers that engage in **bargain hunting** behavior usually explore coupon sites such as wish.com or auction sites such as eBay. Bargain hunting is often combined with browsing and may or may not lead to a purchase.

- **One-time shopping:** Consumers that engage in **one-time shopping** behavior may combine product-focused shopping, browsing, researching, and bargain hunting all at the same time. Such consumers are shopping for a gift or using a gift card and will typically not return to the shop once the purchase is made.

Digital Sellers

Just as there are a variety of buyer segments in the digital retail space, there are a variety of digital sellers. The most common are digital malls, digital marketplaces, auction sites, and buying club sites.

- **Digital malls:** A **digital mall** is where a variety of sellers stock their goods. When you click on an item, you can be directed to a number of options for purchase, each with different prices and terms, such as free or expedited delivery. In a digital mall, consumers are often unaware of which firm is actually selling them the good. For example, if you click "purchase" on a website for a small local business, you may be redirected to Amazon.com, which handles the actual sale. Amazon.com (similar to Alibaba.com) is an example of a digital mall.

- **Digital marketplaces:** A **digital marketplace** is made up of small, independent sellers—Etsy.com is a good, global example of this type of online retailer. Consumers search Etsy for a variety of handcrafted or vintage goods sold by a variety of sellers.

- **Auction sites:** An **auction site** such as eBay lists goods from individuals or firms that can be purchased through an auction bidding process or directly through a "purchase now" feature.

- **Buying club sites:** A **buying club site** such as Boxed.com allows consumers to buy in bulk, similar to warehouse clubs such as Sam's Club or BJ's.

It is important for marketers to remember that multiple consumer segments could be visiting the same digital retail site for a number of different reasons. By studying site visitor behavior, marketers can identify which segments are most common at each site. For example, through user tracking, marketers can determine how many times the average consumer visits a site prior to making a purchase. Studying visitor behavior can also tell firms what types of activities these consumers engage in (are they researching or browsing?) as well as which pages they tend to visit. This information helps firms determine an effective social media marketing strategy that assists consumers in either finding the information they need to make a purchase, or in finding the product that they are looking to purchase. Additionally, if research shows that consumers are increasingly purchasing goods from sites that are not U.S.-based, marketers must also consider how to adjust their social media marketing strategies to fit the global marketplace.

Now that we understand how digital buyers and sellers can influence digital marketing, let's look at how marketers can influence online purchasing behavior, which is discussed in the next section.

The Influence of Marketers on Online Purchasing Behavior

Marketers use analytics to understand consumer purchasing behavior. Once marketers have a better understanding of how consumers make their purchasing decisions, they develop their own strategies to further influence those decisions. When it comes to online purchasing, marketers influence consumers' behavior in both visible and invisible ways.

Visible Marketing Influences

Visible marketing influences on online purchasing behavior include paid search, paid stories, paid display advertising, and sponsorship.

Paid search is online advertising in which a company pays to be a sponsored result of a customer's Web search. Companies usually purchase a given search term—"women's watches," for example—and pay each time a customer clicks on their sponsored search ad. Paid search advertising is usually sold using an auction system with the search terms awarded to the highest bidder.

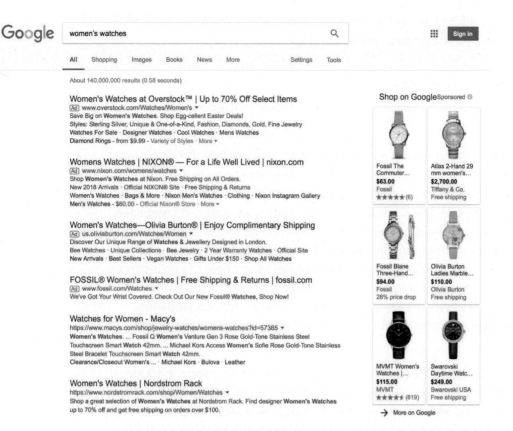

You've no doubt seen paid search ads—the ads listed in this Google Search show the top search results that are often labeled as "sponsored results" or "Ad" to inform consumers that the firm paid to be at the top of the results list.

Paid stories are ads that appear as content designed to look like stories to the viewer. Most advertisers and publishers aspire to deliver paid ads that are cohesive with the page content. For example, on a page that is discussing the top five cruise vacations for Millennials, there may be a "story" about UniWorld, a cruise agency that targets Millennials. In this case, UniWorld would have paid for that story to appear on the page.

Paid display advertising includes everything from banner ads to YouTube video advertising. These ads generate awareness as well as (hopefully) drive traffic to a website. Companies can also purchase advertising on social media sites such as Instagram and Facebook.

Sponsorship firms often sponsor YouTube or Instagram celebrities who in turn endorse the firms' products. These so-called online influencers are often compensated in multiple ways for their endorsements. For example, a celebrity may be paid a fee to discuss the firm's products on his or her YouTube channel and then get paid again when

Felix Kjellberg, aka PewDiePie, started his Swedish gaming/YouTube station in 2010. He now has over 109 million subscribers!

consumers click on the display ad on the YouTube page. Some YouTube celebrities have become millionaires through such advertising fees.

Invisible Marketing Influences

Invisible marketing influences on online purchasing behavior include cookies, geotracking, and bots.

Have you ever searched Google for a product and then noticed that the next time you opened your browser or a social networking site like Facebook the ads that appear on the page were for the product you searched for? How did those ads get there? The answer is with cookies. **Cookies** are small data files stored on websites that can generate a user profile about a consumer; the profile might include browsing history or login information. These cookies are "invisible" to the user. Most of the time, consumers don't know what type of data the cookie is collecting or for what purpose. *Cookie synching* allows cookies to embed in a person's computer and follow that person as he or she moves through various websites. Through cookies, marketers know what consumers are searching for and where they are searching.

Tracking, such as that performed by cookies, can also be used to inform marketers how much consumers are willing to pay for a good based on their previous purchases. This means that as consumers move through websites, the price of the good or service they are searching for may start to be the same regardless of the site they search.

Geotracking is another invisible form of marketing influence on online purchasing behavior. With geotracking, marketers use a consumer's geographic location to determine what goods will come up in a search and at what price points. For example, Staples may offer a 20 percent discount on its products if geotracking shows that competing stores are also in the consumer's local area.

Bots are a third invisible form of marketing influence on online purchasing behavior. A bot is a software application that runs automated tasks over the Internet. On Facebook and other websites, bots look for keywords such as "vacation," "Barbados," and "beach," and then automatically send consumers information related to those keywords. These marketing messages might show up as an ad on Facebook or a sponsored ad on Instagram. Or, consumers might receive an e-mail touting a special deal on a "beach vacation."

Marketers can use your location to offer you special deals and offers.

Through these online tools, marketers are able to influence what consumers see when they search for various products.

In the sections that follow, you will learn specifically about the value of social media and mobile marketing, the differences between social media versus traditional forms of advertising, how marketers use mobile marketing, and how companies rely on customer knowledge and marketing research to build effective social marketing campaigns.

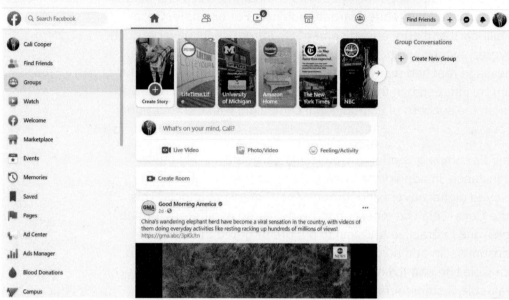

The Value of Social Media Marketing

Marketers have always used media to communicate messages. For example, Coca-Cola has been communicating messages through media since the 1890s, long before Facebook and Snapchat became popular. A new form of media, **social media,** is electronic, two-way communication that allows users to share information, content, ideas, and messages to create a customizable experience.

Social media are described as *customizable* because consumers are able to choose the specific content they wish to interact with or view. For example, when you use Facebook, Instagram, or Pinterest, do you see the exact same content as everyone else using these same apps at the same time? You do not. Instead, the content that shows up in your feed is based on your social media profile, prior usage, friends, likes, and so on. Compare this to the content offered by such broadcast television networks as NBC, CBS, or Fox. Each network offers a traditional program lineup. When you select a traditional television network, you see the same content that everyone else is seeing at that same moment in time.

Social media are a relatively new and still evolving type of media. In the mid-1990s, the Internet was used to communicate commercial messages, leading to e-mail, websites, and blog marketing. Applications more associated with social media did not appear until the mid-2000s (for example, Facebook started in 2004). After rapid growth, social media are just now starting to mature. Current social media platforms, which will keep changing, offer marketers new opportunities for engaging with consumers in meaningful ways. As with other media, marketers have embraced social media as a way to communicate both mass and personalized messages.

Social Media versus Traditional Advertising Media

Traditional mass media such as print and TV ads are static, not easily changeable, and provide one-way communication. In addition, a firm that relies on traditional advertising may receive little direct feedback about its campaigns. Such campaigns may also reach more than just the intended target audience. For example, Coca-Cola's use of 7,000 billboard and print advertisements to introduce its new product, Coca-Cola Life, delivered a static, unchanging message that "Life is a lower calorie cola." The message reached many more people than just calorie-conscious consumers that might be interested in the product. In addition, Coca-Cola received little direct feedback on the success of its traditional campaign, with the exception of sales results.

Conversely, social media campaigns offer customizable, targeted, and two-way communication. For instance, in addition to Coca-Cola Life's traditional media, Coca-Cola also deployed digital advertising through social media channels. This campaign allowed the Coca-Cola Life message to be delivered in different formats (videos, clickable links, and interactive quizzes) to a more targeted audience of health-conscious consumers. In addition, social media users were encouraged to post reviews of Coca-Cola Life and to share their stories with the company and others—providing valuable, instant feedback to, and marketing for, Coca-Cola. These unique qualities of social media compared to that of traditional mass media make social media effective, flexible, and adaptable tools for marketers.

Consumer Interactions

One of the most important differences between traditional and social media is how consumers interact with them. Social media are personal to consumers because every consumer sees different content within their social media feeds. For example, with traditional media, consumers read the same newspapers or watch one of many television channels. Within these media, advertisers place messages that everyone sees at approximately the same time and within the same context (e.g., the same ads, news stories, or programming in proximity to the message). Conversely, for social media, the setting and content around an advertisement is unique for each consumer.

This difference can have many positive and negative implications for advertisers. For the most part, the adaptive and flexible nature of social media makes them an attractive option for marketers that create interesting and creative messages. These messages can be more specifically targeted to specific users and specific content that is complementary to the marketer's product—as consumers view or interact with it in real time. However, some have suggested that social media users are less interested in advertisements because they don't like feeling "forced" to see advertisements for things they didn't select.

An additional difference between traditional and social media is that social media provide more than just network- or newspaper-delivered content. For most traditional media, some form of programming or news is what entices a consumer to tune-in at a specific time or read specific content. Thus, the programming or news establishes an audience that can receive advertising messages. As we will see in this lesson, social media also deliver programming and news, but they do much more.

Interaction with the Company

Social media allow consumers to interact with the company without actually having to visit a store or talk to a company representative. Typical ways to interact with a company include via websites, videos, podcasts, and social media pages. Some interactions include purchasing products, while others include requesting service, asking questions, or simply conducting informational research.

Interaction with Other Consumers

In addition to company interaction, social media also allow consumers to interact with each other. This type of interaction used to occur through personal discussions called "word of mouth," where consumers described to each other both good and bad experiences with a company or product. Today, social media allow this type of interaction to easily occur through blogs, forums, panels, reviews, and comments.

Content Creation

Social media also allow consumers to interact with people indirectly, which can contribute to content creation. Many consumers enjoy voicing their opinion of products, sports teams, companies, and many other things. Typical ways that consumers create content are through blogs, posts, videos, pictures, podcasts, and status updates within social media sites like Instagram, Pinterest, and Facebook.

Consumer Feedback

In addition to the new ways that consumers can interact with social media, social media also allow consumers to easily provide feedback. **Consumer feedback** consists of different ways that customers can report their satisfaction or dissatisfaction with a firm's products. Consumer feedback can take many forms:

- **Consumer reviews:** A common form of consumer feedback is provided through consumer reviews. A **consumer review** is a direct assessment of a product (good, service, or idea) that is expressed through social media for others to see and consider. Choosing what movie to view, where to eat, or what hotel is best is heavily influenced by consumer reviews. For example, **Kayak.com** provides easy-to-see consumer ratings, with a 10 being excellent and a 0 being poor. A hotel's rating is derived by the accumulation of consumer reviews provided by past guests of the hotel that are averaged into an overall rating with comments. Many consumers use these ratings to help in their purchase decision making. Thus, marketers pay close attention to consumer reviews, with an emphasis on responding to negative comments quickly and in a way that can be observed on social media.

- **Chatter (comments, likes, retweets, shares, hashtags, etc.):** Often a marketer's message is further broadcast to other consumers. This practice, called **chatter,** occurs when a consumer shares, forwards, or "retweets" a marketing message. Chatter is another form of consumer feedback. For marketers, the level of chatter represents consumer feedback. The "Share a Coke" campaign, for instance, has consistently been a top trending topic on social media. Consumers across the globe take pictures of their personally labeled Coca-Cola bottles and share them with friends via social media using #ShareaCoke.

- **Influencers:** Consumers who engage with social media at higher levels often become influencers. **Social media influencers** are consumers who have a large following and credibility within a certain market segment. Influencers are often bloggers, celebrities, industry experts, or even people who produce relevant YouTube videos. Often those who have a strong following will be asked to try and report on new products as a high-visibility form of feedback. For example, Manhattan-based beauty start-up Glossier owes its overnight cult status to its ever-growing group of super fans and micro-influencers. Glossier relies on "regular women" to spread brand awareness. A large group of these women closely follow the brand's product announcements, social media posts, and event invitations and forward the brand's messages to their followers on social media and in real life. Building on the support of its influencers, Glossier now has website traffic of more than 1.5 million unique views each month and it has gone from a start-up to a growing beauty brand that often has waiting lists for its products.

Many applications exist that allow companies to monitor all consumer feedback in one consolidated location. Products such as Crimson Hexagon, Facebook Insights, Hootsuite, Klout, and Mention are cloud-based software applications that specialize in bringing together a company's social media feedback in one place.

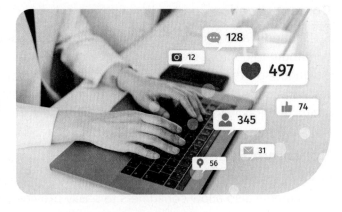

Consider again Coca-Cola: Its marketers (as well as that of most companies) have created a social media command center. Coca-Cola maintains a North American Social Center, which is called the "Hub," that is a real-time newsroom to manage social media marketing for all the Coca-Cola trademarked brands (e.g., Diet Coke, Fanta, and Sprite). The main wall inside the Hub in Atlanta has LCD screens displaying newsfeeds, statistics, and social media chatter. From these up-to-the-minute sources, analysts monitor what people across the globe are saying about one of the world's most well-known brands.

Understanding Mobile Marketing

Mobile marketing is a firm's efforts to communicate with consumers through mobile devices. Specifically, **mobile marketing** is a set of practices that enables organizations to communicate and engage with their audience through any mobile device or network.

Mobile marketing has quickly become one of the most used marketing tools, as more than 77 percent of all Americans own a smartphone. Even more important than ownership, 71 percent of total minutes spent online in the United States are on a mobile device, making mobile marketing a powerful and important marketing tool. Because mobile technology is a delivery tool, mobile marketing works in association with social media but also presents its own unique opportunities and challenges.

The Similarities and Differences between Social Media and Mobile Marketing

Mobile marketing uses many of the same platforms as social media marketing. A **social media platform** is a website-based media channel used to facilitate communication and connection. Social media platforms, such as Facebook, Snapchat, Instagram, and Twitter, are also used in a mobile setting, typically with additional capabilities.

For example, mobile marketing allows for location-based advertising in which mobile users receive notifications and advertising messages based on their physical proximity to a store or an object. Whole Foods has used this location-based mobile marketing with great success. This type of mobile marketing senses when a consumer who has the Whole Foods app is in proximity of a Whole Foods location. Then, a marketing message is sent directly to the consumer through a push notification. Whole Foods even sends customers messages when they are about to enter a competitor's store, hoping to lure the shoppers back to Whole Foods.

Both mobile and social media marketing allow consumers to provide feedback and interact with companies. However, mobile marketing channels allow for more instantaneous feedback compared to other social media marketing channels. For example, diners can review their meal while they are eating it, Uber passengers can rate their ride as they leave the car, and marketing surveys can be pushed to the consumer's phone immediately after a hotel stay.

Mobile technology also makes it easier for customers to make educated last-minute decisions on where to eat. Mobile apps such as Yelp and Google help diners easily see all the options available near them, including reviews, menus, special offers, and prices, all without prior planning. Apps such as **Hookedapp.com** allow college students at many universities to see time- and location-sensitive special offers at nearby restaurants and bars.

As mobile technology matures, marketers are finding new and innovative ways to capitalize on it. For example, marketers can create their own on-the-go apps to add functionality to mobile users' lives while marketing their product. Apps such as BEHR's ColorSmart allow users to upload photos and match colors to BEHR paint colors. Users can take a picture of a room and try out different paint colors on the picture to see what the room could look like. Meanwhile, the use of mobile targeting continues to improve and be used in ways that deliver a message to a consumer at exactly the right time and place where it might be used immediately.

Mobile Applications Used in Social Media

Marketers use different types of mobile applications in their social and mobile marketing campaigns. At the broadest level, traditional social media platforms, such as Facebook and Twitter, reach a broad audience of consumers across a variety of mobile apps and Wi-Fi–based website technologies. Conversely, Snapchat is a mobile-based app popular among younger mobile users. Marketers choose which apps to use and what delivery platforms are most efficient (e.g., mobile, website, etc.) based on which will most appeal to their target audience. For example, marketers create and maintain company mobile sites on social media platforms (Facebook, Snapchat, etc.) and also create their own customized apps.

Company-specific apps can utilize many different types of mobile app functions. Some popular types of apps include geotargeting apps, loyalty program apps, mobile payment apps, scheduling apps, and information apps.

Geotargeting

Many firms incorporate geotargeting within their apps. This technology allows firms to know when customers (who have the related apps) are physically near their stores. Marketers use geo-targeting to push marketing messages to consumers within a store's vicinity in an effort to entice them to it. In terms of services, firms like Uber have used mobile technology's locational ability to revolutionize the taxi industry.

Loyalty Programs

Many companies manage loyalty programs through mobile applications. Starbucks has its own successful app that tracks users' loyalty points and allows users to pay for their morning coffee through the app. In fact, 30 percent of all of Starbucks' sales last year were made through its mobile app.* Starbucks' mobile app also notifies users of deals, promotions, and ways to earn more loyalty.

Mobile Payments

Mobile technology makes it easier for consumers to pay for products. Companies are becoming increasingly willing to accept mobile payments via Apple Pay or Android Pay, which is eliminating the need for consumers to carry cash or credit cards. In addition, many companies offer gift cards that are mobile based, making it easy to track balances and make purchases.

Scheduling

Scheduling apps allow service companies like hair salons, gyms, and restaurants a convenient way for their customers to schedule appointments or manage reoccurring services. For example, at Great Clips hair salon, customers who check-in online are granted priority status over those who do not.

Information

Mobile apps are also quite useful for simple things, such as **information.** Movie theaters, airlines, auto dealers, and almost all businesses are able to provide great deals of information as a consumer desires it, even when they are on the go.

*https://www.pymnts.com/earnings/2018/starbucks-rewards-mobile-app-stocks-loyalty/

Social Media Marketing Campaigns

A **social media marketing campaign** is a coordinated marketing effort to advance marketing goals using one or more social media platforms. Social media marketing campaigns share many similarities with traditional marketing campaigns. Both require setting goals, choosing strategies, and adjusting the plan as events occur. However, in many ways, social media campaigns differ from traditional marketing campaigns. Let's look at the steps involved in developing a social media marketing campaign.

Step 1: Set a Goal.

The first step in a social media marketing campaign (as with any marketing campaign) is setting a goal. For example, the goal may be to increase brand awareness, increase the number of app downloads, or communicate product information. To achieve their goals, marketers develop strategies. To choose the right strategy, marketers start by defining what area of the marketing mix (the four Ps) they will be addressing with their social media and/or mobile campaign.

Social media relate to all four areas of the marketing mix. In different ways, social media can be used as a place (to sell products), product (typically a service), promotion (communication method), or price (communication method).

Place

Social media are integral to the place of the marketing mix by being with consumers wherever they go. In essence, a marketer's "store" can be anywhere. Using social media, consumers can shop, share, and interact with companies from wherever they are. Marketers also capitalize on the strengths of mobile technology and social media to improve traffic at their physical locations. As described earlier for Whole Foods, the ability of marketers to identify a consumers' location in proximity to their store allows marketers to know when customers are nearby and then to push notifications to them.

Product

Social media and mobile technology can be used together as a means to deliver service products. An example is Facebook Marketplace—a convenient way to buy and sell items in a local area. So what is Facebook Marketplace's product? In reality the company's product is a social media app that connects buyers with sellers in a local market. This service, delivered via mobile device and in a social media format, is nothing more than a connector of people. The concept seems simple, but new ways to use social media or mobile technology as a product seem to emerge frequently. Have you used Uber (ride sharing), Airbnb (lodging), DogVacay (pet boarding), or TaskRabbit (work for hire)? These are all mobile applications designed around connecting people who can offer something with people who need the services.

Promotion

Using social media is a powerful way to educate consumers about current promotions, both online and in-store. These promotions are spread by social media users as well as by the company. The previously mentioned "chatter" of social media is one way that a marketer's promotional messages can spread. Chatter often spreads company-distributed materials, known as a campaign "going viral." With a **viral campaign,** promotional messages spread quickly by social media users forwarding them throughout their social networks.

In many cases, consumers don't even realize they are being targeted by social media promotions. For example, a form of promotion called **native advertising** integrates a marketer's advertising message into the content and look of a website or other platform. Native advertising is somewhat controversial because it is paid advertisement for a company, but it can take the look of a blog, social media post, or newsfeed story. Native advertising often matches the editorial content that surrounds it. Many suggest that this type of social media advertising can mislead consumers into thinking that it is an authentic news story or that the social media site endorses a particular product. Think of sponsored ads on Facebook as a form of native advertising. Do you ever have trouble distinguishing between actual user-generated Facebook posts and those paid for by a company? If so, the native advertising was successful.

An example of both native and viral advertising is the way the website BuzzFeed integrates advertisements and products within a quiz. For example, a native advertising "article" (actually a paid advertisement) on BuzzFeed titled "Which Donut Are You?" led readers through a quiz that displayed Dunkin' products while also entertaining readers. This fun "quiz" was actually an advertisement for Dunkin'. Because the quiz was fun and unique, it quickly went viral through blogging, reviews, and social media comments. This native and viral marketing activity expanded the reach of the Dunkin' advertising campaign.

Price

Social media have also changed the way that companies price their products. As mentioned earlier, the concept of showrooming is one way that social media and price interact. Initially, traditional stores were threatened by the ease with which consumers could compare prices. Now, stores are adapting to this increasingly common practice by offering price matching and also keeping a closer watch on how their competitors are pricing similar products.

Consumers have also gained power because of the social nature of social media. Feedback gleaned from social sites can be used by companies to fine-tune the pricing of their products, either by increasing or decreasing prices based on consumer comments related to the value they receive from a product.

Finally, how consumers pay for products is changing based on social and mobile apps that allow for direct payment of products, or the ability to split the cost of a product among many people. Apple's built-in Wallet app allows users to store credit card information on their phone, and more and more retailers are accepting Apple Pay at the register. These easier forms of payment often affect how much consumers are willing to pay for a marketer's product, sometimes increasing the total ticket of grocery store and restaurant purchases.

STEP 2
Select the Correct
Combination of Social
Media Platforms

Step 2: Select the Correct Combination of Social Media Platforms.

The next step is to select the correct combination of social media platforms. To accomplish this, marketers must determine their target audience. Sometimes the most appropriate platform is the company's own website, whereas other times the best platform is an external social media platform that has a larger reach with the target audience. Often a combination of social media and websites is used at the same time. For example, Samsung uses Facebook, Pinterest, and Twitter to promote new Galaxy smartphones. The social media advertising gains the attention of consumers and then directs consumers to **Samsung.com** for complete details.

The choice of which social media platform a marketer chooses for a particular strategy is driven by the goals and target market of the strategic marketing plan. Marketers have a wide variety of choices to implement their plan. For example, they can create their own Facebook pages to generate customer-driven exposure through user likes and shares. Marketers can also use Facebook's paid ad placement services, which allow for more control over who is reached with the ad. Different from just having a Facebook presence, Facebook advertising services target users based on social media content. Parameters selected from past social media use, such as content visited, likes and interests, and social media friends, all can be used as a basis for paid social media selection.

When marketers have a specific target, they select specific social media platforms to more effectively reach their desired audience. For example, Pinterest has become an effective tool for marketers that use engaging photos of their products or services. The magazine *Country Living* is a good example of how Pinterest can be used. *Country Living* generated interest on its Pinterest site when it ran a Dream Bedroom contest. The contest generated user interaction and a larger following of the company's Pinterest presence. The company's marketers were also delighted with the huge increase in the number of times *Country Living*'s products were pinned to users' accounts.

Pinterest is just one example; other social media sites can be used to target specific audiences. Sites such as YouTube, LinkedIn, Snapchat, BuzzFeed, Twitter, and Google+—among many others—all offer different target audience characteristics that marketers can use to achieve different marketing goals.

Where is our target audience spending time on social media?

This is the largest determiner of the social media site choice. Placing marketing messages in the right place for the right audience is a fundamental marketing principle.

What type of content do we want to create within the campaign?

If marketers want to create engaging videos, YouTube may be a good choice, whereas Instagram is well suited to still-image campaigns. If the campaign is designed to share professional articles and news about the company intended for a professional audience, then LinkedIn may be an appropriate channel.

What social media channels are our competitors using?

Often, maintaining a marketing presence in proximity to their competitors is an effective way to highlight product differentiation. Marketers learn a lot about what customers want and how that is changing by observing what their competitors are doing with their social media.

STEP 3

Create Marketing Content

Step 3: Create Marketing Content.

Next, marketers create content for the audience. **Content** is the information, images, videos, and any other delivery method of the marketer's social media message. For example, Coca-Cola's **buy.shareacoke.com** social media website uses images, logos, text, and promotion of nonprofit causes as content to explain the "Share a Coke" campaign. Often marketers take direction of which social media to use based on what their competition is doing. Monitoring the competition's social media can lead a marketer to follow or identify new areas that the competition has missed. Most consumers respond best to original content versus reposted content.

Content as noted earlier, is the information, images, videos, and any other delivery method of the marketer's social media message. Social media provide a unique way of marketing because the content created generates conversation among consumers. The conversation around the product or campaign is almost more important than the message the company is sending. Social media sites are also some of the best places for companies to hear from and engage with their customers. Because of this, a common goal of social media marketing is promoting products through creating customer conversations about a particular brand or

product. These conversations are a form of engagement where a consumer thinks and communicates about a brand.

Additional social media marketing goals revolve around determining the needs of customers and increasing customer service. The outcomes of these goals help create the brand's reputation as consumer needs and service expectations are met and exceeded. However, goals describe what companies are trying to achieve, not the actual steps that need to be taken to achieve them. The process of achieving goals is built on *objectives,* which are the activities that need to be completed in an effort to achieve a goal.

This lesson is a good example of the difference between goals and objectives. The goal of this lesson is to foster student learning about social media. To achieve this goal, several "learning objectives" have been developed to provide easy-to-follow steps that help you as you progress toward the overall learning goal.

The objectives for social media campaigns operate in much the same way—as a roadmap toward achieving goals. In other words, each goal comes with its own set of objectives that can be measured to make sure that consistent progress is being made toward achieving the goal. For example, If the goal of a social media campaign is to increase customer service, an objective could be that all customer comments posted on the company's Facebook page (both positive and negative) will be responded to within four to six hours. If the social media goal is to create conversations about a product, an objective could be set to generate a certain number of shares, likes, or comments within a specific period of time on a social media platform, such as Facebook. As outlined in the strategic planning lesson, the most important thing to consider when setting social media objectives is to make sure the objectives are in line with the overall goal and are specific, measurable, and realistic.

One difficulty with social media marketing campaigns has been that it is sometimes difficult to measure the value of shares and likes in actual dollars for the company. As we will discuss in the next section, as social media matures, the ability to monitor and measure social media success is also improving. Thus, companies are becoming better attuned with setting objectives that can be measured with data that represent the financial returns of social media efforts to firms.

STEP 4
Monitor the Campaign

Step 4: Monitor the Campaign.

Finally, the results of the campaign are monitored. In Step 1, goals were set for the campaign; in this step they are used to monitor how successful (or unsuccessful) a social media campaign is. If the plan isn't working, changes will need to be made. Tourism officials in Egypt, for instance, created a social media marketing campaign with the hashtag #thisisegypt. However, the campaign backfired when people used the hashtag to show the negative realities of living in Egypt. Because this social media campaign delivered results counter to what was intended, it clearly needed adjustment. We'll explore the step of monitoring social media campaigns in more detail later in this lesson.

The goals and objectives of social media marketing campaigns can be measured using many tools that are provided by the social media sites themselves. In addition, other applications have been developed for this purpose, such as Google Analytics or Viralheat, which allow marketers to manage and monitor multiple social media accounts in one central location through the use of a **dashboard.** The dashboard aggregates (counts up) the number of likes, shares, pins, clicks, views, comments, and other variables in one spot, so social media marketers can quickly assess how well they are implementing their objectives and what progress is being made toward their goals. Dashboards can also provide other important information, such as the peak times when people are conversing about their brand. Companies like Tableau provide solutions for businesses to visualize data in real time.

To measure the success of a campaign, companies can determine *bounce rates, click paths,* and *conversion rates* through various online sites:

- A **bounce rate** is the percentage of visitors who enter a website and then quickly depart, or bounce, rather than continuing to view other pages within the same site. In this way, bounce rates measure how many people leave a page immediately after clicking on an ad.

- A **click path** is a sequence of hyperlink clicks that a website visitor follows on a given site, recorded and reviewed in the order the consumer viewed each page after clicking on the hyperlink. Click paths show marketers how customers interact with a page once they click on the ad and can help the company streamline the process from first click to purchase.

- A **conversion rate** is the percentage of users who take a desired action, such as making a purchase. Conversion rates tell marketers how many people who click on an ad actually complete a purchase or subscription or follow the company.

Many companies also try to quantify their likes and shares based on the profit they generate. To make this calculation, companies determine how many consumers they are connecting with (reach) and how often they connect with these consumers (frequency). The reach and frequency statistics are then compared to the expense of all social media activities taken to achieve these results. However, firms cannot perfectly match reach and frequency with dollar amounts because determining reach, frequency, and exact expenditures to produce them is often difficult. Thus, more effective measures of financial return can be gleaned from examining where customers come from. To determine where customers come from, firms can track what social media message immediately preceded a website visit that resulted in a purchase. For example, if you receive a Twitter message that a popular band will be coming to your area and then click a link to purchase a ticket, it is clear that you "came" from a Twitter message. Because most firms use multiple social media platforms to communicate, determining which one drives the most visits and purchases is useful for planning social media campaigns.

Marketers also measure how consumers are responding to their content by using sentiment analysis. **Sentiment analysis** looks at whether people are reacting favorably or unfavorably to products or marketing efforts. Sentiment analysis examines all posts and comments on all streams to come up with sentiment. Different software systems are capable of examining sentiment, with most categorizing sentiments as either positive, negative, or neutral. The advantage of sentiment analysis is that it can be run continually throughout the campaign, allowing a company to make adjustments to its strategic plan as needed. For example, through sentiment analysis, Expedia Canada learned that many people were having a negative reaction to the music in a new video and were able to respond to the negative feedback in a subsequent video, which addressed Expedia's poor music choice in a humorous way.

As mentioned earlier, many companies use technologies like Hootsuite to create social media "war rooms" or hubs where a dedicated social media team can monitor all social media activity in real time. Real-time monitoring is important because things change quickly in social media. For example, Coca-Cola is mentioned 33 times per minute in English alone around the world on social media. Doing the math, every hour almost 2,000 consumers mention Coca-Cola, and every day that number escalates to nearly 48,000 Coca-Cola references. By monitoring this high volume of feedback, Coca-Cola can quickly assess its social media efforts and adjust them as needed.

Social Media Marketing and Ethical Issues

One major concern with social media marketing is customer privacy. Companies have access to private information about many individuals and must protect this information. Violating a customer's privacy by sharing or accidentally leaking customer information can lead to diminished brand reputation and loss of consumer trust. This is both a legal and ethical concern for companies. Various federal regulations around Internet privacy have been proposed, and more are expected in the future. Currently, many in the industry are trying to avoid federal regulations by self-regulating through the Digital Advertising Alliance (DAA), which has set up guidelines for member companies regarding consumer privacy. However, no companies are required to be part of DAA and follow its guidelines.

The honesty of company social media efforts is also important. If companies attempt to deceive customers, there is a high chance that people will find out and the company will be affected. Messages conveyed on social media should be fair and honest and should not be able to be misinterpreted. According to the Federal Trade Commission, if a company is sponsoring content created specifically to endorse its brand or product, it must be disclosed. For example, Walmart paid two bloggers to travel across the United States to write positive stories about Walmart. Walmart did not disclose that the company was supporting the bloggers until an outside party disclosed the information. In the end, the campaign brought a large amount of negative publicity on Walmart.

Employees who work with social media also need to be trained in how to respond to social media posts. Although it is a good practice to respond to negative comments from customers online, employees who write the responses need to make sure they are acting in line with company policies. Most companies try to avoid ethical social media problems by providing clear guidelines about what employees can and cannot post to social media on behalf of the company.

Careers in Social Media Marketing

Social media marketing is an evolving and changing field, and most large companies have marketing positions relating specifically to social media management and strategy. These jobs include management positions, content creators, and analysts:

- A *social media strategist, social media marketing manager,* or *social media specialist* is typically responsible for the overall success of the company's social media efforts. These employees develop social media marketing plans, manage the social media staff, and represent the department's social media efforts to the rest of the company.

- *Online community managers* deal directly with online communities and customers and represent the company to the online world.

- *Influencer relations employees* work with influencers to build and maintain relationships with content creators, YouTubers, bloggers, and other people who are influential to consumers.

- *Social media analysts* and *social media designers* apply their backgrounds in analytics and graphic design, respectively, to the company's social media efforts.

- *Writers* are often hired to create content for the company.

- In addition, if the company maintains its own website there will be a *social media developer* or someone with a background in programming to build and maintain the company's website.

When looking for jobs in social media marketing—or anywhere else—remember the power that social media can provide you to market yourself on sites like LinkedIn. Having an active *and professional* social media presence could help you land a job in social media marketing.

Marketing Analytics: Using Data to Inform Digital Marketing Strategies

Creating Value for Consumers Online

Consumers have very short attention spans, and even shorter while browsing sites online. Therefore, marketers rely on data to inform how they can build attractive and relevant websites to encourage frequent visits and action. It is important that firms study how consumers are using technology and various online platforms in order to determine what content should be presented on that platform. By using data from websites and analytics from online providers such as Google, Facebook, or Twitter, firms can optimize their digital strategies. For example, Taco Bell understands that consumers' reasons for visiting its website change throughout the day; therefore, it changes its content to match what consumers need and want. If you visit its website (**www.tacobell.com**) in the morning, you will see images of breakfast items; around 11 a.m., those images get replaced with lunch items. In the afternoons, the images switch again, this time to snack items, and at dinner time and during late-night hours, the images switch over to more dinner-appropriate items.

Understanding Consumers' Behavior Online

Data analysis allows marketers to determine how to move consumers through the buying cycle. Through the study of consumer behavior, marketers learn a great deal about how and when consumers make market-based decisions. This study remains relevant in the digital environment—the key difference between how consumers behave in a brick-and-mortar store versus how they behave at an online retailer lies in the promptness with which they make a purchase decision. They tend to move more quickly online. Website data and analytics gathered from sites such as Google, Instagram, and LinkedIn not only reveal *how* consumers may be using a firm's website, but also *where they are* in the purchase decision-making process. For example, if you are browsing on a site like Zulily.com, a message may appear notifying you that only two items left of a product you just viewed remain. Or, perhaps you've left some items in your shopping cart. You may receive an e-mail suggesting you "forgot" to make your purchase. These digital marketing strategies are designed to motivate you to make your purchase decision quickly or risk losing out on that product. Using consumers' purchase histories can optimize what items are shown to consumers and remind them of what they purchased last time. For instance, Starbucks sends product recommendations tailored specifically to its customers through its mobile app; these recommendations are based on customers' previous purchase histories and are intended to make it easier for them to order and discover other food or beverage items they might enjoy.

Identifying New Digital Opportunities through Ongoing Research

It seems like every month there is a new platform promising to be the next big social media platform. Marketers must pay attention to these new offerings and determine which, if any, are going to be relevant to their consumers and therefore part of their marketing strategy. Research and data analysis shows that Snapchat and Instagram are popular social media apps among American teenagers. If the firm's target market spends a lot of time on Snapchat, then the firm's marketers should look at Snapchat as a marketing outlet for its products and services. Sometimes an existing social media platform may decide to offer a new feature as a way to attract a new market segment. Instagram launched Instagram Stories, for example, to compete directly against Snapchat. Since its launch, this feature "has emerged as a clear favorite for marketers. . . . Instagram's pure scale, evident in its reach, targeting and retargeting capabilities, has made it an attractive bet for brands." Identifying these new outlets or how to improve the use of existing ones is an ongoing research function of the firm.

Keeping Real-Time Data at the Heart of Digital Marketing

One of the most appealing features of digital marketing is the immediacy of the feedback. When a campaign is launched it can be tracked in real time across all social media platforms. For example, customers can post their queries, reviews, or general comments on social media, which marketers can then immediately gather for their firms to be immediately addressed. Companies who share their promotions in the digital marketplace can get instant responses from customers no matter where they are in the world because they may already be linked to the company through their digital or mobile devices. If a company has a Facebook page set up, it can review real-time data about who is liking their page, who is responding to ads, and how they are responding to them. All of these data allow marketers to quickly adjust their campaigns and fix any underperforming elements. Of course, this wealth of data constantly coming in requires analysis and interpretation, so firms must possess the expertise to generate relevant reports and analyze them for strategy makers. In fact, as digital marketing has become one of the fastest-growing fields in marketing, those seeking careers in digital marketing must possess quantitative analytical skills, content creation proficiency, and written communication skills.

Although social media networking has become one of the swiftest methods for marketers to connect with their customers, call tracking can also provide real-time data that can then be immediately used to modify a digital campaign. Call tracking allows firms to identify which pages are generating the calls to a business, which keywords are inspiring greater click-through in pay-per-click campaigns, and which digital advertising campaign is attracting the most customers (the key here is to use a different phone number for each ad for easier tracking). LA Furniture Store uses call tracking as the deciding factor for evaluating the effectiveness of a marketing channel and attributes call tracking to its improved sales. Call tracking data revealed that 30 percent of its sales were processed over the phone; they also helped the company improve sales even further by identifying the most attractive keywords for its Google and Bing Search Ads.

Regulating Digital Marketing

Digital marketing activities are still classified as advertising and thus are regulated by the FTC, FCC, and individual states. As firms are found to have misused their customers private data, or to have not protected them from misuse, demands for increased regulation in the United States has risen. Users of these sites tend to be unaware of how much data are actually being gathered *from* and *about* them; thus, they are often surprised when data breaches occur or when firms accused of misusing private data are caught. Digital marketers must keep the privacy of their users at the forefront of their data protection efforts and marketing campaigns. They must also ensure that systems are in place to prevent misuse of these data by parties both inside and outside the firm.

Responsible Marketing

UNIQLO: Leveraging Digital Marketing to Its Fullest?

Have you ever heard of UNIQLO? Chances are you have. With over 2,200 stores and a market value of $105 billion, the Japanese casual wear company has quickly surpassed other large fashion retailers such as Inditex (Zara) and H&M.

If you have heard of UNIQLO, you most likely did so via social media. The company's rapid growth can be attributed in part to its digital marketing strategies. Its HEATTECH campaign, for example, received 4 million views, which resulted in 35,000 new customers. This campaign used fast-moving digital technology to encourage customers to take a photo, receive a unique code, learn about the product, and share it on social media.

In addition to its unique marketing campaigns, UNIQLO is using digital platforms to educate customers about its sustainability initiatives. These include reducing its carbon footprint, using less single-use plastics, engaging in responsible manufacturing, and fundraising to protect coastal regions. The company also posts tutorials on its Instagram demonstrating how to repair and re-wear its clothing. UNIQLO seems to be doing its best to leverage digital marketing to its fullest.

Reflection Questions

1. Look into UNIQLO's social media. How is the company using its digital platforms to educate consumers on its commitment to sustainability within the retail industry?

2. Select a brand you are familiar with. After reviewing its social media, discuss how the brand has used digital marketing to create value and build customer relationships.

3. Using that same brand, discuss how can the brand can better leverage digital marketing to improve its connection to its customers.

Lesson 9-3
Advertising and Sales Promotion

This lesson explores how marketers use advertising to accomplish marketing goals, like those of the Old Spice campaign, across various advertising media. In many cases, sales promotion is used to enhance advertising efforts. The concept of sales promotion, and its importance to marketers, is also addressed in this lesson.

By the end of this lesson you will be able to

- Describe the role of advertising in the promotional mix.
- Identify the three objectives of advertising.
- Describe the steps in designing and executing an advertising campaign.
- Describe the major types of advertising appeals.
- Explain ethical challenges related to advertising and ways they are controlled.
- Explain the importance of sales promotion to a firm's success.
- Explain how marketers use data to plan advertising and sales promotion strategies.

Marketing Analytics Implications

- Marketers use data to determine messaging strategy. Messaging strategies for new brands tend to focus on *informing;* for existing brands, *persuading;* and for infrequently purchased or seasonal brands, *reminding.*
- Data are at the heart of advertising campaign planning. The advertising plan is often a subplan of the IMC plan because advertising must work with the other IMC elements. Additionally, data are used to determine which appeal will most effectively deliver a brand's message to its intended audience.
- The FTC requires that firms provide *substantiation* of advertising claims. This means that firms *must* run independent studies, conducted by experts, to validate all claims made. Puffery, for instance, is common in advertising. It is used to exaggerate a product's quality or effectiveness, but consumers will tend not to believe these exaggerations until they see evidence of their truth.
- Data are used to determine an advertising campaign's sales promotion strategy. Multiple promotions are used within a single campaign but are targeted at different places in the channel, such as employees, salespeople, retailers, and consumers.

Old Spice: Using Advertising to Give New Life to an Aging Brand

Old Spice was launched in 1938 as a nautical-themed line of shaving soaps and aftershave lotions for men, and held true to this brand identity for many decades. But by the 2000s, the brand was exhibiting signs of aging ... along with its loyal customers. Old Spice had lost its appeal among younger men.

In the 2000s, the male grooming market began to change as young men aged 18 to 35 began paying greater attention to personal grooming. Products like men's body washes and body sprays became the fastest-growing segment in the personal grooming products market. Procter & Gamble (P&G) recognized the difficulty of repositioning the brand image of Old Spice in the minds of men aged 18 to 35, who associated the brand with their fathers. So instead of focusing exclusively on young men, Chief Marketing Officer Marc Pritchard decided to expand the target audience beyond existing category purchasers to include women, who had significant influence over men's purchases of grooming products, as well as first-time product category buyers.

To start the brand repositioning, P&G launched "The Man Your Man Could Smell Like" campaign, featuring actor and former NFL player Isaiah Mustafa as the new face of Old Spice. The creative team wanted his iconic character to be loved equally by women and men—a ladies' man who was OK for men to love as well. They chose Super Bowl XLIV as their platform to launch the campaign.

This first Old Spice ad was an instant hit, going viral and gaining more traction than any other Super Bowl commercial that year. Keeping trademark touches like a nautical theme and closing whistle, the ad brought fresh messaging, timely humor, and swagger to the brand. In the five months following the first airing, the commercial attracted more than 16 million views on YouTube, and sales of Old Spice body wash more than doubled. By 2019, the commercial had more than 56 million views on YouTube.

The ad was a massive success in terms of viral video, but the creative team was not satisfied with resting on their laurels. They decided to take the Old Spice marketing campaign to real-time Internet. Gathering a team of writers, creative directors, digital agents, and actor Isaiah Mustafa on a film set staged as a bathroom, they then kicked off a "watershed moment for marketing"—the Old Spice Responses campaign.

The group seeded various social networks such as YouTube, Facebook, and Twitter with an invitation to ask questions of Mustafa's character, their dashing, shirtless spokesman with over-the-top humor. All responses were tracked and selected users received direct and personalized responses from Mustafa in the form of short, funny YouTube videos. Embracing the conversational and interactive nature of social media, replying through personalized video was far more phenomenal than merely text responses, and created more buzz than the original television ads.

The Old Spice Responses campaign appealed to people's egos, felt personalized and intimate, and created a significant amount of buzz around the brand in popular culture. It combined unique advertising, creative writing, and the proficient use of social media as a two-way communication channel to activate mainstream brand coverage. Old Spice Responses topped 6 million total views in the first 24 hours on YouTube, receiving more viewer traffic than President Obama's victory speech got in the same timeframe.

Defining Advertising

Advertising is one of the most noticeable activities of marketers because it is everywhere we look, such as during NFL games, before YouTube videos, in your Gmail, or even wrapped on buses. Advertising is a pervasive, recognizable, and important aspect of marketing promotion. When you think of excellence in advertising, companies such as Nike, Capital One, Honda, Subway, or Verizon likely come to mind. This makes sense because these companies rank among the highest on advertising expenditures.

Advertising is important because it helps to keep marketing and brand messages in the minds of consumers. Specifically, marketers use advertising to inform, persuade, and remind consumers about unique features and benefits of their products.

Advertising is nonpersonal promotional communication about goods, services, or ideas that is paid for by the firm identified in the communication.

Two words in this definition—*paid* and *nonpersonal*—are key to understanding how advertising fits into the promotional mix. **Paid advertising** requires a purchased time or space for communicating a message. Because it is paid, advertising has the advantage of control; the purchaser decides how to present the message to the public. Advertising helps marketers control how their message is encoded. **Nonpersonal advertising** uses media to transmit a message to large numbers of individuals rather than marketing to consumers face-to-face.

Firms spend hundreds of billions of dollars on **advertising campaigns** each year in an effort to appeal to large numbers of individuals. An advertising campaign is a collection of coordinated advertisements that share a single theme. For example, the milk industry spends in excess of $50 million annually on its advertising campaigns—the previous "Got Milk?" campaign and the more recent "Milk Life" campaign. Large consumer product companies spend much more. Procter & Gamble (P&G), manufacturer of brands such as Old Spice, Gillette razors, Febreze air freshener, and many others,

Good Health

{starts on the inside!}

Cheerios

Love.

spends in excess of $7 billion annually to advertise its products. P&G uses a variety of advertisement types, including worldwide TV, print, radio, Internet, and in-store signage within the advertising campaigns for each of its product brands.

Advertising Objectives

Marketers use advertising campaigns to achieve three unique objectives: to *inform,* to *persuade,* and to *remind.*

You may not realize it, but you encounter each type of advertising campaign on a regular basis. Marketers choose one type over the other based on the marketing goals for the product they are promoting. So, in any given set of commercial advertisements, you are likely to see one, two, or all three varieties. The question, then, is how do marketers decide which type of advertising to use? One way is to consider where the product falls within the product life cycle (PLC). As a product moves through the PLC, the marketing goals that are set for the product change.

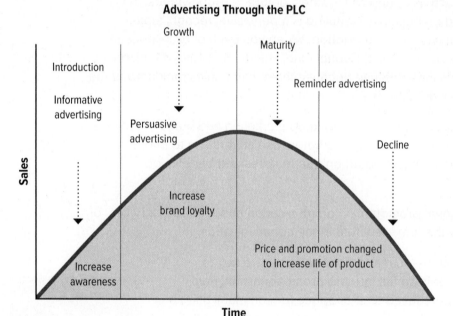

Advertising Through the PLC

To Inform

One objective of advertising is to inform. **Informative advertising** attempts to develop initial demand for a product. This kind of advertising is especially important during the introduction stage of the PLC. The introduction stage is characterized by few or no competitors. In this state, the product is new to the market, and sales are typically slow because customers are not yet accustomed to the product.

The introduction stage is also characterized by high levels of advertising spending. To increase consumer awareness and comfort with a new product, marketers use informative advertising. For example, informative advertising may be used to promote an upcoming meeting of a new club at your school. Things such as the date, time, room number, and a description of the guest speaker help college students (consumers) identify whether the club relates to their interests.

Similarly, marketers of products use the same approach. When the "Got Milk?" campaign started, it wasn't because milk was new. Rather, the campaign was initiated because milk consumption was declining. Initial "Got Milk?" advertisements informed consumers of the health benefits associated with drinking milk. In this case, the "Got Milk?" campaign reintroduced a product that many had forgotten by informing them of existing and newly discovered features and benefits of milk.

These examples describe how marketers use informative advertising to position their product in the consumer's mind, rather than to persuade them to buy the product. In other words, informative advertising is effective at "setting the stage" for a new product, event, or brand.

To Persuade

Another objective of advertising is to persuade. **Persuasive advertising** attempts to increase demand for an existing product. It is common during the growth stage of the PLC as firms attempt to take away market share from competitors. To gain market share from competitors, a company focuses its marketing efforts on showcasing the competitive advantage of its product over others to persuade consumers to try the new product or switch from other products (now that they know about the new product).

Marketers invest heavily in advertising during this stage, but generally spending drops from the high levels of the introduction stage. Consider when a store like Victoria's Secret has its semiannual sale. Victoria's Secret uses persuasive advertising to encourage people to attend the sale while supplies and discounts last!

The marketers of milk transitioned from informative advertising to persuasive advertising as their campaign advanced. Messages turned from informative facts to persuasive calls from celebrity endorsers that were intended to persuade consumers to drink more milk. Persuasive advertising can be used for a long period of time as a product grows from a new product into a mature product.

To Remind

A third objective of advertising is to remind. **Reminder advertising** seeks to keep the product before the public in an effort to reinforce previous promotional activity. This kind of advertising is common in the maturity and decline stages of the PLC. In the maturity stage, the number of customers that can be persuaded to try the product for the first time declines, as most that have an interest have already tried it.

The main objectives of the maturity stage are profitability and maintaining the firm's market share for as long as possible. To accomplish this, firms can follow a couple of different paths. One path attempts to restart the PLC through product modifications and enhancements. The other follows a strategy of maintaining an accepted product (rather than altering it) through reminder advertising. An example of this is an e-mail prompting you to buy season tickets to support one of your school's sports teams. You are most likely already aware of the

team and you may have already attended games, so the e-mail is designed to remind you that the season is approaching and that action is required. Reminder advertising also occurs every year when Anheuser-Busch shows ads with its Clydesdale horses. This is not to inform or persuade the audience, just to make positive brand associations and remind customers of the values of the Anheuser-Busch brand.

The "Got Milk?" campaign utilized a reminder approach for many years before it ended. The slogan "Got Milk?" is largely a reminder. The final implementation of the campaign played on the slogan and reminded consumers to replenish their milk supply. Eventually, the decline stage of the campaign (not milk itself) ended and a new campaign ("Milk Life") was started in an attempt to "re-introduce" milk (again). The "Got Milk?" campaign ran for more than 20 years (from 1993 to 2014) before it was retired for the new campaign.

IF YOUR COFFEE ISN'T PERFECT, WE'LL MAKE IT OVER. IF IT'S STILL NOT PERFECT MAKE SURE YOU'RE IN A STARBUCKS.

IT'S NOT JUST COFFEE. IT'S **STARBUCKS**

To Inform

This ad is for a new product, so the goal is to **inform** the target market in order to make them aware of it and that it's "coming soon."

To Persuade

Almost all advertising has the general objective to **persuade**, but some advertising is *specifically* designed to persuade. In this ad, Starbucks is persuading its target market it makes the perfect cup of coffee.

To Remind

Reminder advertising is typically used for well-established products that have high consumer awareness. Here FTD is reminding people to send flowers for Mother's Day using their established and reliable brand.

It's time to forgive her.

Mother's Day 14ᵗʰ May.

Fleurop. The power of flowers.

Steps in Building an Advertising Campaign

To build an advertising campaign, marketers follow some well-established steps:

- **Identify the target audience.** Marketers conduct research to determine which customer segments will comprise their target markets. Then, they position products to best serve the needs of each target market. These same target markets also define the target audience of advertising campaigns. For marketers to be successful, they must know who will buy their products so they know whom to direct their advertising to. For example, when a university decides to advertise, it starts by determining the intended audience of the campaign. Athletes, parents, high school seniors, or adult students could all be in a university's target audience. Determining which one is the target helps determine what occurs in the following steps.

- **Define objectives.** The next step in creating an advertising campaign is to determine what the campaign needs to accomplish. Recall that advertising can accomplish three different objectives: to inform, persuade, or remind. Deciding which objective to follow is based on the product and needs of the target audience. For example, a university advertising campaign intended to convey application deadlines and course offerings would have the objective of informing consumers.

- **Create advertisements.** Consumers see and hear ads in various formats (visual, print, audio) across many methods (television, online, print, etc.). These advertisements are an outcome of the production and creative services that some marketing specialists perform. Creating advertising messages involves more than just designing ads using technology. Marketers must also create the content of the messages. Most universities have a creative services department that develops marketing materials. Getting the message correct is an important part of advertising campaign success (or failure).

- **Choose advertising tools.** Marketers can choose to deliver their messages using many communication methods. Marketers focus their efforts on choosing the correct method to match the target audience of their advertising campaign. For example, a university campaign designed to reach grandparents would likely utilize a print and television campaign. Adults 65 and older are more likely to read newspapers and watch broadcast television, so these methods would be best suited to reach this audience.

- **Measure the effectiveness of the advertising.** In addition to gauging awareness through pretests (awareness before a campaign) and posttests (awareness after a campaign), marketers also use more advanced metrics to measure advertising effectiveness. Companies often measure print and digital advertising based on whether consumers see the ads and how well they recognize and remember them. A *recognition test* involves showing consumers an advertisement and asking if they recognize the ad. Although recognition tests can help the firm determine whether its ad has caught the attention of the target market, recognition alone may not prompt consumers to buy a product. Marketing professionals often assess what consumers recall about an advertisement using either unaided or aided recall tests. *Unaided recall tests* require consumers to recall ads from memory, without any clues. In *aided recall tests,* respondents receive clues to help stimulate their memory. If consumers don't recognize or recall the firm's print advertisements, marketers may change the size or message of the ad or deliver it through an alternative medium.

Step 1: Identify the target audience.

Snickers has several target markets, including children, teens, and adults. To create an effective advertising campaign, Snickers first needs to identity for which of these target markets the campaign is intended.

Step 2: Define objectives.

Snickers decides to focus on its adult target market. It's next step is to define its objectives. Advertising can perform one of three objectives: to inform, remind, or persuade. For its campaign, Snickers wants to remind adults that Snickers is a good "pick me up" for when they start to get hungry and lose focus at work.

Step 3: Create advertisements.

Before Snickers creates its ad, it will need to decide on an approach. It can use either a rational or an emotional approach, depending on the target market and the objectives of its advertising campaign. In this case, Snickers uses humor, which is an emotional approach, to connect with adults who know that when they get hungry at work, they lose focus.

Step 4: Choose advertising tools.

How will Snickers reach its target audience with its message? It will need to choose a method appropriate for its audience. These tools or methods include online, television, radio, print, outdoor, and other nontraditional channels. To reach working adults ages 25 to 54, Snickers decides to use a mix of outdoor and social media advertising.

Step 5: Measure the effectiveness of the advertising.

Advertising can be fun to design and implement, but it also needs to be effective. To measure advertising effectiveness, companies can use a variety of metrics. Snickers relies on such metrics as sales revenue, click-through rates, and video/page views.

To gauge how often and how many consumers see an advertisement, marketers rely on two types of metrics: frequency and reach. **Frequency** is a count of how often a consumer is exposed to a promotional message (television advertisement, online advertisement, billboard, etc.). **Reach** is the percentage of the target market that has been exposed to the promotional message at least once during a specific time period. These metrics are related but serve different functions. In most cases, advertisers desire to *reach* as many consumers as they can, as *frequently* as they can. However, the constraints of advertising costs and marketing budgets make this balance a difficult one. For example, the Super Bowl is a powerful tool for increasing the reach of a message (that is, the number of people who see it). The game typically garners the highest ratings of any television program of the year and is one of the few televised events during which viewers actively watch the commercials. Marketers, in an attempt to reach as many people as possible, pay more than $5 million per 30-second commercial annually to advertise during the Super Bowl. Unfortunately, no matter how effective a single advertisement might be, it generally takes multiple exposures to move a consumer to make a purchase or to change his or her buying habits.

Although Super Bowl advertising is a great way to reach a lot of consumers, its cost makes it difficult to build much frequency. Most marketing budgets do not allow for multiple $3 million, 30-second advertisements. Today, the balance of reach and frequency is aided by YouTube or social media sharing, on which ads now appear. These online platforms extend the life of the ads, helping to build frequency through repeated viewings. The battle of reach and frequency explains why advertisers strive to be creative and innovative in their Super Bowl advertising—they are gambling that the initial reach will lead to future frequency of viewing of the advertisement. If a Super Bowl ad falls short and is not found to be interesting, the initial reach will not drive future frequency and the large investment will not pay off as expected.

The Major Types of Advertising Appeals

Advertising appeals can be broadly grouped as having either a rational appeal or an emotional appeal.

A **rational appeal** uses logical arguments to attempt to make the viewer *think* about the product and its benefits to better understand why purchasing or using the product is a good decision. An advantage of rational advertising is that it clearly communicates a product's features. However, because rational advertising focuses on product features, it is usually a less entertaining form of advertising than emotional advertising. When marketers decide to use a rational appeal to deliver their message, they typically focus on topics such as savings, health, convenience, or environmental impact. For example, banks and financial services that encourage people to think about their retirement planning often use rational appeals to reach their target audience.

An **emotional appeal** aims for an emotional response from the viewer. The advertiser might use humor, fear, concern, or love to help the viewer feel positive toward the product. An advantage of emotional advertising is that when people feel good after viewing an ad for a product, they become more connected to the product and brand. An example of an emotional appeal would be the Procter & Gamble, "Thank you Mom" campaign during the 2012 and 2016 Olympics.

However, emotional advertising can be risky because sometimes consumers are not able to properly decode the message that the marketer is trying to send. In these cases, advertising that isn't understood by the target audience is ineffective and does not deliver the expected results. If marketers decide to use an emotional appeal in their advertising, they are likely to draw on emotions such as fear, admiration, pleasure, or envy.

RATIONAL GROUP

Savings

With rebates featured prominently, this advertisement for G3 Boats appeals to the consumer's desire to save money.

Health Convenience

This advertisement for an Organic Valley brand protein shake appeals to the consumer's desire for health convenience. The image of a young woman on the go, and the prominent featuring of the protein grams per serving on the label, suggests that this shake is the quick and easy solution to a healthy morning meal.

Convenience

This Walgreens advertisement for flu shots appeals to the consumer's desire for access and convenience. Whether you're a family or a VA patient, you can walk into any Walgreens health clinic to receive a shot.

Environmental Impact

This Heinz advertisement appeals to the consumer's concern for the environment and the general desire to reduce the carbon footprint on the Earth. It features a new recyclable bottle that is 30 percent plant based.

EMOTIONAL GROUP

Fear

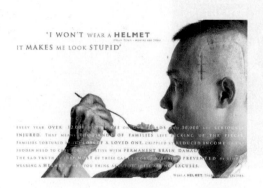

This public service advertisement attempts to reach audiences through fear. It features a young man who now has the mental age of a 2-year-old because of a biking accident in which he did not wear a helmet. The overall message: Wear a helmet, or this could happen to you.

Admiration

This advertisement features tennis champion Serena Williams. In featuring this all-star athlete, Nike is appealing to the consumer's sense of admiration and aspiration. If you want to be a #1 athlete like Serena, wear Nike gear.

Humor

This advertisement for Chick-fil-A uses humor to encourage consumers to visit the restaurant and eat more chicken.

Envy

This advertisement by Omega plays upon human envy. Do you want to drive a boat on the ocean like George Clooney? Too bad you don't own an Omega watch.

Laws, Ethics, and Questionable Practices of Advertising

Advertising is an area where some marketers push ethical boundaries. Because advertisers use rational appeals to communicate about their product, they sometimes inflate or exaggerate claims. Similar problems can arise with emotional appeals when advertisers cross lines of decency and/or appropriateness. Several laws and industry controls exist to reduce ethical problems in advertising. However, despite laws and regulations, it is often difficult to establish when a line has been crossed. For example, laws concerning advertising clearly state that "claims in advertisements must be truthful, cannot be deceptive or unfair, and must be evidence-based." Yet, some advertisers are adept at being technically within the law as they push limits and potentially cross ethical lines.

One of the most common examples of complying with the law while making advertising claims is the use of "puffery" in advertising. **Advertising puffery** is advertising that makes broad exaggerated or boastful statements about a product or service that are subjective (a matter of opinion) rather than objective (measurable facts). In these cases, marketers are able to make claims that cannot technically be proven false, so they are legal—yet the claims may be exaggerations that mislead consumers. Examples of puffery can be found in advertising for food, such as "the best pizza in town" or for a car dealership as "the best prices in the state." These types of claims are difficult to prove as either true or false; they depend on whom you ask or when you measure. Advertising puffery is hard to regulate and even harder to prosecute under the law. But usually consumers can detect puffery, so its impact is often limited.

In the past, some marketers went well beyond puffery by purposefully misrepresenting objective facts. These infractions led to legal and self-regulation of the advertising industry. Deceptive practices such as unsafe products marketed as safe, "bait and switch" advertising (where a marketer would advertise one product and then deliver another), and "subliminal advertising" (where messages not perceptible to the eye were used to send hidden messages) have largely disappeared. The consumerism movement—which began in the 1960s and was led by Ralph Nader—takes credit for the passage of many laws and regulations that help reduce unethical and/or illegal advertising practices. Nader and others in the movement effectively lobbied Congress to pass laws that protect consumers.

Two key federal agencies are responsible for regulating advertising. The **Federal Trade Commission (FTC)** protects consumers and businesses from anticompetitive behavior and unfair practices. The agency seeks to protect consumers from deceptive and unsubstantiated advertising and enforces the provisions of the FTC Act. In addition, the FTC enforces truth-in-advertising laws for all forms of media. The **Federal Communications Commission (FCC)** was created to enforce laws that prohibit obscene, indecent, and profane content from being broadcast on radio or television networks.

Some products and industries fall under the laws administered by additional governmental agencies. For example, food and pharmaceutical products are regulated by the Food and Drug Administration (FDA). Marketers that deliver their products or advertise them through the U.S. mail are regulated by the U.S. Postal Service (USPS). Advertising for products such as intoxicants, cigarettes, and guns are controlled by the laws and regulations of the Bureau of Alcohol, Tobacco, Firearms and Explosives (ATF).

The poor practices of some marketers have led the industry to create and enforce many self-imposed standards that go further than existing laws and regulations. **Self-regulation** involves a structure to settle disagreements in business without having to include government resources. Advertising agencies, media companies, and clients have come together to agree on standards. Associations such as the Advertising Self-Regulatory Council (ASRC) and the Children's Advertising Review Unit (CARU) bring the various groups together to create and enforce ethical practice standards that are more stringent and comprehensive than many government-imposed laws and regulations.

Because of laws, regulations, and self-regulation, advertising today typically represents a fair and honest representation of a marketer's product.

This sign represents the concept of advertising puffery because it claims that it offers "cebu's best."

The Importance of Sales Promotion

Sales promotions are important because they build stronger relationships with retailers and end customers. Furthermore, they build excitement for the product and usually increase short-term sales. **Sales promotion** is a set of nonpersonal communication tools designed to stimulate quicker and more frequent purchases of a product. Firms often use sales promotion to support the other elements of their promotional mix. For example, in recent years McDonald's implemented a sales promotion in which young customers received plastic figures of the cartoon characters The Smurfs in their Happy Meals. The promotion used advertising tools and prompted more consumers to buy Happy Meals in an effort to add to their Smurf collections. Successful sales promotions like this have the potential to build short-term excitement and long-term customer relationships simultaneously. Let's discuss some common sales promotions.

Consumer (Business-to-Consumer) Sales Promotion

There are five main types of consumer sales promotions, which are coupons, rebates, samples, contests/sweepstakes, and loyalty programs. Each is described in greater detail in the remainder of this section.

Coupons

Coupons remain the most common type of sales promotion. **Coupons** are documents that entitle the customers who carry them to a discount on a product. Traditionally, coupons have come in the form of printed vouchers that customers present to claim their discount. However, the expansion of mobile advertising and the popularity of websites like Groupon have allowed companies to distribute coupons digitally to smartphone devices that consumers can redeem at a specific business. Coupons provide firms with an effective method for stimulating sales and encouraging customers to make additional or repeat purchases. In addition, a firm can control the timing and distribution of coupons in a way that does not dilute potential revenue from a customer who is happy paying full price. However, coupons also have disadvantages. Coupon fraud is a serious problem in the form of mis-redemption practices. Mis-redemptions can range from illegal copying of coupons by the consumer to innocent mistakes by an employee who gives too large a discount or forgets to take the coupon from the customer after the purchase. All mis-redemptions have the effect of reducing the firm's profitability, so care must be taken to anticipate and discourage such practices.

Rebates

Rebates allow consumers to recoup a specified amount of money after making a single purchase. Many promotional strategies make use of rebates; marketing professionals offer more than 400 million rebates each year. Although most customers prefer coupons because they're easier to use, most marketers prefer rebates. Rebates provide the incentive of a price decrease; however, because customers often fail to redeem them—statistics show that between 40 percent and 60 percent of all rebates are not redeemed—the firm typically earns greater profits than it would by issuing coupons. Rebates work most effectively when offered in conjunction with a high-cost purchase in which the perceived value of the rebate is magnified. A $200 rebate for a flat-screen television or a $2,000 rebate for a new car typically generates more purchases than small rebates for everyday items that consumers tend to forget about.

Samples

Samples are trial portions of a product used to encourage product purchase. Samples have been an effective sales promotion tool for decades and involve everything from trying a new type of sausage at your local grocery store to getting a free weekend of HBO's programming. Samples can be expensive because marketers are giving away products in many cases, but they can also be a powerful tool for getting customers to actually buy the product. Marketers of baby formula often send samples to new mothers in the hopes that the family will like it and start using that brand of formula during the child's first years of life.

Contests and Sweepstakes

Firms spend approximately $2 billion per year on contests and sweepstakes. The terms *contest* and *sweepstakes* are often used interchangeably, but there is a distinct difference between the two.

- **Contests** are sales promotions in which consumers compete against one another and must demonstrate skill to win. Contests provide marketers with a way to engage consumers and empower them to promote an organization's products and brand. For example, Oreo sponsored an "Oreo and Milk Jingle" video contest that offered cash prizes and trips to the group who did the best job designing and singing the "Oreo and Milk" jingle.

- **Sweepstakes** are sales promotions based on chance. The only requirement to win is that you enter. Every entry has an equal chance of being the winner. An example of a sweepstakes is the HGTV Dream Home Giveaway, in which HGTV randomly selected a viewer to win $500,000 and a cottage in Hawaii. Sweepstakes like HGTV's have the advantage of creating interest and excitement from a broad group of consumers. This particular sweepstakes also helped bring HGTV to a wider audience.

Both contests and sweepstakes have disadvantages. Contests can be expensive to administer because each entry must be judged, meaning firms can't rely on random selection. Multiple federal and state government agencies monitor contests and sweepstakes to make sure they are fair and properly represented to the public.

Loyalty Programs

Designed to strengthen customer relationships, **loyalty programs** allow consumers to accumulate points or other benefits for doing business with the same company. Loyalty programs are especially popular in the airline and hotel businesses. Holiday Inn, for example, has had success with its Priority Club loyalty rewards program. Global membership in the loyalty club has reached more than 65 million, and members are rewarded with points that increase more rapidly the more often they stay at a Holiday Inn. The points can then be turned into free hotel stays, gift cards, and charitable contributions.

Loyalty programs have grown in importance as natural loyalty to products or services has decreased in many industries. Loyalty programs such as Delta's SkyMiles offer incentives that encourage satisfied consumers to go out of their way to fly with Delta.

Trade (Business-to-Business) Sales Promotion

The sales promotion tools discussed up to this point are often directed to individual consumers. **Trade sales promotions** are promotional tools directed at business-to-business (B2B) firms, including wholesalers and retailers, rather than individual consumers. Trade sales promotions often include the same tools discussed earlier (coupons, rebates, contests and sweepstakes, and loyalty programs), plus two other major approaches:

1. **Allowances: Allowances** typically involve paying retailers for financial losses associated with consumer sales promotions or reimbursing a retailer for an in-store or local expense to promote a specific product. For example, a local grocery store might be reimbursed for a radio ad that mentions that Diet Dr Pepper is on sale over the next few days. Typically, the firm will pay the retailer only once it has proof of the financial loss or local promotion costs.

2. **Training:** The other major trade sales promotion is training the reseller's sales force. Training activities such as brochures or on-site demonstrations help retail and wholesale personnel understand the product's benefits. This in turn makes the resellers better equipped to speak with consumers and sell the firm's products. Training also helps ensure that employees at every level understand the features, advantages, and benefits of the products they are trying to sell.

Employee Sales Promotion

Employee sales promotions are typically directed toward front-line employees that deal directly with customers. Employee sales promotions are based on promotional incentives designed to increase sales for a product or brand. Such promotions are common for both salespeople who sell to other businesses (B2B) or those who sell to end consumers (B2C). Because front-line employees interact with customers, they are able to fit products to the needs of their customers. Many manufacturers realize that providing an incentive to front-line employees will provide a reason for these employees to recommend one product over another. Typically, incentives are small and offered by multiple manufacturers so as to reduce unethical selling practices. Employee sales promotions should be used only to direct employees to mention a product, but not as a means to force a specific product on a customer.

Sales promotions reward employees through cash incentives and contests. A *cash incentive* is paid to employees for selling a specific product or brand. In these cases, each front-line employee's sales are tracked and a small amount of money is paid for the sale of certain items that offer an allowance.

A less direct way that front-line employees can be encouraged to sell a particular product is through a sales contest. *Sales contests* are for employees based on their sales volume of a product to customers. Contest prizes can include vacation days, gift cards, travel, or even a new car (for expensive or high-volume sales). These contests encourage sales of a product and also provide a competitive element to the sales environment that can result in higher overall sales of all products, not just those included in the contest.

Marketing Analytics: Using Data for Advertising and Sales Promotion Strategies

Using Data to Determine Messaging Strategy

Consumer knowledge about and preference for a firm's market offerings must be assessed before an advertising campaign can begin. Both consumer knowledge and brand preference can be measured by using established quantitative measures and data analysis. Qualitative methods such as observation, interviews, or focus groups can also be used to understand consumer preference and knowledge. Once the two are known, then the firm will have a better understanding of whether to use a strategy that informs, persuades, or reminds. Launching new products or services often requires that firms inform consumers about the new product but also persuade consumers to try the new offering. These types of messages can be combined within campaigns.

Using Data to Inform Campaign Planning

Once a firm identifies the target market for its products or services, then it uses data to determine who the existing target audiences are within that target market. For example, a university must decide for which target audience it is going to structure its advertising campaign. A campaign directed at parents will use different messages and media than a campaign targeted at high school students. The parents might receive messages that stress campus safety and faculty credentials whereas the high school students might receive messages that highlight campus social life and community. Once the audience is identified, the rest of the planning process can begin.

Objective setting is also a data-driven process. Data are needed to understand what level of product knowledge exists among a target audience as well as at which stage the target audience is currently in the purchase process. A target audience comprising new users of a product (such as high school students in the college market) will receive different messages than would existing users (such as college seniors in the market for graduate school). Data regarding purchase rates are also examined. Infrequently purchased products or products with seasonality of demand require special attention because consumers are not always in the market for these goods.

Ad placement is determined by media usage of the target audience. Each media outlet provides data about a product's target market. These data include demographic information such as age, gender, income, occupation, and salary, as well as psychographic data related to interests, values, opinions, and so forth. Using these data allows firms to best align the placement of the messages in the media that the target audience is most likely to consume. Thus the media must match the audience and then the creative elements can be designed around the chosen media. For example, ads that appear on YouTube are similar to those that appear on television. However, because YouTube allows for longer videos, ads on YouTube tend to be more story based. Likewise, newspaper ads comprise different creative elements than those developed for radio or digital media.

Regardless of the media chosen, all ads must be tested prior to launch. Data gathered from target audiences reveal how an ad is rated. The subjects in these studies are asked to rate the ad in a number of categories including likability, impact on their desire to purchase or to learn more about the product or service, and ad execution. If the subjects in the study like the ad overall but feel the music, for instance, is not right, the advertiser has the chance to make adjustments to the campaign prior to launch. Once the ad is running, then the campaign must be measured. For digital campaigns, these measurements can be taken hourly because the data are being collected in real time. For example, if our university posts an ad on Instagram to a target audience of high school seniors, it will know in real time how many members of this audience viewed the ad and acted on it. These numbers show up in the number of times the "Learn More" link was clicked or by the number of times the post was liked. For traditional media, where such real-time measurements are not possible, measurement is often taken before, during, and after the campaign to assess effectiveness.

Larger companies will hire data firms to monitor and report data regarding campaign reach. Nielsen and other data companies can provide data on total audience size as well as target audience size. This allows the advertiser to calculate the total audience size for the campaign as well as how many target audience members have been exposed to the ad.

Using Data to Identify and Justify Advertising Claims

Each advertising campaign is crafted around a carefully researched messaging strategy that delivers on its defined objectives. The claims made in the ads must be relevant and compelling to the target audience lest the message not be received—research helps identify what is considered relevant and compelling to consumers. For example, if data gathered by our earlier-mentioned university note that its target market is most interested in aspects of campus life, then all messages it crafts must consistently address campus life. Some of these messages will be rational appeals based on objective information gathered through research; for example, perhaps prospective students care about the size of dorms, library hours, access to shopping, student academic clubs, and so forth. Other messages will be emotional appeals based on subjective information, such as how fun and exciting the campus culture is, how cool the building designs are, or how hot the music scene is in the local town.

Now, because advertising is a regulated form of marketing communications, companies must ensure that all advertising messages are *truthful*. If an advertisement presents a claim that is later proven to be untruthful, then the company that released the ad could be accused of either deceptive advertising or misleading the consumers—especially if consumers have relied on these false claims to make their purchase decision.

Let's consider the case of Skechers. In 2009 Skechers launched a new shoe line called Shape Ups. The advertisements for these shoes claimed they would help people lose weight and improve muscle tone. Many consumers purchased Shape Ups based on this claim; unfortunately, no evidence existed to support it. The FTC found that these claims were deceptive and had harmed consumers. As a result, Skechers was required to reimburse all consumers who purchased the shoes.

To avoid making false claims, the FTC requires *substantiation* for any advertised claim. This means that the firm must provide the FTC with research conducted by an independent expert. The claim must be based on evidence that is clear. Obviously, this is easier to do for rational claims. For emotional claims, substantiation still must be provided but is more difficult. What constitutes as *better?* What does *more fun* mean? Further research is conducted to inform companies of what consumers consider "better" or "more fun."

It is important to remember that all of the federal and state agencies involved in the regulation of advertising must adhere to certain standards. For example, the FDA regulates direct-to-consumer drug advertising. Drug companies are obligated to not only inform consumers of the positive effects of a drug, but they must also warn them about possible negative side effects as well—even if they sound really bad. The goal is to ensure that consumers base their purchase decisions on factual information.

Using Data to Determine the Sales Promotion Mix

As with the other IMC elements, the choice of which sales promotion to use in a campaign must be driven by data. A recent survey completed by Mobile Commerce Daily, for instance, noted that "96% of mobile-device users will use smart devices to search for digital coupons. . . . Online coupons have 10 times the redemption rate of conventional coupons." Such information is invaluable to companies as they consider which sales promotion mix will prove most effective for promoting their brands.

Each of the promotional mix tools can achieve different objectives. For example, data have shown that rebates and coupons can change consumers' perception about the price of a good or service. This can positively influence consumers to choose one firm's brand over that of its competitors. Samples have also shown to be effective at encouraging consumers to try a product, especially newly launched ones. And although data about contests and sweepstakes indicate they can be highly effective, they are also highly regulated. How the sweepstakes or contest is run and how prizes are awarded are regulated by each state. Companies that utilize direct-mail promotions must adhere to the requirements enumerated in the Deceptive Mail Prevention and Enforcement Act. These include specific mandated disclosures such as statements indicating that no purchase is necessary and that purchase will not improve an individual's chances of winning. Statements must also outline the terms and conditions of the promotion (name of the sponsoring company, details about the prizes offered, and their approximate retail value).

Again, firms must carefully study their target market consumers to understand which promotional tool will be most effective, and if applicable, what the regulations are at both the federal and state levels.

Responsible Marketing

The Blurry Areas of Advertising and Sales Promotion

There are many ethical grey areas in advertising. Some of them have been discussed in this lesson and in other Responsible Marketing cases. In addition to puffery or misleading language, one of the biggest issues in advertising involves deceptive advertising. This could involve tactics from questionable pricing strategies or misleading or inappropriate messaging.

Sometimes retailers employ a deceptive pricing strategy called "bait and switch" in their ads. Companies advertise a product at a really low price just to get people to come into the store; this is usually an item they have very limited quantities of. But once the customer is in the store, they try to switch them to a higher price (and higher profit) item. Another tactic of deceptive pricing occurs when brands exclude hidden fees or terms and conditions, elevating the price significantly.

Sometimes ads can be more subtle in their mistakes, such as crossing a line that is inappropriate or upsetting to an audience. Consider the 2019 Peloton ad, entitled the "Gift that Gives Back." In this ad, a woman receives a Peloton for Christmas, then records herself riding it every day for the next year, and finally curates a video for her partner to show how much the bike changed her. To many, this seemed like a message that was telling someone that she needed to be thinner or better or in some way superior to what she was. Many consumers blasted the ad on social media and in other media sources.

Reflection Questions

1. What makes an advertisement or sales promotion deceptive?

2. How much regulation should the FTC (Federal Trade Commission) have on deceptive advertising?

Lesson 9-4
Public Relations

Public relations (PR) is an important component of a firm's integrated market communications program. PR strategies provide information and build a firm's image with the public, including customers, employees, stockholders, and the community. In this lesson you will learn more about the role of PR within the promotional mix as well as different types of PR activities.

By the end of this lesson you will be able to

- Describe the role of public relations within the promotional mix.
- Identify the different types of public relations activities.
- Explain the importance of different types of publicity.

Young Drivers Demonstrate Their #StreetTread: Michelin's PR Campaign to Reduce Teen Accidents

What do sneakers and tires have in common? An innovative PR campaign from Michelin North America thinks teens will want to find out. In fact, the tire company's "Teens Prove Their #StreetTread" campaign won the Best of Silver Anvil Award from the Public Relations Society of America in 2019.

How did the campaign come about? Michelin's public relations firm, Ketchum, discovered that poorly maintained tires are a major contributor to car accidents and that car accidents are the #1 cause of teen fatalities in the United States. Research also indicated that few teens knew how to properly maintain their tires.

And here's where the sneakers–tire connection comes in: Ketchum realized that for many teenagers, sneakers are an important part of their social identity. It also realized that tires and sneakers have something in common: tread! Ketchum used

this connection between sneakers and tires to inspire young people to learn more about proper tire maintenance.

First, Ketchum worked with sneaker brand Vans to create a limited-edition Vans sneaker modeled after the famous Michelin man.

Next, it had a giant "sneakermobile" drive across the country to create social buzz and build awareness about the campaign. Teens could come to the "sneakermobile" and buy the special sneakers for a penny if they could demonstrate that they could check their tire pressure as well as proper tire tread depth.

According to Ketchum, the campaign generated 10,000 stories, 1.5 billion media impressions, and reached about 37 million teens. Most importantly, though, the campaign may help save lives and reduce the number of automobile-related fatalities among teens.

Defining Public Relations

Public relations (PR) is nonpersonal communication focused on promoting positive relations between a firm and its stakeholders. Similar to advertising, PR is passive in that the customer is not always involved in the exchange. Firms use PR to achieve different objectives. For example, it can be used to build a positive image for the company, handle unfavorable stories or events, and maintain positive relationships with the media.

As noted in the branding lesson, consumers often associate certain attributes with a firm. PR is often used as a way to facilitate those associations. By aligning with a cause or a sentiment, marketers are hoping to better communicate the story of their brand. For example, Red Bull sponsors the X Games to signal to its consumers that it is part of the "extreme" lifestyle. Recently, Heineken launched the "Worlds Apart" campaign to encourage people to break down their barriers and chat with one another over a beer. This type of campaign is designed to convey a positive image about a brand through its associations and what it stands for. The hope is that this type of activity allows customers to feel more connected to the brand.

PR can be especially important when an organization faces a crisis, as BP did during and after the 2010 *Deepwater Horizon* oil spill in the Gulf of Mexico. In response to the negative publicity BP received due to the spill, the company filled its website with technical briefings with BP officials and maps and charts detailing the company's efforts to contain the leak. BP also produced and posted short films featuring BP officials, representatives of government agencies, and area residents helping in the Gulf cleanup effort. In addition, the company publicized other positive actions it was taking, such as donating $1 million to help a food bank feed people whose incomes were washed away by the spill and actively hiring laid-off workers to clean beaches and animals contaminated with oil. Finally, BP gave the Gulf States money to promote the region's seafood industry in an effort to improve economic conditions in the area.

BP had to work hard to clean up its image in the wake of the massive *Deepwater Horizon* oil spill in 2010. Locals in the region, however, felt that BP's response was insufficient and insincere, and just a gimmick for the cameras.

Public relations today is a 24-hour-a-day, 7-day-a-week job for marketers across government entities, for-profit industries, and nonprofit organizations. Advances in technology cause information to spread faster than ever before and enhance the influence of that information on potential customers. Marketers should use all of the tools they can to make sure the public perceives the company in the most positive way possible. A simple truth in marketing is that people want to do business with organizations they like and respect. It is incumbent upon marketers to share the organization's actions with the public in a way that entices customers to buy from the company, give money to the charitable organization, or support a specific person or idea.

Public Relations Activities

Organizations use a variety of tools for public relations. While there are many types of public relations activities, some of the most common include the following:

- **Annual reports:** Annual reports provide a forum for the organization to share with its stakeholders what it has achieved over the past year. They present the firm with an opportunity to highlight financial successes as well as charitable and philanthropic work that portray the organization in a positive light. Most publicly traded companies create an annual report.

- **Speeches:** Speeches provide an avenue for members of an organization to market their message directly to a group in a longer-form speech. These speeches, usually given by key members of the organization, can generate positive press and help to develop relationships with stakeholders and the public at large.

Steve Jobs's 2005 speech "How to Live Before You Die" is considered one of the best CEO speeches ever and generated positive press for Apple.

- **Blogs:** Blogs are a social media tool through which individuals can share their thoughts and knowledge with the public. Blogs can be a mix of insights, humor, or other personal interests and have become increasingly popular with marketers focused on PR. Established industry executives also make use of blogs. The *Disney Parks Blog* does a great job of taking fans "behind the scenes" at Disney parks. This allows fans to feel like they are part of the magic.

- **Brochures:** Brochures typically are intended to inform and/or engage the public. A modern PR brochure can be on paper or online and provides a forum for educating the public about a firm, its mission, or a specific cause. Brochures often present information similar to that found in an annual report, but in a shorter, more accessible way. Marketers should ensure that the brochure contains useful information, is visually engaging, and is consistent with the rest of the company's promotional mix elements.

- **Event sponsorship:** Event sponsorship involves firms paying to sponsor, host, or support a cultural, sport, or charitable activity. For example, many firms partner with the Susan G. Komen foundation during October's Breast Cancer Awareness Month. Adidas sponsors the Boston Marathon every year to show its association with the famed race. Coca-Cola has sponsored the Olympic Games since 1928.

Annual Reports

Annual Reports:

- Pros: Able to provide a lot of information about strategy, direction, and growth for shareholders.
- Cons: Consumers do not often read annual reports.

Speeches

Speeches:

- Pros: Provide a human interaction with customers. Allow customers to feel more connected to a firm's leadership.
- Cons: If something in the speech is offensive, customers will talk about it forever.

Blogs

Blogs:

- Pros: Company completely controls the message to the audience. Firm can respond to complaints or issues with customers directly.
- Cons: Customers don't often access blogs. Customers might not read all of the content on a blog.

Brochures

Brochures:

- Pros: Provide a "take-away" for customers that they can keep with them to reference.
- Cons: Sometimes end up in the junk pile.

Event Sponsorships

Event Sponsorships:

- Pros: Many customers like for firms to be associated with a cause.
- Cons: Some customers might not like the event a firm is associated with.

Publicity

Publicity involves disseminating unpaid news items through some form of media (e.g., television story, newspaper article, etc.) to gain attention or support. Publicity is part of public relations and focuses primarily on how a firm is represented through the media.

During the most recent economic recession, firms often worked with the media and used publicity as a way to lower the cost of their public relations efforts. Firms can often write press releases or content to manage how information is presented using publicity. For example, if Tesla is introducing a new product, the company might write a press release to ensure that the media (1) know the event is happening and (2) announce the key components that Tesla wants presented. However, sometimes publicity is word of mouth or comes from external sources.

There are two types of publicity, online and offline. Online publicity is disseminated through any Internet-based platform. A primary example of online publicity is the use of social media. When consumers post about an organization on social media, this is considered publicity as the firm has no control over what the consumers post. After Pepsi's unsuccessful marketing campaign featuring Kendall Jenner, thousands of people posted on Twitter to reprimand Pepsi for its insensitive advertisement, causing the company to remove it immediately.

Offline publicity includes more traditional print media such as magazines, journals, and newspapers. Marketers can cultivate relationships with the media to help manage how messages show up in offline materials. However, journalists have an obligation to report the facts. When videos emerged of United Airlines forcibly

Red Bull's Mini Cooper draws attention wherever it goes. It's like a moving billboard. Fans of the brand know that where the Red Bull car goes, adventure will follow.

removing a passenger from one of its airplanes in 2017, newspapers and news outlets around the globe reported on the story and United Airlines's stock plummeted.

The major advantage of publicity is that, when done well, it allows marketers to communicate with consumers at an extremely low cost. The major disadvantage of publicity is that organizations have less control over how the information is presented because they don't pay for it. Bad publicity and negative news can harm an organization's reputation and image.

Responsible Marketing

Public Relations Nightmares

Public relations professionals help organizations maintain a favorable image in the mind of the public. Yet public relations activities have broader societal implications than targeted communications campaigns because PR activities go beyond just interactions with a firm's target customers. Rather, public relations can impact how communities or societies as a whole view the organization, which can greatly affect a company's overall brand image.

For example, in late 2021, online mortgage company Better.com's founder and CEO Vishal Garg fired almost 900 employees via a Zoom call and then accused some of the recently fired employees of stealing from the company by working fewer hours than required. Even if this were true, the fact that the firing was so impersonal left many customers—even those who had never previously heard of Better.com—with an unfavorable image of the company. The video then went viral, with many Twitter users claiming #thisisevil. Better.com then had to struggle to determine the best way to respond to such public backlash.

Sometimes, PR scandals can arise from poorly executed attempts at positive public relations. In 2017, for example, sportswear manufacturer adidas sent an e-mail to finishers of the Boston Marathon with a subject line that read "Congratulations, you survived!" Although the message was meant to highlight the achievement of completing a marathon, it was met with consternation because of the lack of sensitivity with respect to the Boston Marathon bombing in 2013.

Reflection Questions

1. What should firms do when they make an honest or careless mistake?

2. What kind of actions should a firm engage in to maintain a positive image with the public?

3. What should firms do when one employee or a group of employees reflect poorly on the firm?

Personal Selling

How Personal Selling Puts the Roof over Your Head

There's an interesting stigma to personal selling in North America that doesn't really exist anywhere else in the world. Why? Maybe because the term *salesperson* triggers for some an image of a fast-talking person in a cheap suit selling used cars—not an uncommon character to imagine, but it's a fictional one at best.

Personal selling is the purest form of marketing: the art of persuading, convincing, and influencing people to create sufficient value that they willingly engage in an exchange with us. One out of every nine people in the United States makes their living by directly trying to get people to make a purchase, and it's a safe bet that most of the other eight engage in similar activities without the title. From the shoes you're wearing to the antibiotics you took when you were last sick to the school you're taking this class at, it's *all* most likely the result of personal selling.

It's also a safe bet that you don't have to look much farther than your own social media feed to see several of your friends or relatives in one of the most ubiquitous personal selling businesses: real estate. There are about 2 million licensed real estate agents in the United States, which is about 1 percent of the total working population of the country!

Buying a home is a high-involvement, high-risk, very emotional purchase, and the thought of making such a massive purchase *without* using the guidance and advice of an agent wouldn't even cross most people's minds. RE/MAX is one of the largest and most recognizable real estate agencies on Earth, and it wouldn't exist without its network of over 123,000 agents working in more than 100 countries around the world. RE/MAX sells almost a quarter-*billion* dollars' worth of real estate every year, with the average agent making $120,000 in salary.

RE/MAX agents are perfect examples of professional salespeople who need to be experts in marketing, promotions, social media, negotiations, the market environment, technology, and customer psychology all at the same time. RE/MAX even runs its own award-winning RE/MAX University, which offers online courses and training programs for new and experienced agents alike. The company also runs a global training center called Gateway where agents can learn everything they need to know to be good salespeople: from licensing legalities to luxury branding to self-defense.

As you can see, real estate is a great example of a relationship-based business model that's almost entirely customer-driven and enabled by professional salespeople.

Lesson 9-5
Personal Selling

In this lesson, you will learn the fundamentals of selling and sales management, as well as building customer relationships. These skills enable marketers to sell effectively and to manage and motivate a sales force.

By the end of this lesson you will be able to

- Describe the importance and nature of personal selling.
- Describe the personal-selling environment.
- Describe characteristics of successful salespeople.
- Explain the steps in the personal-selling process.
- Describe the major functions of sales management.
- Articulate how sales managers recruit, select, and train salespeople.
- Describe salesperson compensation, motivation, and evaluation.

Understanding Personal Selling

Personal selling is the two-way flow of communication between a salesperson and a customer that seeks to influence the customer's purchase decision; it is paid for by the firm. Personal selling is an integrated marketing communications tool that takes many forms and can include someone trying to sell you insurance or a new car. It can even take the form of the person behind the fast-food counter trying to get you to upsize your order. In all sales situations, the goal of selling is to develop good customer relationships. For relationships that occur over time, salespeople engage in relationship selling. **Relationship selling** involves building a trusting relationship with a customer over multiple sales interactions.

The sales function can drive the economic success of most firms. In today's information-rich world, the role of personal selling is more important than ever. Some suggest that consumer access to information spells the end to the need for salespeople, but the increasing complexity of products and technologies suggest the opposite. The main reason that selling is more important than ever is because it is flexible. Personal selling differs from the other tools of the promotional mix because messages flow directly from the salesperson to the customer, often face-to-face. In settings where a product or its use are complex and varied from customer to customer, salespeople are the best way to describe a product's benefits.

For RE/MAX, real estate agents serve as salespeople who work hand in hand with customers (home buyers) to locate products those customers may be interested in purchasing. RE/MAX brokers need to be knowledgeable, for example, of the towns and markets in which they do business so that they can provide the most value for their customers. Without well-trained and professional sales agents, RE/MAX would not be able to meet the demands of its customers.

However, personal selling isn't right for every product or service; there are many pros and cons of personal selling, so marketers have to be certain to use this form of promotion in the correct settings.

Pro: Relationships

Personal selling enables firms to develop personal relationships with customers. Relationship selling involves building a trusting relationship with a customer over a long period of time. Relationship selling is important because few firms can survive on the profits generated from one-time transactional sales.

Pro: Immediate Feedback

Verbal feedback allows the salesperson to listen to the customer's likes, objections, and concerns. Nonverbal feedback is physical and visual communication that gives insight into the customer's mindset and the likelihood that he or she will buy. Both forms of feedback allow the salesperson to adjust the sales presentation to provide a customized solution that best fits the customer's needs.

Con: Inconsistent Messaging

One challenge for personal selling is ensuring that each salesperson communicates a message that is consistent with the marketing communications strategy. Inconsistent messaging is the equivalent of a firm having as many marketing strategies as it does salespeople. Firms can overcome this challenge through training of the sales force.

Con: Expense

A major challenge of personal selling is the cost involved. The average cost of a sales call varies across industries but usually averages several hundred dollars per visit. Costs are even higher when you consider that, for most products, one sales call will not result directly in an order.

The Personal-Selling Environment

Salespeople work in fast-changing, competitive, and complex environments. Every product that is sold has different characteristics. But, for most products, the selling environment includes some elements of rapid change, threats from competitors, and increasingly complex relationships and sales processes. Let's look at each of these factors.

Changing Environment

For salespeople, few things remain constant. New products and solutions are common, requiring that salespeople invest much of their time learning and building product knowledge. Also, the wants and needs of customers are always changing, so salespeople act as a voice of customers within their own firm, reporting back on new and different requirements of customers. Think about a salesperson who works at Verizon. Every year, almost every phone from companies like Apple, Samsung, HTC, LG, Motorola, and many more are updated and replaced. It is the salesperson's job to understand all of these changes and be able to explain why they offer benefits to consumers.

Competitive Environmen

One reason that the sales environment changes so fast is because almost all sales situations are characterized by competitors vying for the same business from customers. This competition drives product innovation and advancement of a salesperson's portfolio of products, services, and solutions for customer needs. Think again of the salesperson at Verizon. If only one phone provider was available to customers, the salesperson's job would be much simpler. If there were no Sprint, AT&T, or T-Mobile, customers would not have choices and would be required to accept any level of service that the salesperson offered. But competition does exist, so to best serve customers, salespeople must keep up with what their company is doing while also monitoring what their competitors are doing.

Complex Environment

Sales environments are also increasingly complex. Complexity is driven by both change and competition. In addition, complexity is added to markets as products, services, and solutions become more comprehensive. Think once more of the Verizon salesperson. Not only does the salesperson have to keep track of product changes and competitive offers, he or she also has to be able to explain coverage areas, service plans, wireless contracts, and warranties (to name only a few). Most firms that sell products have become "all-inclusive" solution providers, offering the physical product and all of the service required to maintain the product. The addition of extra levels of service has added a great deal of complexity to personal-selling environments.

Technology is attributed to the increased pace of personal-selling environments. Advances in communications and the mobility of technology have been prevalent and helpful in sales environments. However, the portability and 24/7 aspect of technology has also made the sales job somewhat of a no-boundaries position. For example, it is not uncommon to see a salesperson at a coffee shop with e-mail, phone, and instant messaging all operating at the same time. There are pros and cons of this scenario. On one hand, the salesperson has more opportunity and is less constrained to an office. On the other hand, the salesperson has fewer distinctions between his or her work and personal life.

The question is, given these difficult environmental factors, why would an individual decide to take a sales job? There are many reasons, but some of the more prominent ones include:

Availability of Jobs

In many cases, a graduating college senior may face a job market that offers more sales jobs than any other type of positions. Because learning to sell may be the only option, many decide to give it a try to see if they like it and can be successful in the career. Not everyone enjoys a sales job, but most who try learn something that helps them land in a different job as a result of their sales experience.

Autonomy

Many salespeople work from home or in large territories that allow them to choose what schedule and priorities to pursue on a daily basis. This freedom allows a very flexible way for many to earn a living while also pursuing family, community, or personal objectives.

Earning Potential

A sales job is usually compensated through fixed pay (salary, base pay, etc.) and variable pay (commission, bonus, etc.). For motivated salespeople, variable pay is viewed as a benefit because it relates to the salesperson's hard work. A Bureau of Labor Statistics survey reports that average sales wages range from $39,000 (bottom 10 percent) to $165,000 (top 10 percent), with a median average wage of $75,000. Despite the range, the median wages are higher than most careers, indicating that salespeople typically earn a higher wage than similar skill level non-sales jobs.

Engaging Atmosphere

Many salespeople report that one of the main benefits of a sales career is the engaging atmosphere. Salespeople perform a variety of activities in the course of their day. In addition, most salespeople work with customers, which expands their workplace to areas outside their own firm. Because of this, many people who appreciate learning, interaction, challenge, and creativity find a selling career to be a good choice.

Characteristics of Successful Salespeople

Successful salespeople are those who remain motivated. **Motivation** is the basis of a continued commitment to working toward a goal. Think of a salesperson as similar to an NFL quarterback. The quarterback decides which plays to call, communicates and implements the plays, and is responsible for the success or failure of each play. Throughout the game, some plays are successful and some are not, but what counts is how the quarterback responds. If the quarterback becomes demotivated by a series of poorly executed plays, the game will suffer. However, if the quarterback maintains a motivation to win, chances are the results will improve. The same is true in sales. Maintaining motivation throughout both the highs and lows of a sales career is what drives long-term success.

Characteristics such as a good personality, an optimistic outlook, resilience, a customer focus, an ability to communicate, and adaptiveness also contribute to success in sales.

Good Personality

A "good personality" in sales is one that exhibits humility, conscientiousness (reliability and dependability), achievement orientation, curiosity, perseverance (not easily discouraged), and lack of self-consciousness.

Optimism

Optimistic people are said to see the glass as half full, as opposed to half empty. This saying represents a general sense of hopefulness and confidence that optimistic people possess. Professional salespeople must continually look for the positive even when their sales efforts are not successful in the short term. Optimism is helpful in overcoming failure.

Resilience

Similar to optimism, resilience relates to a response to failure. Resilient people have the capacity to quickly recover from difficulty and stress. This helps salespeople move on from lost sales and opportunities quickly, which refocuses them on finding the next sale as opposed to remembering the last failed sales attempt.

Customer Focus

Today's sales professionals should have a general service orientation that results in them being focused on serving customer needs. Salespeople should attend to the needs of other people and their businesses; finding enjoyment from serving others helps salespeople succeed.

Communication

Communication is the first step in serving customer needs. Professional salespeople are therefore good communicators. Salespeople must be able to write, speak, and act appropriately in diverse situations. In addition, salespeople must be good listeners. Salespeople who listen actively can process what is said before responding.

Adaptiveness

Salespeople rarely encounter the same situation twice, so to be successful they must be adaptive to many situations. Every customer and selling situation poses new challenges and opportunities. Salespeople who can adapt and alter their message in response to customers are typically more successful than those who cannot.

The Personal-Selling Process

As shown in the figure, the personal-selling process follows a relatively straightforward seven-step model, which includes (1) prospecting and qualifying, (2) the pre-approach, (3) approach, (4) presentation, (5) handling objections,

(6) closing the sale, and (7) follow-up. Let's explore each of these steps.

Step 1: Prospecting and Qualifying

Prospecting and qualifying are the lifeblood of the personal-selling process as firms are constantly seeking new customers for their business. **Prospecting** involves the search for potential customers—those who need or want a product and fit into a firm's target market. The personal-selling process is often compared to a large funnel: at the top we find all the customers. As prospecting begins, so does the narrowing of the funnel. Prospecting limits the set of potential customers to *prospects*— that is, the customers most likely to use a salesperson's product. Potential customers can be found in a variety of ways, such as through customer referrals, trade shows, industry directories, websites, social media, and networking.

While merely finding potential customers is important, the main goal of this step is to find *qualified* prospects. **Qualifying** prospects involves identifying which customers within the firm's target market have not only a desire for the product but also the authority to purchase it and the resources to pay for it. Qualifying prospects continues the funnel process, as an even smaller subset of qualified prospects emerges. Salespeople qualify prospects using a number of tools, including talking with the potential customer and conducting marketing research to better understand the target customer's needs, wants, and ability to pay. Once a salesperson qualifies a prospect, the prospect is further targeted based on his or her sales potential.

It is important to note that not every qualified prospect is immediately targeted by a salesperson. Some prospects may be qualified but are lower priority than other qualified prospects. Salespeople maximize their effectiveness by carefully prioritizing which qualified prospects are the best fit with personal and organizational goals.

Step 1	Prospecting and qualifying
Step 2	The pre-approach
Step 3	The approach
Step 4	The presentation
Step 5	Handling objections
Step 6	Closing the sale
Step 7	Follow-up

Steps of the Personal-Selling Process

Step 2: The Pre-Approach

For the prioritized and qualified prospects, salespeople move closer to initiating a sales call by conducting research. This step is called the **pre-approach** because it occurs before the actual personal-selling process. At this step, sales professionals analyze all the information available to them about a prioritized and qualified prospect. The pre-approach goes beyond the basic information that is considered in the qualifying stage to include very specific information, such as key decision makers, account histories, product needs, growth potential, and industry trends. By preparing in the pre-approach, salespeople signal to prospects that they are professionals who can likely be trusted. Quality pre-approach research focuses on a consumer or organization and the environmental forces at work in the consumer's life or the organization's industry. Developing a deep understanding of the prospect's situation, before ever meeting with the prospect, typically results in success because it allows the salesperson to enter the next stage armed with insightful questions and ideas.

Step 3: The Approach

The **approach** step includes the initial meeting between the salesperson and the prospect. During the approach, the sales professional meets and greets the prospect, provides an introduction, establishes a rapport that sets a foundation for the relationship, and asks open-ended questions to learn more about the prospect and his or her needs and wants. Ultimately, the customer must be convinced that the salesperson is offering something of value. Again, the success of the approach depends largely on the level of preparation the salesperson has achieved in the pre-approach step.

Step 4: The Presentation

After the approach, the salesperson should be prepared to present the product's major features, describe its strengths, and detail how it will improve the business or life of the potential customer. In any situation, the **sales presentation** provides a forum to convey the organization's marketing message to the prospect by doing the following:

1. **Explain the value proposition.** The sales presentation should make it clear what value the product holds for the customer.

2. **Assert the advantages and benefits of the product.** Each sales presentation should clearly state the advantages and benefits of the product relative to competing products or not purchasing a product at all.

3. **Enhance the customer's knowledge of the company and product.** Customers want to do business with organizations they like and trust. In addition to providing important information about the product, the sales presentation should also reinforce why the organization will be a good partner to the customer.

4. **Create a memorable experience.** Salespeople should spend time thinking about what they want the customer to remember about the presentation. Customers will use these memories as they make purchase decisions, so focusing on key words, phrases, or images during the presentation can be critical.

Planning and preparation lead to successful sales presentations; however, salespeople should not underestimate the value of active listening during the presentation. **Active listening** is a skill that salespeople can develop to improve how they fully concentrate, understand, respond, and remember what is said during a sales meeting. Talking too much can signal a lack of interest in the prospect's needs and wants.

In addition, presentations must be prepared with an understanding of the value of the customer's time. They should quickly and efficiently link the firm's goods, services, and ideas to solutions that help the customer.

Step 5: Handling Objections

Objections are the concerns or reasons potential customers offer for not buying a product. Objections can be an opportunity to clarify and reassure the customer about pricing, features, and other potential issues. Handling objections requires professionalism, strong communication skills, and a sincere respect for the prospect's concerns. Salespeople should make sure to validate the prospect's objection, no matter how trivial it might seem, because the prospect finds it important. The ability to answer specific objections is one of the major advantages of personal selling compared to other elements of the promotional mix.

Common techniques for overcoming objections include:

- **Acknowledging the objection.** "Yes, our prices are higher because our product is better." This allows the salesperson an opportunity to stress the benefits of the product. Consumers are willing to pay higher prices provided they've been given a clear reason (higher quality, better safety, more efficient, etc.) for spending more.

- **Postponing.** "We'll discuss the delivery option in a few minutes, but first let me ask about your needs in this area. . . ." Salespeople should postpone addressing objections if the full context of an appropriate answer has not been developed. This strategy works best if the salesperson plans to address an objection shortly. Postponing for too long will frustrate customers and reduce their level of trust.

- **Denial.** "That is not accurate. The truth is. . . ." If a customer mentions something that is completely false, the salesperson should strongly deny the point, but only in a way that is not offensive or insulting to the customer.

Too Expensive

Your Product Is Too Expensive

Often a customer will present a price objection. In these cases, it is important for a salesperson to acknowledge the objection and then offer reasons why the price is exceeded by the value that the product provides.

It Won't Work

Your Product Won't Work

Sometimes an objection relates to an answer later in the sales presentation. In these cases, it usually makes sense to ask for customer agreement to postpone discussing the objection until later, if it is still a concern. However, the salesperson should make sure to address the concern as soon as possible.

Your Competitor Is Better

Your Competitor Is Better

Some customer objections are simply not true. If a firm actually offers better products than competitors, the salesperson needs to deny a claim that his/her competitors are better. A good way to deny (in a nice way) is to provide testimonials and reviews from other customers.

Step 6: Closing the Sale

Closing the sale occurs at the point when the salesperson asks the prospect for the sale. The "close" is often the most difficult part of the personal-selling process because it requires the salesperson to overcome the basic human fear of being rejected. The majority of customers will not take the initiative to close the sale, so the act of asking for the sale is very important to securing it.

Salespeople generally use one of the three major closing strategies that follow:

1. **Summarization close.** "As you have stated, you are seeking a product that will do the following. . . . Our product can sufficiently cover the requirements that you had mentioned. Would you like to finalize an order with us?" The salesperson summarizes the product's benefits and how it meets the customer's needs before asking for the sale.

2. **Trial method close.** "Would you be interested in trying the product for a few days before making a final decision?" The salesperson solicits customer reaction without asking for the sale directly.

3. **Assumptive close.** "What date do you want those products delivered?" The salesperson asks the customer about the parameters of the sale knowing that if the customer responds with a specific date, he or she has decided to make the purchase.

No salesperson can rely on only one closing strategy. Each can be used effectively, depending on the customer and the specific situation. Salespeople who listen closely to the clues given by the customer during the earlier steps of the personal-selling process will be best prepared to select the appropriate closing strategy.

Step 7: Follow-Up

Conventional wisdom says that it costs five times as much to acquire a new customer as to keep an existing one. Also, it is important to follow up with customers to minimize postpurchase dissonance or regret. Because of this, the follow-up stage is critical to creating customer satisfaction and building long-term relationships with customers. It includes the following:

- **Improving customer satisfaction:** If the customer experienced problems with the firm's product, the salesperson can intervene and become a customer advocate to ensure satisfaction.

- **Promoting positive reviews:** A tremendous amount of activity happens after the sale. Most importantly, customers often share their experiences with others. They may talk with a few friends and family, or they may share their thoughts with many through social media. If a salesperson thinks that a customer is satisfied, he or she is wise to ask for referrals of other customers that may be in the market for a similar product or service. Also, many salespeople ask customers to recommend them through social media, which can lead to new customers.

- **Enhancing sales opportunities:** Diligent follow-up can also lead to uncovering new customer needs or wants and securing additional purchases.

Prospecting and Qualifying

Real estate agents at RE/MAX use social media, face-to-face networking, referrals, cold calls, and leads to identify potential customers who might be interested in buying or selling their home.

Pre-Approach

Before meeting with potential customers, real estate agents at RE/MAX may analyze the market in which they will be working in order to provide knowledge and expertise to potential customers.

Approach

A RE/MAX agent typically uses the first meeting with customers to conduct a needs assessment to better understand the customers' wants, needs, and financial parameters.

Presentation

RE/MAX agents next perform a comparative market analysis in which they present customers with options that fit the parameters they discussed in the approach. This may include showing customers different properties within a given market.

Handling Objections

For a real estate agent, objections are part of the process, as it helps agents understand what customers like and don't like. Agents often spend considerable time with clients showing them different homes until they refine the search and no longer encounter as many objections.

Closing the Sale

For RE/MAX agents, closing the sale may occur quickly or slowly, largely depending on the situation of the customer and the type of market in which they are operating (buyer vs. seller market, for example). In all cases, a sale is unlikely to close until all customer concerns are answered and only after a customer decides that the property meets their wants and needs within their financial parameters.

Follow-Up

Follow-up for real estate agents would usually consist of checking in to be sure the customers are happy in their home and possibly presenting their customers with some token of appreciation for their business.

Major Functions of Sales Management

Sales management refers to a manager's responsibility for formulation and implementation of a sales plan used to deploy salespeople to interact with customers. Functions include the sales force type and size, the type of salesperson, sales territory management, and sales goal setting. The elements of the sales formulation should be considered across the needs of the entire sales organization because altering one or more element may have ramifications on the remaining elements. Let's look at each function in more detail.

Sales Force Type

Most sales organizations structure their sales force around **captive salespeople** who work directly for the firm. This structure is similar to almost all other work environments, where employees are "on the payroll" and are directly subject to the rules, procedures, and policies of the company. However, sales managers can also structure their sales force around **independent sales representatives** who do not work directly for the firm. These reps, also known as **manufacturer's reps,** sell products for many companies on a contractual basis.

Sales managers typically prefer to utilize captive salespeople because the salesperson has a sole focus on selling the company's products, and it is also easier to control how the salesperson conducts his or her sales activities. However, in some cases, for example where a firm sells a small amount of expensive products across a wide geography, the use of a manufacturer's rep makes more sense as a sales solution.

Salesperson Types

There are three main types of sales roles within a firm's sales force:

1. **Inside salespeople** sell to customers from within the organization. Firms often use inside salespeople to process repeat business, especially when products are relatively simple. Typically inside salespeople work from an office, or call center, and make outbound calls to customers or receive inbound calls from customers

2. **Outside salespeople** visit customers in the field for selling purposes. This type of sales role is most common because it involves a salesperson taking the product and message to a customer. Because salespeople are oriented to customer needs, most salespeople travel to the offices of their clients rather than having their clients come to them. In the case of RE/MAX, agents often visit customers in their homes and make sales by showing them different properties that are available to them.

3. A **sales team** consists of a group of salespeople and other employees who act collectively to sell to a customer. In many cases, creating custom product solutions for customers can be complex. Selling situations may require technical, logistical, or financial expertise that salespeople cannot provide. In these cases, a sales team can be formed to bring together the required expertise to develop solutions to customer problems.

Many companies utilize all three types of salespeople.

Defining Sales Goals

Sales goals are important to sales managers and the overall organization. The revenue generated by selling products is the engine that drives the activities of most firms. Thus, setting realistic sales expectations, and then attaining them, has an impact on all functions of the firm. Sales goals are the basis of firm planning, which includes such things as forecasting production levels, employment needs, logistics systems, and capital investments. Specific to sales, revenue and growth goals help sales managers determine many elements of their sales plan, including salesperson quotas, compensation, and sales force size.

Sales Force Size

The size of the sales force depends on the formulated goals of the sales organization. For example, if a company that sells janitorial products (cleaners, paper products, etc.) sets a sales goal of $15 million for a new sales area, it can work backward to determine how many salespeople it requires. Based on historical sales data, the company can estimate that each account will buy an average of $100,000 of supplies a year and each salesperson can effectively serve an average of 20 customer accounts. In this scenario, the firm would need to employ seven to eight salespeople (7.5 to be exact). This is a simple example, but it represents how managers use sales forecasts based on past performance to determine how many salespeople should be needed to meet the sales office's goals.

Sales Territory Management

Once a sales manager determines the number of salespeople to employ, the next step is to determine an effective way to utilize those salespeople. Sales managers typically organize salespeople based on some sort of territory strategy. Defining what territories the salespeople will serve is often accomplished in one of four ways.

Geographic Approach

A **geographic approach** typically places salespeople within an exclusive territory. Expenses due to travel are minimized with this strategy, but salespeople are expected to be experts in all products and are required to sell to all different types of customers within a particular territory.

Product Approach

A **product approach** involves salespeople who sell a particular product or set of products. With a product strategy, salespeople often have demonstrated expertise in the products that they sell though they do not have individual territories. As a result, it is possible for the same customer to receive sales calls from different salespeople when the customer is interested in purchasing different products from the same company.

Customer Approach

A **customer approach** organizes the territory around the customer. The sales manager may divide customers based on size, amount spent, or type (e.g., consumer or business; end user or distributor; government or institution). Key accounts, which are those accounts that have been identified by the company as critical to the organization's success, often arise within a customer-based strategy.

Combination Approach

A **combination approach** utilizes multiple types of territory strategies to organize the sales force. For instance, companies may organize salespeople via a customer strategy with some members dedicated to key accounts and others to standard accounts. Then, within standard accounts, the sales force may be organized by territory or by product. This approach is more complex, but it can also aid companies to extract the main benefits of the three territory approaches.

Recruiting, Selecting, and Training Salespeople

The recruitment of salespeople is a vital part of implementing the planned sales program. This is especially true because the turnover rate in sales is higher than in many other areas and can cost companies a lot in terms of money, lost opportunity, and damaged relationships. Thus, sales managers are almost always recruiting salespeople. Most sales managers keep an active file of potential job candidates and use networking events to evaluate potential salesperson candidates who may be available now or in the future (perhaps especially if they are currently employed by a competitor).

While losing salespeople to turnover is expensive, hiring and training new salespeople is also expensive. Some estimates suggest that the costs to hire and train a salesperson can exceed $100,000 in the first year, which is typically a time when the salesperson doesn't produce much revenue. So it is important for sales managers to hire and train the right salesperson. To improve the likelihood that the salesperson is the right fit, sales managers consider several key characteristics that have been shown to relate to success in sales. As described earlier, characteristics such as a good personality, an optimistic outlook, resilience, a customer focus, an ability to communicate, and adaptiveness all contribute to success in sales.

For newly hired and experienced salespeople, training is a key element of sustained success. Because customer demands, company goals, and products change frequently, training is used to keep salespeople current and well informed. According to a report by *Brainshark,* most sales companies spend from $1,500 to $5,000 in training per existing salesperson every year. This investment is only a start because training takes salespeople away from selling—often a week of training may result in short-term decreases in sales. However, training is a priority for successful firms because, in the long term, keeping salespeople up to date on new products, market conditions, and technology pays dividends in sustainable sales growth.

Motivating, Compensating, and Evaluating Salespeople

Once a strategic sales program has been determined, the sales manager must effectively implement the program. **Sales implementation** involves designing a compensation and motivation system for salespeople that drives firm goal attainment.

Compensation design is important because salespeople often strive to achieve financial rewards for their effort. The components of a sales compensation plan include salary, incentives, and a sales expense account. **Salary** is the fixed portion of a salesperson's income whereas **incentives** are the variable portion. Because different sales efforts drive different levels of sales success, in most cases salesperson income is variable. The variable portion of income includes commissions, quota/bonus programs, and sales contest rewards. These variable components of sales force pay are used to incentivize and reward above-average sales, while limiting pay for those who do not devote as much effort to selling the firm's products.

Sales managers are responsible for motivating their sales force. Motivation can be **intrinsic** (an innate desire or interest in a task) or **extrinsic** (external rewards or consequences associated with a task). Sales managers motivate by encouraging activities that fulfill the intrinsic desires of salespeople. For example, allowing salespeople freedom and autonomy to manage their time allows them to be creative and engaged in how they sell. Extrinsic rewards also motivate salespeople. To stimulate motivation through extrinsic means, sales managers utilize incentive programs, promotions, and social recognition.

Sales Control and Assessment

The process of ensuring that sales achievements are consistent with the objectives of the organization is called **sales control.** To understand what areas need to be controlled, sales managers engage in **sales assessment,** which is the act of evaluating the sales performance of an organization and its sales force.

For example, sales managers may need to reassess their sales plan or implementation, redistribute the sales force, or engage in sales training. As a result, sales managers should be comfortable in understanding quantitative and qualitative assessments of sales performance. At a higher level, sales managers should understand how specific sales numbers affect the company's bottom line, and at a lower level, sales managers need to know which members of the sales force are and are not performing according to expectations. Sales managers can try to improve the performance of those who are performing below expectations through training or motivational practices.

Responsible Marketing

Personal Selling and Ethics

Unfortunately, salespeople and the sales profession often have a reputation as being pushy or, even worse, unethical. In reality, personal selling is a critical function for many businesses because it is one of the only ways to truly build personal relationships with customers. Although sales is a noble profession, and a great revenue generator for a firm, some salespeople still engage in unethical practices to make a sale.

One example involves offering a free trial and charging customers at the end of the trial, without being explicit about these intentions. In 2018, employees of a firm accused their employer of using this practice. The company representatives were directed to tell customers that they could try a product free for 30 days and that the company would cancel their trials before they converted to paid subscriptions. However, in many instances, the trials were never cancelled, and customers had to pay a subscription fee.

Another unfortunate behavior that sometimes occurs in sales is stereotyping. Stereotypes are usually related to a person's gender identity, race and ethnicity, socioeconomic status, language, age, and other factors. Some salespeople may use stereotypes to predict who might be a good (or bad) prospect. While this may not be malicious or intentional, it can still be damaging to customers.

Reflection Questions

1. Have you ever felt like a salesperson was behaving unethically toward you? How did you respond?

2. What steps can sales professionals take to ensure they behave ethically?

3. What steps can companies take to train sales professionals to behave ethically? What consequences could companies implement if employees fail to behave ethically?

Customer Relationship Management

Retaining—keeping—customers is one of the main objectives of all marketers. It is well known that gaining a new customer is much costlier than keeping an existing customer. Thus, marketers focus a lot of energy on keeping customers purchasing existing products and expanding customer purchases into new, or frequent, product purchases. As we discuss in this lesson, because of its importance, marketers use customer relationship management processes and technologies in a proactive and strategic way to make sure that customer relationships endure.

By the end of this lesson you will be able to

- Define customer relationship management.
- Identify strategies for creating value for customers and capturing value from customers in return.
- Describe the customer relationship management process.
- Describe the security and ethical issues involved in using customer relationship management systems.
- Describe how companies can judge the effectiveness of their customer relationship management efforts.
- Explain the importance of data in customer relationship management.

Marketing Analytics Implications

- The success of any customer relationship management (CRM) system is the quality of the data entered into it and the quality of the firm's data capture and collection strategy.
- Once the data capture and collection strategy is in place, the firm must then develop a strategy for how it will use the data to address important research questions.
- Data security must be ensured at the collection, maintenance, and analysis phases of the CRM strategy.

A Better Shave?: How the Dollar Shave Club Used CRM to Create a New Industry

Have you seen the Dollar Shave Club (DSC) ads on your social media? If you're a man (women are welcome, too), chances are that you have because DSC uses social media as a basis of its customer acquisition strategy.

It all began in 2012, when the start-up created a short YouTube video (**https://www.youtube.com/watch?v=ZUG9qYTJMsI**) that went viral and quickly drew new customers to the "club." But earning customers isn't where the story ends. The

In a 2012 YouTube video that went viral, Dollar Shave Club founder Michael Durbin introduced a new way for men to buy grooming products. The "club's" product is home delivery of quality grooming products purchased via a subscription.

real secret of DSC's success is how it *keeps* its customers. Keeping customers has fueled the club's amazing growth in a previously brick-and-mortar-dominated market. In fact, in four short years, DSC went from a start-up business to a $1 billion all-cash acquisition for consumer product giant Unilever.

The strength behind DSC's customer retention strategy is technology. DSC is a big believer in using customer relationship management (CRM) technology and customer data to keep customers happy, engaged, and purchasing. DSC uses a powerful system to integrate data from several sources, known as *touchpoints,* including CRM, customer support, and data analytics to ensure that it has a rich understanding of its members.

Armed with a deep customer understanding, DSC is able to deliver an outstanding and personalized customer experience through products and other value-added features, such as news and information blogs. "We don't respond to situations; we respond to people" is the philosophy that drives DSC's member engagement. It is estimated that DSC has more than 3 million active subscribers, who not only enjoy the brand but also participate in a great relationship with the company. In the end, it is managing relationships and using data that are key to DSC's success.

The concept of building strong customer relationships is not new. We are all familiar with loyalty clubs, such as air miles, hotel points, coffee stars, and more. However, DSC is different. DSC is a subscription service, where the customer is "somewhat" committed to be a repeat customer yet has many options to discontinue the subscription. Marketers of firms that utilize a subscription model, like DSC, have realized that managing relationships is much more important than just encouraging repeat business. For subscription firms, retaining customers is arguably more important than finding new ones (especially over the long term). Thus, knowing exactly what customers want, creating value, and offering an experience that enhances customer retention is critical to a firm's survival.

What's next for DSC? Dollar Shave Club wants to be an online shop where "guys—a term more relatable and less clinical than 'men'—can buy everything they need in a bathroom to look, smell, and feel their best."

Customer Relationship Management

Retaining—keeping—customers is one of the main objectives of all marketers. It is well known that gaining a new customer is much costlier than keeping an existing customer. Thus, marketers focus a lot of energy on keeping customers purchasing existing products and expanding customer purchases into new, or frequent, product purchases. As we discuss in this lesson, because of its importance, marketers use customer relationship management processes and technologies in a proactive and strategic way to make sure that customer relationships endure.

By the end of this lesson you will be able to

- Define customer relationship management.
- Identify strategies for creating value for customers and capturing value from customers in return.
- Describe the customer relationship management process.
- Describe the security and ethical issues involved in using customer relationship management systems.
- Describe how companies can judge the effectiveness of their customer relationship management efforts.
- Explain the importance of data in customer relationship management.

Marketing Analytics Implications

- The success of any customer relationship management (CRM) system is the quality of the data entered into it and the quality of the firm's data capture and collection strategy.
- Once the data capture and collection strategy is in place, the firm must then develop a strategy for how it will use the data to address important research questions.
- Data security must be ensured at the collection, maintenance, and analysis phases of the CRM strategy.

Defining Customer Relationship Management

Companies interested in improving customer relationships and empowering their employees to support that effort often formalize the process by making customer relationship management a large part of their marketing strategy. **Customer relationship management (CRM)** is the process by which companies get new customers, keep the customers they already have, and grow the business by increasing their share of customers' purchases.

CRM is an overall strategy that unifies all of a company's activities under the overarching goal of achieving customer satisfaction through the right actions, attitudes, and systems. Companies that adopt CRM use data to understand customer needs and wants. Based on that understanding, they respond to and anticipate customer expectations in a way that delivers value to the customer. These activities, if done well, can foster favorable impressions of the company and its goods or services.

Creating and Capturing Value

The goal of CRM is to facilitate the creation of value for customers and to capture value from customers. At its essence, the first strategic way that CRM is used is as a formalized process that helps marketers understand customer needs and wants. As you may recall from the lesson on value creation, customer value starts with an understanding of consumer wants and needs. CRM software aids in this process by capturing consumer data, storing those data, and predictively using those data to understand customer needs.

As consumers increasingly engage with firms across mobile, Internet, and in-person interactions, companies are able to track touchpoints. A **touchpoint** is any point at which a customer and the company come into contact. Touchpoints are a basis for understanding customer needs at the individual customer level. These various touchpoints help analysts to create and maintain customer profiles, including buying habits and purchasing patterns. This enables the firm to develop tactics for building a long-term, mutually beneficial relationship with the customer. CRM technologies support this effort by allowing marketers to perform the following value creating strategies:

- Track consumer behavior over time.
- Take the information gathered and tailor goods and services accordingly.
- Capture data that allow the firm to identify customers who are likely to be profitable.
- Interact with customers to learn what they need and want.

Salesforce.com is a leading CRM software system that is used to track all aspects of a customer's relationship with a firm. Salesforce compiles data on the buying habits, potential future needs, and even wants that may not yet be needs.

The Internet, in particular, has proven invaluable in allowing companies to collect individualized data to use to send personalized messages that market directly to individual consumers. For example, HelloFresh delivers its customers personalized weekly recipes and all the ingredients for those recipes based on customers' preferences. These targeted recipes are derived by sophisticated analysis of CRM data and improve as customers continue to build touchpoints with the firm.

HelloFresh strategically uses customer shopping preferences to help suggest future product purchases—in this case, healthy recipe options—across demographics.

In another example, retail giant Target tracks its customers' shopping patterns and sends them sales offers that complement their previous purchases. This can even allow Target to promote baby supplies to women based on purchases of unscented lotion and vitamin supplements. In this case, sometimes CRM can have a "creepy" side, as Target can often guess things about its customers that its customers may not be sharing with others, like being pregnant. In similar fashion, companies such as Amazon, Facebook, Google, and Spotify use CRM strategically to create value by knowing what consumers want or need—even before the consumers realize they want or need it.

It follows that companies that strategically use CRM also capture value. CRM systems allow marketers to capture value by offering superior service and attention to customer desires, which allows them to establish and maintain valuable relationships.

Relationships based on customer value creation should capture value for the firm by being profitable and enduring in the long term. Using CRM, marketers can readily develop a customer orientation because they focus on gathering data that allows them to understand the needs of individual customers rather than an entire market segment, as in a traditional marketing approach. This narrower focus allows companies to obtain maximum profits from customers they already have rather than spending time and money prospecting for new customers.

In-Person

A traditional touchpoint is an in-store interaction. Stores like Abercrombie & Fitch (AF) keep good records of what their customers buy and are shopping for in their stores.

Website

The AF website encourages users to log in to their AF account. Once logged in, AF's CRM system records touchpoints for all items that the consumer looks at.

Mobile

AF's mobile app creates many touchpoints, including location, usage, product views, and frequency of use.

Internal Information

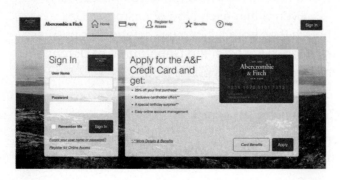

There are many more touchpoints that AF tracks. Payment status, timeliness, and usage are all touchpoints that help to complete a picture of consumer buying habits.

The Customer Relationship Management Process

The CRM process revolves around a cycle of activities. CRM is an iterative rather than a linear process: the firm repeats the sequence of steps as necessary to reach the desired result.

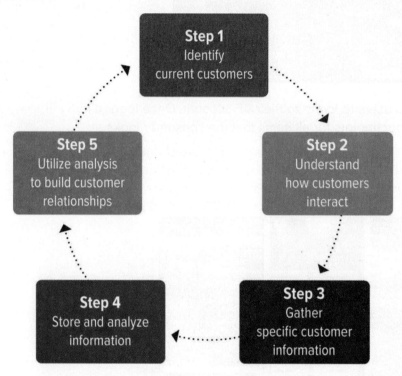

Let's take a closer look at each step in the process.

Step 1: Identify Current Customers

The company initiates the CRM process by identifying its current customers. For example, a company like Dell that has both business customers and individual customers would identify its customers by finding out their locations, breaking down computer purchases by customer type, quantifying the frequency of both individual and business purchases, and determining how many computers each type of customer typically purchases. Even within each type of customer, Dell would want to note distinctions. Though they both fall into the category of business customer, Arkansas State University might differ from the Department of Defense when it comes to the type and number of computers purchased.

Step 2: Understand How Customers Interact

Next, the company seeks to understand how customers interact with the firm—that is, how they purchase and communicate with the firm (e.g., via the Internet, in brick-

and-mortar stores, through a salesperson). Interaction between the company and its customers serves as the basis for a CRM system.

For example, Dell has several channels through which it sells computers to individuals. It would probably want to know which types of customers are more likely to order a computer on the company's website, and which would prefer to go to a retail store to buy one. Interactions can take many forms: phone calls and e-mails to customer service, conversations with salespeople, purchases, questionnaire responses, coupon redemption, requests for information, and repair or product return requests.

Step 3: Gather Specific Customer Information

Previous steps involved gathering general customer information. The next step is to gather specific information on each individual customer's touchpoints with the company, such as website visits, purchase history, use of coupons or promotional codes when purchasing a good or service, warranty card submissions, point-of-sale data, customer inquiries, or any other time the customer has had contact with the company in any way.

A salesperson serves as an excellent source of information about in-person customer interactions. He or she can record information, including the customer's contact information and good or service preferences into a CRM system.

Internet interactions in which a customer goes on a company's website for information, purchases goods or services, or provides feedback on a good or service are an increasingly popular way for companies to gather information. For example, Dell uses software called Dell Premier to track purchase information. In a consumer example, Starbucks's app tracks customers' purchases and rewards customers for repeat purchasing. The app allows in-store purchase information to be collected at checkout and tagged directly to the customer.

Most recently, advances in facial recognition are being harnessed as a CRM tool that identifies and tracks individual customer in-store movement. It is currently unknown how customers will view intrusive technologies, like facial recognition, but for marketers, they promise to help improve CRM data on individual customers.

Step 4: Store and Analyze Information

After the company has gathered the appropriate data, it must store the data so that CRM data analysis applications can access them. CRM databases store the information for individual or business customers, depending on the needs of the firm. The information input into the CRM database must be accurate if the company hopes to use it later to take action that will create and maintain satisfied, profitable, long-term customers.

The analysis that leads to such action is accomplished through data mining techniques within the CRM system. **Data mining** is a process that involves the computerized search for meaningful trends in a large amount of data. Using data mining software such as ProClarity, marketers at Dell, for example, can search for relevant data, organize the data based on select criteria, and create customer profiles that can be used to analyze customers.

Companies rely on four different techniques to analyze data: **customer segmentation analysis, recency-frequency-monetary analysis, lifetime value (LTV) analysis,** and **predictive modeling.**

Customer Segmentation

Customer segmentation analysis involves creating customer profiles based on demographic characteristics, purchase patterns, and other criteria and placing them into various categories.

Recency-Frequency-Monetary

Recency-frequency-monetary analysis involves categorizing customers by their buying patterns, such as how recently they have purchased a good or service, how often they purchase from the company, and how much money they spend on the company's products. Based on this analysis, the system ranks customers according to how profitable they are (or their profitability potential) so the firm can target them for marketing efforts.

Lifetime Value (LTV)

Lifetime value analysis allows the company to monitor the actual costs of doing business with customers to ensure that it is focusing on the most profitable customers. In lifetime value analysis, the system compares the costs of retaining customers with the costs of acquiring new customers to determine how much money each type of customer requires. With this analysis in hand, a company can predict how valuable a customer will be over time. The system can also help a company identify potential customers on whom it may be worth spending money to develop a relationship.

Predictive Modeling

Predictive modeling uses algorithms to try to determine the future actions of customers. Based on patterns of previous buying behavior, the system attempts to predict how customers will act. For example, Norwegian Cruise Lines might use its CRM system to predict whether a customer will purchase a cruise in the future based on the timing and frequency of previous cruise purchases. The company could use the results of its predictive modeling to identify the customers who are likely to purchase a cruise in the near future, enabling marketers to focus on them for promotional activities.

Step 5: Utilize Analysis to Build Customer Relationships

The fifth step in the CRM process is to utilize the information gathered and analyzed in the previous steps to build customer relationships. At this stage, information is sent to functional areas within the company, like sales and marketing, that then use it to customize their activities to target specific customers.

Norwegian Cruise Lines, for example, may offer a discount or cabin upgrade to customers who have frequented its cruises. We'll discuss these types of tailored examples, as well as additional ways a firm can leverage the information obtained through the CRM process, later in this lesson.

Tailored Customer Promotions

Perhaps the most obvious use of CRM data is to **tailor promotions** to match customer profiles.

For example, Kroger offers discounts on gasoline purchases for members of its loyalty card program. Based on past purchases and the customer's receipt, Kroger generates discount coupons to send to customers, creating an incentive for the customer to return to the store.

While repeat customers might be targeted for such loyalty programs, infrequent buyers could receive different incentives, such as a coupon good for their next purchase.

Reduce Cognitive Dissonance

Firms also can use CRM information to **reduce cognitive dissonance** (buyer's remorse) by congratulating the buyer on his or her choice and reinforcing the best aspects of the good or service.

For example, a car manufacturer could send an email to customers congratulating them for purchasing a car that was rated the highest in its class for initial customer satisfaction by J.D. Power. These efforts help to reassure buyers that they made a good purchase decision.

Improve B2B Relationships

Firms that sell to other businesses can use CRM systems to **improve business-to-business relationships**. They may cross-sell their products (promoting other products they sell that might be purchased by an existing business customer), track customer service complaints and returned goods, and tailor promotional programs to specific business customers based on classification.

Beyond this, CRM systems offer several unique advantages to B2B firms. Suppliers can generate sales forecasts from information in CRM systems. The system also can provide product availability information to manufacturers as they establish production runs or to retailers as they plan sales.

The Ethics of Customer Relationship Management Systems

Not all customers feel comfortable having their information accumulated and stored in a company's computer system. The issue of privacy is becoming increasingly acute as instances of hacking into computers to steal personal information have been made public and customers realize that any computer security system that stores birthdates, credit card numbers, addresses, and other personal information can be breached. For example, firms such as Target, Equifax, Uber, and Yahoo were all recently hacked, exposing data on millions of customer accounts to potential criminal activity.

Additionally, because companies can capitalize on information about a customer's buying habits and product preferences, among other things, a general discomfort with invasion of privacy has become widespread. Such information could be sold to or traded between companies. Because good customer relationships are built in part on trust, any doubt about the security of personal data can lead customers to take their business elsewhere.

To guard against a breach of security, CRM systems must have robust firewalls to protect the privacy of customers. While all CRM applications are vulnerable to security breaches, companies using cloud computing CRM applications should understand the increased risks involved in entrusting data to a third party and require the third-party vendor to employ protective measures to discourage and prevent data hacking. Many systems require a multi-step login system using a system such as DUO, which is a two-way verification system.

The U.S. government has put in place laws for protecting the financial, health, telephone, and e-mail information of citizens. Still, companies should supplement these laws with policies that govern how the company can collect information, how it can use the information it collects, whether it may share the information with other companies, and how it will protect the information. Companies interested in protecting the privacy of their customer information must therefore make a concerted effort not only to develop such policies but also to make sure they are followed. Failure to do so could lead consumers to lose trust in the company and, as a result, switch to a competitor.

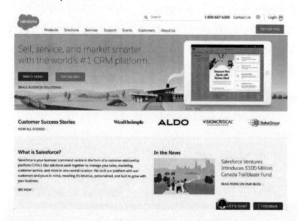

Safeguarding the privacy and trust of a company's customers requires constant vigilance and regular upgrading of security systems. Salesforce.com is a cloud-based CRM system that stores incredible amounts of sensitive data regarding customers.

Effective Customer Relationship Management

How can a company determine whether its CRM strategy is working? There are four basic criteria a company can use to judge the effectiveness of its CRM program:

1. **Share of customer. Share of customer** differs from market share in that it measures the quantity of purchase dollars each customer spends on the

company's products rather than the number of customers. If a company's CRM efforts lead to an increase in the number of goods or services purchased by a consumer, then it has been successful in increasing its share of customer. As share of customer increases, so do the company's profits. If the CRM efforts do not increase the company's share of customer, the company must evaluate them to determine why and apply corrective measures. For example, if a customer buys a Nissan Altima, the company can use CRM data to guide contacts that lead to service appointment, warranty extensions, and future purchases.

2. **Customer equity. Customer equity** is a ratio that compares the financial investments a company puts into gaining and keeping customers to the financial return on those investments. A company can determine the value of its CRM program from this ratio. If a company determines that it is spending more on CRM than it is getting back in profit, it needs to evaluate the program and correct the problems. For example, if a salesperson neglects to follow up with B2B customers to offer postsales service that other companies do not offer, the company has given up the chance to increase the financial return on its investment in those customers.

3. **Customer focus. Customer focus** measures how well a CRM program prioritizes customers based on each customer's profitability. A CRM program that enables marketers to identify and focus on highly profitable customers will likely yield more profit for a company than one that only allows a company to communicate equally to all customers. If a CRM system can properly identify highly profitable customers, the company can direct its selling efforts only to those customers who are or are likely to become profitable in the future.

4. **Lifetime value. Lifetime value (LTV)** is measured by the total profit a customer brings to a company during the time that the individual or firm is a customer. The CRM efforts of a company, if done right, should be able to maximize the LTV of customers large and small. Companies that can predict which customers will generate the most profit over a long period of time can eliminate or reduce services to customers with low LTV, thereby reducing customer service costs while maintaining or even increasing services to customers with high LTV.

Understanding the criteria for measuring the effectiveness of a CRM strategy is important because, unless a firm knows the extent to which any aspect of its business is succeeding, it cannot identify and address problems that may affect its profitability. Successful CRM programs require monitoring and assessment to adequately support the firm's overall marketing efforts.

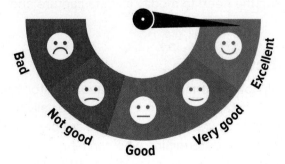

CRM systems offer many indicators of how well CRM systems are working. This simple "health score" scale pulls together CRM information from a variety of measurement areas to provide firms a quick way to check if an individual customer is happy or at risk.

Marketing Analytics: The Importance of Data in Customer Relationship Management

Gathering Quality Data for Successful CRM

Successful customer relationship management (CRM) depends on quality data, and the type of data collected will vary with each firm. Often the task of inputting customer data will be assigned to the sales force because it has the most direct contact with customers. However, inputting the data is just the first step in successful CRM; those data also need to be regularly refreshed to ensure they are accurate and as up-to-date as possible. CRM systems will hold both qualitative data, such as personal and demographic information, as well as quantitative data, such as order history, value of sales, and average order. For the firm to ensure that it has the data necessary for analysis, it must have a data capture and collection strategy in place. Quality data combined with a quality data collection strategy are the key to a high-performing CRM system.

Using CRM Data to Address Research Questions

Once the data capture and collection process is complete and the data are entered into the CRM system, marketers can then use the data to address a wide variety of research questions. For example, if a restaurant chain wants to know which of its servers are most successful at upselling, its marketers could query the CRM system to search across customer bills. They could then organize those search results by restaurant and by server, making it easier for marketers to identify upselling patterns among the top upselling restaurants and servers. Perhaps the data reveal that high-profit items such as bottles of wine or dessert are upsold most often; this information would help the restaurant chain better understand how its top-upselling servers were trained, which would then potentially inform the training strategy of other servers in the chain that are not upselling as well. The use of CRM data to address research questions enables firms to identify what, if any, adjustments might be necessary for improving a product or service; but, if the data do not yield useful answers, this could signal to a firm the need to modify its data capture and collection process.

Ensuring the Security of CRM Data

Given that CRM systems contain a wealth of personal information about customers and employees, it is imperative that these data be protected. This is especially true in sensitive industries such as pharmaceuticals where health records are protected by law. Firms must exercise extreme caution in determining who can be permitted access to these data; they must also monitor who is accessing the most sensitive data. In firms where clients are competitors, CRM data must be "firewalled," meaning that employees working on one client's account cannot access data about another client. Security protocols must be in place and employees must be trained about the importance of data security and client privacy.

Responsible Marketing

Protecting Data: Who Should Be Responsible?

As you learned in this lesson, customer relationship management (CRM) is the process by which companies attract and retain customers. Customer data are critical to the successful implementation of CRM strategies; the more data a company has, the more it can learn about its customers. Robust data yield stronger relationships and can enable more personalized communication. In a world where every customer wants customization, data are key.

While obtaining and interpreting customer data is critical to success in today's competitive environment, companies must also consider the privacy concerns of its customers. In January 2021, men's clothing retailer Bonobos was a victim of a cyberattack. A hacker downloaded all of Bonobos's customer data, including sales history, addresses, phone numbers, and partial credit card information. The attacker then posted the full Bonobos database on a free hacker forum. Bonobos alerted consumers via e-mail that their information may have been compromised. Bonobos also worked with its security provider to stop the attack as soon as it was aware of it, and it worked to tighten security measures in the future.

Reflection Thoughts

1. How much responsibility do consumers have in protecting their own data?

2. What additional measures should Bonobos have undertaken to ease consumer concerns?

3. Do the benefits of aggregating customer data and developing customization strategies outweigh the consequences of data breaches and violations? Consider both the brand and consumer perspectives.

Communicating Value: Test

1. All the activities that communicate the value of a product and persuade customers to buy it is referred to as
   A. marketing.
   B. promotion.
   C. positioning.
   D. advertising.
   E. salesmanship.

2. The marketing mix element where most of an organization's communications with the marketplace occurs is
   A. price.
   B. place.
   C. product.
   D. promotion.
   E. position.

3. Which of the following is not an element of the promotion mix?
   A. advertising
   B. sales promotion
   C. social media
   D. personal selling
   E. public relations

4. As a promotion mix element, public relations refers to
   A. nonpersonal promotional communication about a firm's goods and services.
   B. a set of nonpersonal communication tools designed to stimulate more frequent purchases of a product.
   C. advertising focused on promoting the company's image with its customers.
   D. communication focused on promoting positive relations between a firm and its stakeholders.
   E. the two-way flow of personal communication between the firm and its customers.

5. Mars, Inc., makers of Snickers candy bars launched its You're Not You When You're Hungry promotion campaign. The company's television commercials portrayed regular people disguised as celebrities behaving badly because they were hungry. Along with the commercials, print ads were released featuring inversion illusions showing the difference between someone whose hunger is satisfied and someone whose hunger is clearly not. In addition, the print ads included a coupon for a Snickers candy bar. This type of promotional strategy is referred to as
   A. an integrated promotion mix.
   B. an integrated marketing concept.
   C. an integrated advertising mix.
   D. integrated promotion communications.
   E. integrated marketing communications.

6. Which of the following is a reason for using an integrated marketing communications strategy?
   A. Marketers using a coordinated message make it easy for customers to recognize their messages, regardless of medium.
   B. Marketers have discontinued the use of newspaper advertising because of its lack of popularity.
   C. Marketers have evolved from segmented strategies to a single concentrated message focusing on only one element of production.
   D. Marketers have increased their advertisement spending for promotional techniques that generate gradual sales responses.
   E. Marketers rely solely on the social media to integrate their marketing communication.

7. Which of the following promotional tools is a part of advertising?

A. coupon

B. contest

C. rebate

D. billboard banner

E. allowance

8. Mr. Thompson is a professor of Marketing at the local university. In the student orientation program, he addressed the freshmen regarding the advantages of taking Marketing as a major course. Which of the following types of advertising did Mr. Thompson use?

A. informative advertising

B. persuasive advertising

C. personal advertising

D. product placement

E. reminder advertising

9. In the context of Internet advertising, which of the following is true of click-through rate (CTR)?

A. It is the amount a firm pays for a thousand views of its ad.

B. It is the amount a firm pays each time a customer clicks on an advertisement.

C. It is a ratio showing how often people who see an advertisement end up clicking on it.

D. It is a measure of the number of cookies that a firm tracks for each user who visits its website.

E. It is an ineffective indicator of how well a firm's advertisements perform.

10. _____ allow consumers to recoup a specified amount of money after making a single purchase.

A. Allowances

B. Rebates

C. Coupons

D. Contests

E. Sweepstakes

11. Chicago Inc., a leading retail store, has a chain of department stores across the United States. The products available at its outlets include cosmetics, jewelry, clothes, footwear, and home accessories. Chicago Inc. offers its customers a Gold Card to which points are added every time the customer's bill amount exceeds $500. These points can be accumulated and used for purchases at any of its outlets. Which of the following promotion mix elements is the cosmetic company using?

A. allowances

B. rebates

C. coupons

D. contests

E. loyalty programs

12. Which of the following promotion tools commonly involve problems due to misredemption practices?

A. allowances

B. rebates

C. coupons

D. contests

E. loyalty programs

13. Digital marketing can appear as all of the following forms *except*

A. search engine optimization.

B. email marketing.

C. guerrilla marketing.

D. inbound marketing.

E. social media marketing.

14. When you type in "sustainable travel," the website for G Adventures appears first in the list. This is an example of

   A. search engine optimization.

   B. inbound marketing.

   C. social media marketing.

   D. native advertising.

   E. a cookie.

15. Which of the following technologies enables digital marketers to build customer relationships through exceptional customer experiences?

   A. the use of cookies

   B. excellent website design

   C. inbound marketing

   D. geotracking

   E. search engine optimization

16. A digital buyer who is engaged in product-focused shopping is someone who

   A. is searching for the best deal on a pair of running shoes.

   B. is checking out the specifications of different camping equipment.

   C. is trying to locate discounts on groceries.

   D. is looking at guitars on eBay while waiting for an appointment.

   E. has repurchased cat litter from Amazon.

17. Each of the following categories is a type of online purchasing behavior *except*

   A. researching.

   B. product-focused shopping.

   C. bargain hunting.

   D. browsing.

   E. auctioning.

18. All of the following examples are types of invisible marketing influences *except*

   A. bots.

   B. cookies.

   C. cookie synching.

   D. geotracking.

   E. sponsorship.

19. When Eva types "makeup" into Google, the first search item is a sponsored ad for Almay. This is *best* described as

   A. a paid story.

   B. a paid search.

   C. cookie synching.

   D. geotracking.

   E. a sponsorship.

20. Today's advertising fails to represent marketers' products fairly and honestly despite the laws, regulations, and self-regulation put in place enforcing them to do so.

   A. TRUE

   B. FALSE

Appendix:
The Marketing Plan

What To Expect

Often the simplicity of a new idea put into practice, like Uber, makes us think, "I could have come up with that!" It isn't usually the novelty of the idea that makes a product or service successful. Rather, it's the ability to operationalize an idea, in a profitable and repeatable way, that makes the difference between an interesting insight and the next start-up success story. Let's take a look at what a marketing plan should look like.

Chapter Topics

The Marketing Plan

The Marketing Plan

In this lesson, you will learn about the purpose and the components of the marketing plan. The marketing plan is an essential, operational document for almost all firms. Start-up businesses create a marketing plan to set goals and determine the direction of their business. Established businesses use the marketing plan to continually assess and reassess their markets and strategy accordingly.

By the end of this lesson you will be able to

- Explain the purpose of the marketing plan.
- Identify the five key components of an effective marketing plan.
- Describe the three key characteristics of an effective mission statement.
- Define the executive summary.
- Identify the elements of a situational analysis.
- Explain the role of the marketing strategy in a marketing plan.
- Explain the purpose of the financials section of the marketing plan.
- Describe how firms implement and adapt the marketing strategy.

Ever Need a Quick Recharge?: Turning an Idea into a Marketing Plan

The ability to turn a cool idea into a marketable business requires strategic planning. Most ideas progress from a thought into a real business through a marketing plan. For example, Recharge is a new app that focuses on the human need for rest. Have you ever been traveling and thought how great it would be if you could find some place to take a nap? Finding that place can be a challenge, however, especially if you're waiting for a flight at the airport or actually driving in your car. Recharge's app helps users locate and reserve a vacant hotel room that can be rented by the minute at any time of the day or night. Recharge's current target audience is businesspeople traveling to major cities looking for high-end, luxury hotels. However, if the app gains acceptance, as it seems to be doing based on financial backing by the airline JetBlue, the app could be the next Uber—but for hotels.

How is Recharge taking the simple idea of renting vacant hotel space from a concept to a business? Founded in 2015, Recharge started by developing a marketing plan that answered questions such as:

- What market exists?
- What internal, external, and competitive forces exist?
- Who should be included in the target market?
- What marketing strategy should be followed based on the target audience?
- What financial return is expected?
- What areas should be monitored and adapted as the plan progresses?

These questions, and others like them, are the foundation of a marketing plan. Take a moment to review Recharge's FAQ section on its website (**https://recharge.co/faq/**).

Answers to questions concerning the target market, strengths, and strategy are all readily available in the FAQs, which suggests that Recharge is following a well-developed marketing plan.

Will Recharge be the "next big thing"? Only time will tell, but having a solid marketing plan is definitely a step in the right direction toward success.

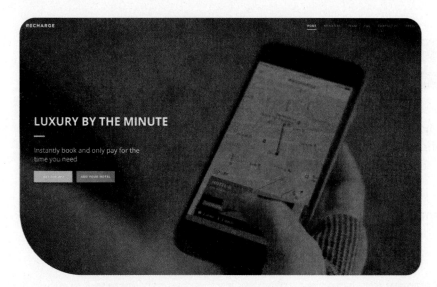

The Recharge app allows weary travelers to find a nearby hotel room, available for rent by the minute, to use for a nap, shower, or meeting.

The Components of a Marketing Plan

The **marketing plan** is a formal document that outlines many aspects of a firm's mission, market position, strategy, financial position, and plan to implement and monitor the marketing strategy in a defined period of time. Creating a marketing plan requires the input, guidance, and review of employees throughout various departments of a firm, not just the marketing department, so it is important that everyone in a firm understand the plan's components and purpose.

The specific format of the marketing plan differs from organization to organization, but most plans include the five sections shown in the figure below:

Key Components of the Marketing Plan

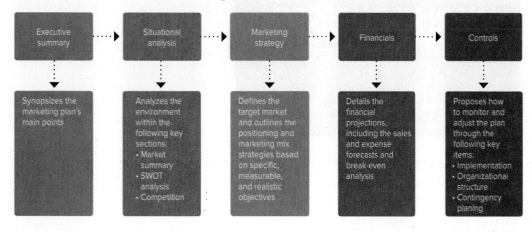

These five sections communicate what the organization wants to accomplish and how it plans to achieve its goals. Each of the components should be grounded in the firm's overall mission, which is ideally defined in a clear and succinct mission statement. Because the mission statement drives marketing strategy, we'll first discuss the characteristics of an effective mission statement and then go into further detail about the specifics of each section of the marketing plan.

The Mission Statement

A **mission statement** is a concise affirmation of the firm's long-term purpose and should appear as early as possible in the marketing plan, usually before the executive summary. An effective mission statement provides employees with a shared sense of ambition, direction, and opportunity.

Mission statements come in many forms. Some are more detailed than others, but each one should communicate what the company thinks is its reason for doing business. Click on the Click & Learn interactive below to read some diverse examples of different mission statements.

Chipotle's Mission Statement

". . . our devotion to seeking out the very best ingredients we can—raised with respect for animals, farmers, and the environment—remains at the core of our commitment to Food With Integrity. And as we've grown, our mission has expanded to ensuring that better food is accessible to everyone."*

Harley-Davidson's Mission Statement

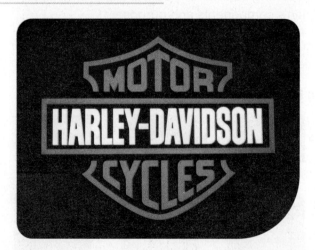

"We fulfill dreams through the experience of motorcycling, by providing to motorcyclists and to the general public an expanding line of motorcycles and branded products and services in selected market segments."

Sherpani's Mission Statement

"Sherpani is a product leader in lifestyle bags for women who define their own meaning of success and who do not conform to just one way of life. Inspired by modern, alpine culture, we are focused on creating unique designs, with smart features; using high quality materials.

We believe that simplicity always works; that nature restores the human spirit and that pure distinctive silhouettes positively affect a women's self confidence."*

A firm should begin the process of developing a mission statement by considering the following classic questions posed by Peter Drucker, who is considered the father of modern management:

1. What is the current state of the business?
   - What is our business?
   - Who is our customer?
   - What is our value to the customer?

2. What is the expected and desired future state of the business?
   - What *will* our business be (where are we headed given our current plans)?
   - What *should* our business be (where should we be headed, if we change our current plans)?

These basic, yet challenging, questions are the most essential a firm will ever have to answer. From there, the firm should focus on following these three guidelines to create an effective mission statement:

1. The mission statement should focus on a limited number of goals.

2. The mission statement should be customer oriented and focused on satisfying basic customer needs and wants.

3. The mission statement should capture a shared purpose and provide motivation for the employees of the firm.

Focus on a limited number of goals.

Companies whose mission statements contain 10 or more goals are typically focusing too much on small, less meaningful objectives rather than creating a broader statement that provides purpose and direction to the entire organization.

Be customer oriented and focus on satisfying customer needs and wants.

Apple has been one of the most successful companies of the past decade because it has designed innovative new products such as the iPod, iPhone, and iPad. These products have been successful because they are designed around the consumer's desire to watch movies, listen to music, and communicate.

Capture a shared purpose and provide motivation for the employees of the firm.

Typically, a well-designed mission statement will be simple and clearly capture the firm's strengths, as Google's does: "Google's mission is to organize the world's information and make it universally accessible and useful." This broad yet simple statement is a powerful example that a mission statement does not need to be complex to be effective.

The Executive Summary

The **executive summary** provides a one- to two-page synopsis of the marketing plan's main points. Because it is so brief, every line of the executive summary should convey the most valuable information contained in the marketing plan.

Depending on the organization's size and objectives, the marketing plan may be viewed by dozens or even hundreds of people. Some will take the time to read each line in the marketing plan, but most are looking for a quick way to understand the basic ideas and strategies behind the plan. The executive summary provides this resource.

Note that although the executive summary is listed first, firms typically draft this portion of the marketing plan last because it is a summary of the overall plan.

The Situational Analysis

A **situational analysis** is the systematic collection of data to identify the trends, conditions, and competitive forces that have the potential to influence the performance of the firm and the choice of appropriate strategies. The situational analysis section is often considered the foundation of a marketing plan because organizations must clearly understand their current situation to make strategic decisions about how to best move forward.

The situational analysis consists of

- A summary of the firm's markets
- A review of the firm's strengths, weaknesses, opportunities, and threats.

To review the firm's markets and market positions, a categorization matrix, such as the *Boston Consulting Group (BCG) Grid,* is used. A formal process of reviewing, defining, and matching strengths, weaknesses, opportunities, and threats, called a *SWOT analysis*, is a systematic approach to understanding internal and external factors important to the firm. We'll examine the BCG Grid and SWOT analysis next.

The Market Summary

The **market summary** describes the current state of the firm's overall market and specific target market(s). The firm's **market** is a broad group of consumers or organizations that is interested in and able to buy a particular product. In addition, the market summary looks at the firm's specific target markets—the group of customers toward which an organization has decided to direct its marketing efforts.

For example, marketers of the Recharge app, described earlier in this chapter, would conduct a market analysis at both a high level (the general travel market) and a specific level (the business and luxury and travel market). For the travel market, information such as air travel, rental car, and hotel occupancy trends would inform a market summary. A quality market summary should provide a perspective on important changes in specific target markets. Recharge would benefit from information on how many business travelers are experiencing long layovers due to changes in the airline industry. Historical trends and future projections of a firm's markets are the foundation for the marketing plan of any firm, as they allow firms to plan with insight into what is expected to be happening when the marketing plan is being implemented.

Business Portfolio Analysis

Firms typically market more than one product or operate in more than one target market. To effectively plan marketing efforts for each, a firm needs to know how each product is doing in relation to the other products the firm sells. Most firms conduct some form of business portfolio analysis to determine how their products are doing currently and to provide guidance for marketing planning. A **business portfolio analysis** is a method of categorizing a firm's products for strategic planning based on competitive position and expected growth rate. A business portfolio analysis is more useful for an existing business with multiple products, and not as effective for a start-up. One popular tool used to conduct a portfolio analysis is the Boston Consulting Group (BCG) matrix. This two-by-two matrix, shown in the figure below, graphically depicts the strength and attractiveness of a market. The vertical axis measures market growth while the horizontal axis measures relative market share. Each quadrant describes a certain type of business condition (star, cash cow, question mark, and dog) for different products of the firm's product portfolio.

The Boston Consulting Group (BCG) Matrix

The quadrants of the BCG matrix can be described as follows:

- *Star products* combine large market share with a high growth rate. Firms with star products generally have to invest heavily in marketing to communicate and deliver value as the industry continues to grow. Marketing efforts around star products focus on maintaining the product's market position as a leader in a growing industry for as long as possible.

- *Cash cows* are products that have a large market share in an industry with low growth rates. A company may decide to allocate only enough marketing resources to keep sales strong without increasing costs or negatively affecting profits.

Relative Market Share

	High	Low
High Market Growth	**STAR** Large market share AND high growth. Marketing efforts focus on maintaining the product's market position as a leader.	**QUESTION MARK** Small market share AND high growth Marketing efforts include significant investment in promotion, product management, and distribution.
Low	**CASH COW** Large market share AND low growth. Marketing efforts allocate only enough marketing resources to keep sales strong without increasing costs or negatively affecting profits.	**DOG** Small market share AND low growth. Marketing efforts should be to discontinue the product.

- *Question marks* have small market share in a high-growth industry. Products in this quadrant are typically new to the market and require significant marketing investment in promotion, product management, and distribution. Question marks have an uncertain future, and marketers must monitor the product's position in the matrix to determine whether they should continue allocating resources to it.

- *Dogs* are products that have small market share in industries with low growth rates. Products that fall into this category typically should be discontinued so the firm can reallocate marketing resources to products with more profit potential.

As part of the market summary, the BCG matrix allows a company to determine where its product will fall in the marketplace and serves as a starting point for developing marketing strategies to address that market position.

STAR: Amazon Echo

Smarthome devices are on the rise and demonstrating high growth rate. These devices are used to listen to music, order products, control home settings, and even order your dinner.

Amazon's Alexa-enabled smarthome devices currently have about 31 percent penetration of the U.S. market with the Google Home behind at about 23 percent of the market.

QUESTION MARK: AmazonFresh

Companies battle for share in the grocery delivery market, which is expected to see growth. Although Amazon initially had a small market share in this area with its AmazonFresh service, its purchase of Whole Foods in 2017 could potentially move it to the "star" product category.

CASH COW: Book Sales

Amazon started business in 1994 as an online bookseller.

Although this market is not growing, the company retains the largest market share of all booksellers, making book sales its cash cow in terms of e-commerce.

DOG: The Kindle Fire

The tablet market has slowed down and shown −5 percent growth from 2018 to 2019.

Apple is the leader in this market with about 30 percent market share; Amazon is a smaller player with less than 9 percent share, which places its Kindle Fire in the "dog" product category.

Amazon Echo: travelbetter.co.uk/Alamy Stock Photo; AmazonFresh: Andriy Blokhin/Shutterstock; AmazonBooks: SEASTOCK/Shutterstock; Amazon Kindle Fire: AlexanderC111/Shutterstock

SWOT Analysis

A **SWOT analysis** is an evaluation of a firm's strengths, weaknesses, opportunities, and threats. A SWOT analysis can be a valuable tool in the development of a marketing plan. Firms use a SWOT analysis to identify internal and external factors that the firm needs to consider in its planning. The ultimate goal of a SWOT analysis is not just to identify factors of the firm and its market, but rather to use this knowledge to match strengths with opportunities in an effort to gain **competitive advantage,** which is the superior position a product enjoys over competing products if consumers believe it has more value than other products in its category. In addition, SWOT identifies weaknesses that can be improved and threats that should be proactively managed.

Internal Considerations

Strengths and weaknesses are items under the control of a firm. Strengths underlie a firm's competitive advantage, whereas weaknesses can deter from a firm's success. Specifically:

- **Strengths** are internal capabilities that help the company achieve its objectives. Strengths are internal considerations because the firm has control over its strengths. Items such as manufacturing capacity, innovation, market share, and customer loyalty are potential strengths that firms can use as they form their marketing strategy.

- **Weaknesses** are internal limitations that may prevent or disrupt the firm's ability to meet its stated objectives. A firm's weaknesses are also in the control of the firm, to some extent. Of course, no firm desires to have weaknesses, but often they are unavoidable. For example, a start-up firm often has financially related weaknesses, such as a small budget for promotion, lack of critical employees, and/or an ineffective manufacturing capacity. Some of these weaknesses can be resolved quickly, whereas others may be a continuing problem as the firm grows. However, identifying weaknesses is important because it keeps the firm focused on areas that require special strategic marketing attention in the immediate to long term.

External Considerations

As described in the lessons on marketing strategy and business environments, many factors facing firms are outside the immediate control of a firm. Competition is a major external force, as well as other factors, such as economic, demographic, and cultural trends. Some external factors are positive and offer opportunity, whereas others are negative and are a threat to the firm. Because external factors can greatly affect a firm, both opportunities and threats need to be identified and analyzed in the marketing plan:

- **Opportunities** are external factors that the firm may be able to capitalize on to meet or exceed its stated objectives. Marketers are very interested in opportunities because they form the foundation of competitive advantage. Firms that can recognize a market opportunity and then match an internal strength to the external opportunity are often the most successful firms. For example, Apple realized the opportunity for an intuitive mobile device that would serve the market initially created by the BlackBerry mobile phone. Apple's strength in touchscreen technology and intuitive operating systems were matched with opportunity, and the iPhone quickly seized market share.

- **Threats** are external factors that may challenge the firm's short- and long-term performance. Threats are very important to marketers. In many cases, identifying and dealing with threats may be the biggest and most important task that marketers perform. In the iPhone example, think about the opposite side, the threat that Research in Motion Technologies, maker of the BlackBerry mobile phone, should have been prepared for. BlackBerry was the pioneer of mobile technology, virtually becoming synonymous with mobile e-mail, texting, and surfing in the early- to mid-2000s. In 2006, BlackBerry was aware of the touchscreen threat but simply did not realize the size and scope of the threat. In 2006, BlackBerry led the smartphone market by a wide margin; but by 2009, BlackBerry's hasty and poorly planned response to the iPhone had failed, and the BlackBerry had fallen from dominant market leader to nearly forgotten—in just a few short years. Thus, accurately identifying threats and explaining how they will be minimized is an important part of the marketing plan.

Foundational Principles of SWOT Analysis

Most important when considering SWOT information are the following foundational principles of SWOT analysis:

1. Strengths and opportunities should be aligned to capitalize on the development of competitive advantages.

2. Weaknesses should be clearly defined and the firm should either work to convert the weakness into a strength or change its focus to reduce the importance of the weakness.

3. Threats should be honestly acknowledged, and the firm should either invest to convert the threat into an opportunity or divest to leave a market that should no longer be served.

Once the SWOT analysis is completed, firms use the foundational principles of SWOT to guide their marketing strategy, which we describe in the next section.

Strengths

- Brand recognition
- Effective supply chain strategy
- Rigorous food safety standards
- Affordable prices and high-quality products
- Decentralized yet connected system
- Innovative excellence program
- Promotes ethical conduct
- Profitable

Weaknesses

- Inflexible to changes in market trends
- Difficult to find and retain employees
- Drive to achieve shareholder value may conflict with corporate social responsibility
- Promotes unhealthy food

Opportunities

- International expansion
- Positive environmental commitments
- Declining economy could increase demand for lower-priced restaurants
- Corporate social responsibility committee
- Honest and real brand image

Threats

- Weak economy could lead to fewer people dining out
- Consumer trend to choose perceived healthier restaurants
- Health concerns surrounding beef, poultry, and fish in some markets
- Potential labor exploitation in some countries
- Contributes to global warming

The Marketing Strategy

Once the situational analysis is complete, marketers focus on defining their marketing strategy. A **strategy** is the set of actions taken to accomplish organizational objectives. A successful marketing strategy can lead to higher profits, stronger brands, larger market share, and a number of other desired outcomes for stakeholders of the organization. The marketing strategy requires firms to consider the best way to allocate resources to achieve the goals of their mission statement. This includes identifying and creating the best products, identifying the proper pricing strategies, executing the most meaningful promotion plan, and distributing products efficiently to reach the intended audiences. All of this is based on the needs of the intended target market.

In 2017, yogurt maker Chobani implemented a strategy to continue its sales growth by introducing a new product, Chobani Smooth.

The marketing strategy section appears after the situational analysis section because findings from market surveys and a careful SWOT analysis inform a firm's short-, medium-, and long-term marketing strategies.

Marketers spend a great deal of time defining the firm's mission, analyzing and describing the market summary, and designing and detailing their marketing strategy. However, the remainder of the marketing plan is of almost greater importance than the initial sections.

The Financials

In the financials section, the marketing plan outlines the financial situation of the firm and how marketing efforts will affect it, hopefully to the positive. The overall profitability of both the product and the firm can be found in the financials section of the marketing plan. **Financial projections** provide those reading the plan with a bottom-line estimate of the organization's profitability. Financial projections can include numerous items, but all should contain a sales forecast, an expense forecast, and a break-even analysis for the period covered by the marketing plan.

The specific calculation and reporting of sales/expense forecasts and break-even analysis is the subject of advanced marketing, finance, and accounting courses. However, each is described in general detail as follows.

Sales Forecast

A **sales forecast** projects how many units of a product the company expects to sell during a specific period. Companies can use their sales for the past several years to predict sales for the upcoming year. Firms factor in changes to the market, compiled in preparing their marketing plan, to estimate how sales may change in the coming year or over the period covered by the plan.

Expense Forecast

The **expense forecast** is an estimate of the costs the company will incur to create, communicate, and deliver the product. Without an expense forecast, marketers will have a difficult time allocating resources and predicting when the product will become profitable. Both fixed and variable expenses are outlined in the expense forecast. *Fixed expenses* include things that can't be changed easily, such as store rent, utilities, and cost of technology. *Variable expenses* include items that are used in greater or lesser frequency as products are sold, such as shift labor expenses, production materials, and sales commissions.

Break-Even Analysis

Break-even analysis combines the data provided in the sales and expense forecasts to estimate how much the company needs to sell to cover its expenses. The break-even analysis of the financial section is designed to provide an accurate picture of exactly how much the firm needs to sell to reach a point where revenue equals expenses. Ideally, the firm projects that its marketing activities should expect to profit based on its investment in marketing efforts.

The Controls

The final section in most marketing plans outlines the controls the firm will put in place to monitor and adjust the plan as the firm executes on the strategy laid out in it. The controls section should cover implementation, organizational structure, and contingency planning.

Implementation

The **implementation section** provides a detailed account of how the specific actions of the marketing plan will be carried out and who will be responsible for carrying them out. Each step of the implementation of a marketing plan, such as buying advertising on a specific television channel or utilizing a new Twitter hashtag, should tie back to the marketing strategy and the specific objectives laid out during the strategic planning process. Marketers should carefully monitor each marketing strategy and expect to make adjustments depending on results or as market conditions change over time.

Organizational Structure

An outline of the organizational structure helps hold specific departments and individuals responsible for the parts of the marketing plan that fall under their control. When elements of the marketing strategy are not implemented correctly, it's often because the plan does not clearly indicate who is responsible for carrying out each activity. By clearly outlining who is accountable for which tasks, the marketing plan can help clarify ownership and drive positive results.

Contingency Planning

Contingency planning defines the actions the company will take if the initial marketing strategy does not achieve results. Contingency planning is also important as you market yourself throughout your career. Even after completing this course, graduating from college, and gaining more work experience, there will be numerous times in your professional career when strategies you've developed don't work out as planned. To prepare for such eventualities, it will be important for you to consider how you will monitor your progress and change course, if necessary.

Coca-Cola famously changed direction after its New Coke product failed to meet company objectives. The firm reintroduced the old Coke formula as Coca-Cola Classic and began to add profits and market share again.

The marketing plan typically concludes with the controls section. However, figures, tables, charts, or any type of appendix may be included at the end of the marketing plan to support information described in the plan.

When the management of a firm accepts a marketing plan, its true purpose is just beginning. A marketing plan is an action-oriented document that should be reviewed regularly by employees of the firm to help keep their marketing efforts on track and in line with the established plan.

Glossary

A

80/20 rule A behavioral segmentation rule that suggests that 20 percent of very loyal customers account for 80 percent of their total demand.

Accounting the process of measuring, classifying, analyzing, and communicating financial information

active listening A skill that salespeople can develop to improve how they fully concentrate, understand, respond, and remember what is said during a sales meeting.

actual product Component of the product that involves the form of the product itself such as the brand, features, quality, style, size, color, and packaging.

Actual Product Component of the product that involves the form of the product itself such as the brand, features, quality, style, size, color, and packaging.

advertising Nonpersonal promotional communication about goods, services, or ideas that is paid for by the firm identified in the communication.

advertising Non-personal promotional communication about goods, services, or ideas that is paid for by the firm identified in the communication.

advertising campaign A collection of coordinated advertisements that share a single theme.

advertising campaign A collection of coordinated advertisements that share a single theme.

advertising effectiveness studies A type of research that measures how well an advertising campaign meets marketing objectives.

advertising puffery Advertising that makes broad exaggerated or boastful statements about a product or service that are subjective (a matter of opinion) rather than objective fact.

advertising puffery Advertising that makes broad exaggerated or boastful statements about a product or service that are subjective (a matter of opinion) rather than objective fact.

allowances Trade promotions that typically involve paying retailers for financial losses associated with consumer sales promotions or reimbursing a retailer for an in-store or local expense to promote a specific product.

Allowances Trade promotions that typically involve paying retailers for financial losses associated with consumer sales promotions or reimbursing a retailer for an in-store or local expense to promote a specific product.

approach A part of the personal-selling process that involves meeting the prospect and learning more about his or her needs and wants.

aspirational reference group Individuals a consumer would like to emulate.

assurance The knowledge and courtesy of employees and their ability to convey trust and confidence.

attitude A person's overall evaluation of an object involving general feelings of like or dislike.

auction site A digital retail site that lists goods from individuals or firms that can be purchased through an auction bidding process or directly through a "purchase now" feature.

augmented product Component of the product that contains the services, experiences, warranties, and financing that enhance the product value.

Augmented Product Component of the product that contains the services, experiences, warranties, and financing that enhance the product value.

B

bargain hunting Digital purchasing behavior that involves coupon sites such as wish.com or auction sites such as eBay. Bargain hunting is often combined with browsing and may or may not lead to a purchase.

behavioral segmentation Segmentation method that categorizes consumers based on *what they actually do* or *how they act* toward products.

beyond compliance leadership An environmental marketing strategy that focuses on communicating to stakeholders the company's attempts to adopt environmentally friendly practices.

bots A software application that runs automated tasks over the Internet.

bounce rate The percentage of visitors who enter a website and then quickly depart, or bounce, rather than continuing to view other pages within the same site.

brand The name, term, symbol, design, or any combination of these that identifies and differentiates a firm's products.

brand associations Consumer thoughts connected to the customer's memory about the brand.

brand equity The value the firm derives from consumers' positive perception of its products.

brand equity The value the firm derives from consumers' positive perception of its products.

brand extension The process of broadening the use of an organization's current brand to include new products.

brand loyalty When a consumer displays a steadfast allegiance to a brand by repeatedly purchasing it.

brand mark Non-word elements that visually represent the brand.

brand recognition The degree to which customers can identify the brand under a variety of circumstances.

brand revitalization A strategy to recapture lost sources of brand equity and identify and establish new sources of brand equity. Also known as *rebranding*.

branding Developing and maintaining the name, term, symbol, design, or any combination of these that identifies and differentiates a firm's products.

break-even analysis The process of calculating the break-even point, which equals the sales volume needed to achieve a profit of zero.

break-even analysis Combination of the data provided in the sales and expense forecasts to estimate how much the company needs to sell to cover its expenses.

break-even point The point at which the costs of producing a product equal the revenue made from selling the product.

browsing Digital purchasing behavior wherein the consumer is not really looking to make a purchase.

business portfolio analysis A method of categorizing a firm's products for strategic planning based on competitive position and expected growth rate.

business-to-business (B2B) marketing Marketing to organizations that acquire goods and services in the production of other goods and services that are then sold or supplied to others.

buyer–seller relationship A connection between a firm and/or its employees intended to result in mutually beneficial outcomes.

buyers Those who submit the purchase to the salesperson in a B2B buying center. This role is often more formal, such as a purchasing manager.

buying center The group of people responsible for strategically obtaining products needed by the firm.

buying club site A digital retail site that allows consumers to buy in bulk.

C

cannibalization When new products take sales away from the firm's existing products rather than generating additional revenues or profits through new sales.

captive salespeople Salespeople who work directly for a sales firm.

capturing value The result of marketing efforts that return value to the firm, typically in the form of profit.

causal research A type of research used to understand the cause-and-effect relationships among variables.

chatter Another form of consumer feedback that occurs when a consumer shares, forwards, or "retweets" a marketing message. For marketers, the level of chatter represents consumer feedback.

click path A sequence of hyperlink clicks that a website visitor follows on a given site, recorded and reviewed in the order the consumer viewed each page after clicking on the hyperlink.

closing The point at which the salesperson asks the prospect for the sale.

co-branding A strategy in which two or more companies issue a single product in an effort to capitalize on the equity of each company's brand.

coding The process of assigning a word, phrase, or number to a selected portion of data, so that it can later be easily sorted and summarized.

cognitive dissonance The mental conflict that people undergo when they acquire new information that contradicts their beliefs or assumptions

collaborative relationship When both parties share resources (e.g., financial risk, knowledge, and employees) in an effort to attain a common goal that provides beneficial outcomes to both parties.

combination approach Utilizes multiple types of territory strategies to organize the sales force.

communication gap The gap between service delivery and the company's external communications.

communication method A means for marketers to get a message to consumers.

communication process How a message is transmitted from (encoded by) a sender and decoded by a receiver.

compensatory decision making A complex decision-making model that considers multiple decision-making criteria and is typically applied to purchase decisions that involve some level of risk.

competitive advantage The superior position a product enjoys over competing products if consumers believe it has more value than other products in its category.

competitive advantage The superior position a product enjoys over competing products if consumers believe it has more value than other products in its category.

competitive intelligence Involves gathering data about what strategies direct and indirect competitors are pursuing in terms of new-product development and the marketing mix.

concept test A procedure in which marketing professionals ask consumers for their reactions to verbal descriptions and rough visual models of a potential product.

consumer behavior The way in which individuals and organizations make decisions to spend their available resources, such as time or money.

consumer confidence A measure of how optimistic consumers are about the overall state of the economy and their own personal finances.

consumer ethnocentrism A belief by residents of a country that it is inappropriate or immoral to purchase foreign-made goods and services.

consumer feedback Different ways that customers can report their satisfaction or dissatisfaction with a firm's products.

Consumer feedback Different ways that customers can report their satisfaction or dissatisfaction with a firm's products.

consumer review A direct assessment of a product (good, service, or idea) that is expressed through social media for others to see and consider.

consumer willingness to pay A consumer's ability and attitude toward the price of a product when making a product decision.

content The information, images, videos, and any other delivery method of the marketer's social media message.

contests Sales promotions in which consumers compete against one another and must demonstrate skill to win.

Contests Sales promotions in which consumers compete against one another and must demonstrate skill to win.

contingency planning The actions the company will take if the initial marketing strategy does not achieve results.

convenience products Inexpensive products that are purchased frequently and require very little consumer involvement in the purchase process.

convenience retailers Stores that offer a limited variety and assortment of merchandise, usually snack foods and minor essentials, at an easily accessible location.

conversion rate The percentage of users who take a desired action, such as making a purchase.

cookies Data files stored on websites that can generate a profile or other data about consumers.

core product Component of the product that relates to the basic benefit obtained by customers of the product.

Core Product Component of the product that relates to the basic benefit obtained by customers of the product.

corporate social responsibility (CSR) An organization's obligation to maximize its positive impact and minimize its negative impact on society.

cost-plus pricing A pricing method in which a certain amount is added to the cost of the product to set the final price. Also known as *markup pricing*.

coupons Documents that entitle the customers who possess them to a discount on a product.

coupons Documents that entitle the customers who possess them to a discount on a product.

cultural fit How well marketing efforts are designed to meet the needs of a specific market.

currency fluctuation How the value of one country's currency changes in relation to the value of other currencies.

currency manipulation When the government of a country artificially controls the value of its currency relative to the currencies of other countries.

customer approach Organizes the sales territory around the customer.

customer equity A ratio that compares the financial investments a company puts into gaining and keeping customers to the financial return on those investments.

customer excellence A strategy designed to put the customer at the center of all marketing activities by focusing on value, satisfaction, and quality of service.

customer focus Measures how well a CRM program prioritizes customers based on each customer's profitability.

customer relationship management (CRM) The process by which companies get new customers, keep the customers they already have, and grow the business by increasing their share of customers' purchases.

customer satisfaction A state that is achieved when companies meet the needs and expectations customers have for their goods or services.

customer segmentation analysis A data analysis model that involves creating customer profiles based on demographic characteristics, purchase patterns, and other criteria and placing them into various categories.

customer value The perceived benefits, both monetary and nonmonetary, that customers receive from a product compared to the costs associated with obtaining it.

D

dashboard A central location where all social media activity can be easily monitored.

data Facts or measurements of things or events.

data mining A process that involves the computerized search for meaningful trends in a large amount of data.

deceptive pricing An illegal practice that involves intentionally misleading customers with price promotions.

decider The person who chooses the good or service that the company is going to buy.

decline stage The stage of the PLC that is preceded by declining sales and profits. Depending on the product, the decline in sales may occur over a long period of time. During the decline stage, competitors drop out of the market as the product becomes unprofitable.

decoding In the communication process, how the receiver perceives and interprets the sender's message.

delivery gap The difference between what the company thinks the customer expects and how it actually delivers that service.

demand analysis A type of research used to estimate how much customer demand there is for a particular product and understand the factors driving that demand.

demographic segmentation Segmentation method that divides markets using characteristics such as age, gender, income, education, and family size.

department stores Stores that offer a wide range of products displayed as a collection of smaller "departments" within the store.

dependent variable The test or outcome variable in a study that is influenced by changes in another variable.

derived demand When demand for one product occurs because of demand for a related product.

descriptive research A type of research that seeks to understand consumer behavior by answering the questions who, what, when, where, and how.

descriptive statistics Statistics that describe the characteristics, features, and properties of the data in a study sample.

design and standards gap The difference between the company's plan and the company's execution of that plan.

differentiated targeting A targeting strategy used when an organization simultaneously pursues several different market segments, usually with a different strategy for each.

diffusion How new products are likely to be adopted, the rate at which they will be adopted, and the process through which their products will spread into markets.

digital mall A digital retail site where a variety of sellers stock their goods.

digital marketing Online marketing that can deliver content immediately to consumers through digital channels, devices, and platforms to build or promote a company's marketing message.

digital marketplace A digital retail site made up of small, independent sellers.

direct competitors Firms that compete with products designed around the same, or similar characteristics.

direct competitors Firms that compete with products designed around the same or similar characteristics.

direct marketing The total number of activities a seller engages in to encourage the exchange of goods and services with the buyer. The seller communicates with the target audience using one or more different forms of media to solicit a response from a prospect or customer.

direct marketing channel marketing channel in which products travel directly from the producer to the consumer

direct ownership Requires a domestic firm to actively manage an overseas facility.

direct response advertising A form of marketing that requires an immediate response or call to action.

direct selling A non-store retail method that involves salespeople interacting with customers directly, usually at home, work, or at an organized "party."

dissociative reference group People that the consumer would *not* like to emulate.

distribution When a company ships its goods to its customers.

distribution center (DC) A type of warehouse used specifically to store and ship finished goods to customers.

distribution intensity The number of outlets a manufacturer chooses to sell through.

distributor A firm that buys noncompeting products, warehouses them, and resells them to retailers or directly to end users.

diversification A marketing strategy that seeks to attract new customers by offering new products that are unrelated to the existing products produced by the organization.

Dominican Republic–Central America Free Trade Agreement (DR-CAFTA) Focuses on eliminating tariffs, reducing nontariff barriers, and facilitating investment among the member states.

downstream flow The movement of goods (and other things, like information, promotion, etc.) from a source toward intermediaries and the final customer in a supply chain.

drugstores Stores that primarily sell pharmaceuticals, health and wellness products, over-the-counter medicines, and beauty products, as well as a limited assortment of food and beverages.

dumping A protectionist strategy in which a company

sells its exports to another country at a lower price than it sells the same product in its domestic market.

E

e-mail marketing A cost-effective form of digital marketing used to retain, nurture, or attract a new customer base.

E-tailers Stores that allow customers to shop for and buy products online (or via a mobile device) for home delivery.

early adopters Consumers that purchase and use a product soon after it has been introduced.

early majority Consumers that are careful in their purchase approach, gathering more information and spending more time thinking about the purchasing decision than earlier adopters.

eco-branding An environmental marketing strategy that focuses on creating a credible green brand. For this strategy to be effective, consumers must recognize a noticeable benefit from their purchase.

eco-efficiency An environmental marketing strategy that involves identifying environmentally friendly practices that also have the effect of creating cost savings and driving efficiencies throughout the organization.

elastic demand A scenario in which demand changes significantly due to a small change in price.

emotional appeal Advertising appeal that attempts to create

an emotional response to the advertising.

emotional appeal Advertising appeal that attempts to create an emotional response to the advertising.

empathy The provision of caring, individualized attention to customers.

employee sales promotions Promotional incentives designed to increase sales for a product or brand.

employee sales promotions Promotional incentives designed to increase sales for a product or brand.

enablement tools Applications that streamline buyer–seller engagement in an effort to improve the customer's experience.

encoding In the communication process, the transformation of the sender's ideas and information into a message that usually includes words, symbols, and/or pictures.

environmental cost leadership An environmental marketing strategy in which firms seek a price premium for their environmentally friendly products.

environmental scanning The act of monitoring developments outside of the firm's control with the goal of detecting and responding to threats and opportunities.

escalator clause A section in a contract that ensures that providers of goods and services do not encounter unreasonable financial hardship as a result of uncontrollable increases in the costs of or decreases in

the availability of something required to deliver products to customers.

ethics Moral standards expected by society.

European Union (EU) A single European market formed to reduce barriers to the free trade of goods, services, and finances among member countries.

evaluative criteria The attributes a consumer considers important about a certain product.

exchange An activity that occurs when a buyer and seller trade things of value so that each is better off as a result.

exchange rate The price of one country's currency in terms of another country's currency.

exclusive distribution A low-volume distribution strategy in which products are sold through very limited channels.

executive summary A one-to two-page synopsis of the marketing plan's main points.

expense forecast An estimate of the costs the company will incur to create, communicate, and deliver the product.

experimental conditions The set of inputs (independent variables) presented to different groups of participants. In an experiment, typically only one or two inputs will be changed in each condition.

experimental manipulation The intentional change that the researcher makes to the independent variable.

experiments Procedures undertaken to test a hypothesis.

exploratory research A type of research that seeks to discover new insights that will help the firm better understand the problem or consumer thoughts, needs, and behavior.

exporting Selling domestically produced products to foreign markets.

external environment Changes that occur outside a firm's immediate environment and control. Includes such external factors as demographic, economic, sociocultural, political/legal, competitive, and technological changes.

external information search Information beyond consumers' personal knowledge and experience that can support them in their buying decision.

external validity The extent to which the results of the experiment can be generalized beyond the study's sample of subjects.

extraneous variable A variable that is not intentionally part of the experiment but that could influence the results.

extrinsic motivation Stems from positive or negative external factors such as financial or social rewards, embarrassment or termination.

F

family life cycle The distinct family-related phases that an individual progresses through over the course of his or her life.

Federal Communication Commission Federal agency that enforces the Federal laws that prohibit obscene, indecent, and profane content from being broadcast on the radio or television.

Federal Communications Commission (FCC) Federal agency that enforces the federal laws that prohibit obscene, indecent, and profane content from being broadcast on the radio or television.

Federal Trade Commission (FTC) Federal agency that protects consumers and businesses from anticompetitive behavior and unfair and deceptive practices.

Federal Trade Commission (FTC) Federal agency that protects consumers and businesses from anticompetitive behavior and unfair and deceptive practices.

feedback The receiver's response to the sender's message that closes the communication process.

field experiments Experiments performed in natural settings like stores or malls.

financial projections A bottom-line estimate of the organization's profitability

fixed costs Costs that remain constant and do not vary based on the number of units produced or sold.

focus groups Data collection tool in which a moderator engages a small group of people as they discuss a particular topic or concept

with each other in a spontaneous way.

franchising A contractual arrangement in which the franchisor (known brand) provides a franchisee (local owner operator) the right to use its name, marketing, and operational support in exchange for a fee and, typically, a share of the profits.

frequency A count of how often a consumer is exposed to a promotional message (television advertisement, online advertisement, billboard, etc.).

functional quality The quality of the service process.

G

gatekeepers Individuals who control the flow of information into the company that all other users will review in making a purchasing decision.

geographic approach Places salespeople within an exclusive territory.

geographic segmentation Segmentation method that divides markets into groups such as nations, regions, states, and neighborhoods.

geotracking Use of a consumer's geographic location to determine what goods will come up in a search and at what price.

global marketing A marketing strategy that consciously addresses customers, markets, and competition throughout the world in an effort to sell more products in more markets.

government markets Federal, state, and local entities that purchase everything from heavy equipment to paperclips.

gray market The sale of branded products through legal but unauthorized distribution channels.

gray market Branded products sold through legal but unauthorized distribution channels.

gross domestic product (GDP) A measure of the market value of all officially recognized final goods and services produced within a country in a given period.

growth stage The stage of the product life cycle characterized by increases in sales, profits, and competition.

H

heterogeneity The quality that services possess in which they are inherently variable because they cannot be mass produced.

high-involvement products Significant purchases that carry a greater risk to consumers if they fail.

hypothesis An educated guess based on previous knowledge or research about the cause of the problem under investigation.

I

idea generation The stage of new product development in which a set of product concepts is generated from which to identify potentially viable new products.

idea screening The stage of new product development in which the firm evaluates an idea to determine whether it fits into the new product strategy.

immediate environment Internal and external factors that impact a firm. Comprises the internal structure of the firm as well as a firm's customers and corporate partners.

immediate environment Internal and external factors that impact a firm. Comprises the internal structure of the firm as well as a firm's customers and corporate partners.

implementation section A detailed account of how the specific actions of the marketing plan will be carried out and who will be responsible for carrying them out.

inbound marketing A form of digital marketing that utilizes such tools as blogs, webinars, or follow-up e-mails to entice consumers to a product or service offer without forcing an interaction or a purchase.

incentives The component of compensation that is variable.

income distribution How wealth is allocated across the population of a country.

independent sales representatives Salespeople who do not work directly for the firm but rather represent various firms and products on a contractual basis. Also known as *manufacturer's reps.*

independent variable The causal variable in a study controlled by the researcher used to influence changes in another variable.

indirect competitors Firms that compete with products that have different characteristics but serve a similar function.

indirect competitors Firms that compete with products that have different characteristics but serve a similar function.

indirect marketing channel marketing channel in which supply chain intermediaries, such as wholesalers, distributors, and retailers, are a part of the product's sale and delivery

inelastic demand A situation in which a specific change in price causes only a small change in the amount purchased.

inferential statistics Statistics that make inferences about a larger population from the data in a study sample.

inflation An increase in the general level of prices of products in an economy over a period of time.

influencers Individuals who affect the buying decision by giving opinions or setting buying specifications.

information The result of marketing research, where data are analyzed and presented to support decision making or improve understanding of a defined problem.

information search When consumers seek information to support them in their buying decision.

informative advertising A type of advertising that attempts to develop initial demand for a product.

informative advertising A type of advertising that attempts to develop initial demand for a product.

innovation The creation of a new or significantly improved product offering.

innovators Consumers that adopt a product almost immediately after the product is launched.

inseparability The quality that services possess in which both the service provider and the customer must be present during the service, as production and consumption are simultaneous activities.

inside salespeople Salespeople who sell to external customers from within the organization.

institutional markets A wide variety of organizations, including hospitals, schools, churches, and nonprofit organizations.

intangibility The quality that services possess in having no physical substance, and therefore cannot be touched, stored, or possessed like goods.

integrated marketing communications (IMC) strategy A promotional strategy that involves coordinating the various promotional mix elements to provide customers with a clear and consistent message about a firm's products.

intensive distribution A distribution strategy that involves placing a product in as many outlets as possible.

intermodal transportation Moving of freight (consumer goods) in a container using multiple modes of transportation (e.g., ship, rail, semi-truck).

internal information search When consumers use their past experiences with items from the same brand or product class as sources of information.

internal validity The extent to which changes in the dependent variable were actually caused by manipulations of the independent variable.

International Monetary Fund (IMF) Works to foster international monetary cooperation, secure financial stability, facilitate international trade, promote high employment and sustainable economic growth, and reduce poverty around the world.

interview A data collection tool in which the researcher works with one participant at a time, asking open-ended questions about how the individual perceives and uses various products or brands.

intrinsic motivation The innate desire to complete a task or engage in an activity.

Introduction stage The stage of the product life cycle when a product is new and sales are low, while expenses are high.

involvement The personal, financial, and social significance of the decision being made.

J

joint venture A domestic firm partners with a foreign company to create a new entity.

K

kiosks A non-store retailing method that offers a temporary, inexpensive, movable format. Kiosk retailers can move their carts or pods to go to areas of high foot traffic.

knowledge gap The difference between actual customer expectations and the company's perceptions of customer expectations.

L

laggards Consumers that tend to not like change; they may remain loyal to a product until it is no longer available for sale.

late majority Consumers that tend to be cautious about new things and ideas.

learning The modification of behavior that occurs over time due to experiences and other external stimuli.

licensing A legal process in which one firm (the licensee) pays to use or distribute the resources—including products, trademarks, patents, intellectual property, or other proprietary knowledge—of another firm (the licensor).

lifestyle A person's typical way of life as expressed by his or her activities, interests, and opinions.

lifestyle segmentation Psychographic segmentation strategy that divides people into groups based on their opinions and the interests and activities they pursue.

lifetime value (LTV) The total profit a customer brings to a company during the time that the individual or firm is a customer.

lifetime value (LTV) analysis A data analysis model that allows the company to monitor the actual costs of doing business with customers to ensure that it is focusing on the most profitable customers.

locational excellence A focus on having a strong physical location and/or Internet presence.

logistics The part of supply chain management that plans, implements, and controls the flow of goods, services, and information between the point of origin and the final customer.

logistics The part of supply chain management that plans, implements, and controls the flow of goods, services, and information between the point of origin and the final customer.

low-involvement products Inexpensive products that can be purchased without much forethought and that are purchased with some frequency.

loyalty An accrued satisfaction over time that results in repeat purchases.

loyalty programs Sales promotions that allow consumers to accumulate points or other benefits for doing business with the same company or a group of companies.

loyalty programs Sales promotions that allow consumers to accumulate points or other benefits for

doing business with the same company or a group of companies.

M

manufacturer brands Brands that are managed and owned by the manufacturer rather than a reseller. Also known as *national brands*.

manufacturer's reps Salespeople who do not work directly for the firm but rather represent various firms and products on a contractual basis. Also known as *independent sales representatives*.

marginal cost The change in total cost that results from producing one additional unit of product.

marginal revenue The change in total revenue that results from selling one additional unit of product.

market A broad group of consumers or organizations that is interested in and able to buy a particular product.

market development A marketing strategy that focuses on selling existing goods and services to new customers.

market entry strategies Different operational ways that firms use to enter the international marketplace, including exporting, licensing, franchising, joint venture, or direct ownership

market penetration A marketing strategy that emphasizes selling more of existing goods and services to existing customers.

market segmentation A process of dividing a larger market into smaller groups, or market segments, based on shared characteristics.

market segments Relatively homogeneous groups of consumers that result from the segmentation process.

market summary Section of the situational analysis in the marketing plan that describes the current state of the firm's overall market and target market(s).

marketing An organizational function and set of processes for creating, communicating, and delivering value to customers and managing customer relationships in ways that benefit the organization and its employees, customers, investors, and society as a whole.

Marketing the activity, set of institutions, and processes for creating, communicating, delivering, and exchanging offerings that have value for customers, clients, partners, and society at large.

marketing environment Internal and external factors that affect a firm's ability to succeed.

marketing environment Comprises internal and external factors that affect firms and may help or hurt a product in the marketplace.

marketing manager Manages the day-to-day marketing activities of the organization and long-term marketing strategy for the company.

marketing mix A combination of activities related to product, price, place, and promotion that represent areas a firm can adjust to influence demand for its good, service, or idea; often referred to as the four Ps of marketing.

marketing mix A combination of activities related to product, price, place, and promotion that represent areas a firm can adjust to influence demand for its good, service, or idea; often referred to as the four Ps of marketing.

marketing plan A formal document that outlines many aspects of a firm's mission, market position, strategy, financial position, and plan to implement and monitor the marketing strategy in a defined period of time.

marketing research The act of collecting, interpreting, and reporting information concerning a clearly defined marketing problem.

marketing strategy The set of actions taken to accomplish organizational objectives.

marketing strategy The set of marketing actions taken to accomplish marketing objectives.

markup pricing A pricing method in which a certain amount is added to the cost of the product to set the final price. Also known as *cost-plus pricing*.

materials management The inbound movement and storage of materials in preparation for those materials to enter and flow through the manufacturing process.

mathematical modeling A type of causal research that involves using equations to model the relationships between variables.

maturity stage The stage of the PLC during which late majority and repeat buyers make up an increasing percentage of the customer base. The main objectives of the maturity stage are earning profits and maintaining the firm's market share for as long as possible.

membership reference group The group to which a consumer actually belongs.

mission statement A concise affirmation of the firm's long-term purpose.

mobile marketing A set of practices that enables organizations to communicate and engage with their audience in an interactive and relevant manner through and with any mobile device or network.

modified rebuy Occurs when the customers' needs change slightly or they are not completely satisfied with the product they purchased.

motivation The inward drive people have to get what they need or want.

motivation The basis of a continued commitment to working toward a goal.

N

national brands Brands that are managed and owned by the manufacturer rather than a reseller. Also known as *manufacturer brands*.

native advertising Form of promotion that integrates a marketer's advertising message into the content and look of a website or other platform.

needs States of felt deprivation. Consumers feel deprived when they lack something useful or desirable, such as food, clothing, shelter, transportation, and safety.

new buy A business customer purchasing a product for the first time.

new category entries Products that are new to a company but are not new to the marketplace.

new product development (NPD) process The process of conceiving, testing, and launching a new product in the marketplace.

new product strategy development The stage of new product development in which the company determines the direction it will take when it develops a new product.

new-to-the-market products Products that have never been seen before and create a new market.

niche marketing A targeting strategy that concentrates on a large share of a small market segment.

noise Distractions in the communication process that disrupt the reception of a marketing message.

non-personal advertising Advertising that uses media to transmit a message to large numbers of individuals rather than marketing to consumers face-to-face.

non-store retailing The selling of goods and services outside the limitations of a retail facility.

noncompensatory decision making A form of consumer

decision making wherein consumers do not evaluate other alternatives at their disposal; rather, they purchase products based on what they've purchased in the past to solve their market problems.

nonpersonal advertising Advertising that uses media to transmit a message to large numbers of individuals rather than marketing to consumers face-to-face.

nonprobability sampling A type of sampling that does not attempt to ensure that every member of the target population has a chance of being selected.

North American Free Trade Agreement (NAFTA) A free trade zone among the United States, Canada, and Mexico.

O

objections The concerns or reasons customers offer for not buying a product.

objective A specific result that marketers aim to achieve within a time frame and with available resources.

observation A data collection tool that involves watching how people behave and recording anything about that behavior that might be relevant to the research objective.

odd pricing A pricing tactic in which a firm prices products a few cents below the next dollar amount.

off-price retailers Stores that offer an inconsistent variety and assortment of branded products from an array of suppliers by reselling products that did not sell at another retailer.

omnichannel retailing A multi-channel retailing approach that allows the customer to have an integrated customer experience across all of a retailer's distribution platforms.

one-time shopping Digital purchasing behavior that may combine product-focused shopping, browsing, researching, and bargain hunting all at the same time. Consumers are shopping for a gift or using a gift card and will not return to the shop once the purchase is made.

operational excellence A strategy of focusing operational principles, systems, and tools toward improving customer satisfaction with the firm's products and services.

opinion leaders Individuals who exert an unequal amount of influence on the decisions of others because they are considered knowledgeable about particular products.

opportunities External factors that the firm may be able to capitalize on to meet or exceed its stated objectives.

outside salespeople Salespeople who visit customers in the field for selling purposes

outsource To procure goods, services, or ideas from a third-party supplier rather than from an internal source.

P

packaging All of the activities of designing and producing the container for a product.

paid advertising Advertising that requires a purchased time or space for communicating a message.

paid advertising Advertising that requires a purchased time or space for communicating a message.

paid display advertising Includes everything from banner ads to YouTube video advertising. These ads generate awareness as well as (hopefully) drive traffic to a website.

paid search Online advertising in which a company pays to be a sponsored result of a customer's Web search.

paid stories Ads that appear as content designed to look like stories to the viewer.

participants People who are subjected to experimental conditions that are controlled by the researchers.

penetration pricing A pricing strategy that involves setting prices low to encourage a greater volume of purchases. Also known as *volume maximization*.

perceived quality A consumer's perception of the overall quality of a brand.

perceptual map A valuable tool for firms in helping them understand their competitors' positions in the marketplace. It creates a visual picture of where products are located in consumers' minds.

perishability The quality that services possess in which they cannot be stored, inventoried, reused, or returned.

personal selling Two-way flow of communication between a salesperson and a customer that is paid for by the firm and seeks to influence the customer's purchase decision.

personality The set of distinctive characteristics that lead an individual to respond in a consistent way to certain situations.

persuasive advertising A type of advertising that attempts to increase demand for an existing product.

persuasive advertising A type of advertising that attempts to increase demand for an existing product.

place The activities a firm undertakes to make its product available to potential consumers.

planned obsolescence A marketing criticism in which companies frequently come out with new models of a product that make existing models obsolete. It can also involve limiting the product lifespan rather than the product availability.

positioning A company's efforts to influence the customer's perception about its products, services, and even ideas. These efforts assume that consumers compare the benefits of products and brands against those of the competitors.

post-purchase evaluation Consumer feelings and perceptions related to the process and product that are evaluated and assessed as either positive or negative.

pre-approach A part of the personal-selling process that involves identifying specific information about a prospect.

predatory pricing The practice of first setting prices low with the intention of pushing competitors out of the market or keeping new competitors from entering the market, and then raising prices to normal levels.

predictive modeling A data analysis model that uses algorithms to try to determine the future actions of customers.

prestige pricing A pricing strategy that involves pricing a product higher than competitors to signal that it is of higher quality.

price The amount of something—money, time, or effort—that a buyer exchanges with a seller to obtain a product.

price The amount of something—money, time, or effort—that a buyer exchanges with a seller to obtain a product.

price bundling A strategy in which two or more products are packaged together and sold at a single price.

price discrimination The practice of charging different customers different prices for the same product.

price elasticity of demand A measure of price sensitivity that gives the percentage change in quantity demanded in response to a percentage change in price (holding constant all the other determinants of demand, such as income).

price fixing When two or more companies collude to set a product's price.

price gouge To overcharge a customer in times of great need, often when no other option exists

price sensitivity The degree to which the price of a product affects consumers' purchasing behavior.

price skimming A pricing strategy that involves setting a relatively high price for a period of time after the product launches. Also known as *profit maximization.*

primary data Data that are collected specifically for the research problem at hand

primary data collection When researchers collect data specifically for the research problem at hand.

private label brands Products developed by a retailer and sold only by that specific retailer. Also known as *store brands.*

probability sampling A type of sampling in which every person in the target population has a chance of being selected, and the probability of each person being selected is known.

problem recognition The stage of the buying process in which consumers recognize they have a need to satisfy.

product A specific combination of goods, services, or ideas that a firm offers to consumers.

product The specific combination of goods, services, or ideas that a firm offers to its target market.

product The specific combination of goods, services, or ideas that a firm offers to its target market.

product accumulation Involves receiving goods from various suppliers, storing the goods until they're ordered by a customer or other company-owned facility, and consolidating orders to achieve transportation economies of scale.

product allocation Picking available goods to fill customer orders; distribution centers can be set up to pick full pallets of goods, full cases of goods, or individual goods.

product approach Involves salespeople who sell a particular product or set of products.

product assortment Occurs when the distribution center mixes goods coming from multiple suppliers into outgoing orders so that each order includes a variety of goods rather than just one type of good.

product development A marketing strategy that involves creating new goods and services for existing markets.

product development The stage of new product development at which a firm determines that the good can be produced or the service can be offered in a way that meets customer needs and generates profits.

product dumping When a company sells its exports to another country at a lower price than it sells the same product in its domestic market.

product excellence A strategy that focuses on the importance of high-quality and value-adding products.

product item A single, specific product such as an iPhone.

product launch The final preparations for making the fully tested product available to the market

product life cycle (PLC) A series of stages that happen during a product's life.

product line Variations of a product within a product category.

product line A group of related products marketed by the same firm.

product line depth The number of products within each of the company's product lines.

product Line extensions Products that extend and supplement a company's established product line.

product mix All of the products that a company offers to its customers.

product mix breadth The number of product lines that the company offers.

product sortation Gathering goods with similar characteristics in one area of the distribution channel to facilitate proper inventory controls and effectively provide customer service.

product-focused shopping Digital purchasing behavior that involves replacing an existing product or purchasing a product that has been pre-chosen.

professional purchasers or buyers Employees of companies who make purchase decisions in the best interest of their organizations.

profit margin The amount a product sells for above the total cost of the product itself.

profit maximization A pricing strategy that involves setting a relatively high price for a period of time after the product launches. Also known as *price skimming*.

profits Revenue minus total costs.

promotion All the activities that communicate the value of a product and persuade customers to buy it.

promotional mix A combination of tools used by marketers to promote goods, services, and ideas and to accomplish their communication objectives.

proportional relationship Two quantities are in a proportional relationship if they have a constant ratio, or if the graph of the quantities on a coordinate plane is a straight line through the origin.

prospecting The search for potential customers—those who need or want a product and fit into a firm's target market.

prototype A mockup of a good, often created individually with the materials the firm expects to use in the final product.

psychographic segmentation Segmentation method that relies on the science of using psychology and demographics to segment consumers.

public relations (PR)
Nonpersonal communication focused on promoting positive relations between a firm and its stakeholders.

publicity Disseminating unpaid news items through some form of media (e.g., television story, newspaper article, etc.) to gain attention or support.

pull strategy A supply chain strategy in which customer orders drive manufacturing and distribution operations.

purchasing power
A measure of the amount of goods and services that can be purchased for a specific amount of money.

push strategy A supply chain strategy in which a company builds goods based on a sales forecast, puts those goods into storage, and waits for a customer to order the product.

push-pull strategy A supply chain strategy in which the initial stages of the supply chain operate on a push system, but completion of the product is based on a pull system.

Q

quadratic equation
an equation containing a single variable of degree 2. Its general form is $ax^2 + bx + c = 0$, where x is the variable and a, b, and c are constants ($a \neq 0$).

qualifying A part of the personal-selling process that involves identifying which potential customers within the firm's target market have not only a desire for the product but also the authority to

purchase it and the resources to pay for it.

qualitative research
Research that studies the qualities of things; characterized by in-depth, open-ended examination of a small sample size, like in-depth interviews or focus groups.

quantitative research
Research that studies the quantity of things; characterized by asking a smaller number of specific and measurable questions to a significantly larger sample size.

quasi-strategic greening
A sustainable marketing strategy that usually involves more substantive changes in marketing actions; for example, redesigning the firm's logo or overhauling a product's packaging.

quota sampling A type of sampling in which a certain number of participants are picked based on selection criteria such as demographics.

rational appeal Advertising appeal that uses logical arguments, which lead to the consumer thinking about the product and its benefits.

rational appeal Advertising appeal that uses logical arguments, which lead to the consumer thinking about the product and its benefits.

reach The percentage of the target market that has been exposed to a promotional message (television advertisement, online advertisement, billboard, etc.) at least once during a specific time period.

reach The percentage of the target market that has been exposed to a promotional message (television advertisement, online advertisement, billboard, etc.) at least once during a specific time period.

rebates Sales promotions that allow consumers to recoup a specified amount of money after making a single purchase.

rebates Sales promotions that allow consumers to recoup a specified amount of money after making a single purchase.

rebranding A strategy to recapture lost sources of brand equity and identify and establish new sources of brand equity. Also known as *brand revitalization*.

receiver Interprets the message transmitted by the sender in the communication process.

recency-frequency-monetary analysis A data analysis model that involves categorizing customers by their buying patterns, such as how recently they have purchased a good or service, how often they purchase from the company, and how much money they spend on the company's products.

recession A period of time during which overall gross domestic product (GDP) declines for two or more consecutive quarters.

reference group
A collection of people to whom a consumer compares themselves.

reference prices The prices that consumers consider reasonable and fair for a product.

relationship selling Building a trusting relationship with a customer over multiple sales interactions.

reliability The ability to perform the promised service dependably and accurately, at the promised time.

reminder advertising A type of advertising that seeks to keep the product before the public in an effort to reinforce previous promotional activity.

reminder advertising A type of advertising that seeks to keep the product before the public in an effort to reinforce previous promotional activity.

Repositioning Involves reestablishing a product's position to respond to changes in the marketplace.

request for proposal (RFP) A formal listing of specifications for a needed product that is sent to a supplier firm asked to bid on supplying the product

researching Digital purchasing behavior wherein the consumer is purchasing a product for the first time. Unlike browsing, which has no expected outcome, researching is more deliberate and will likely result in a purchase either online or offline.

resellers Retailers and wholesalers that buy finished goods and resell them for a profit.

respondents Sample participants who respond to questions that are asked by the researchers.

responsiveness The willingness to help customers and provide prompt service.

retailer A company that purchases and resells products to consumers for their personal or family use.

retailers Companies that purchase and resell products to consumers for their personal or family use.

revamped product A product that has new packaging, different features, and updated designs and functions.

revenue The result of the price charged to customers multiplied by the number of units sold.

S

salary The component of compensation that is fixed.

sales assessment The act of evaluating the sales performance of an organization and its sales force.

sales control The practice of aligning sales results with the greater objectives of the organization.

sales forecast Projection of how many units of a product the company expects to sell during a specific time period.

sales forecasting A form of research that estimates how much of a product will sell over a given period of time.

sales implementation How the sales plan is practiced by the sales force through recruitment, compensation design, and sales force motivation.

sales management A manager's responsibility for formulation and implementation of a sales plan used to deploy salespeople to interact with customers.

sales presentation A forum to convey the organization's marketing message to the prospect.

sales promotion A set of nonpersonal communication tools designed to stimulate quicker and more frequent purchases of a product.

sales promotion A set of non-personal communication tools designed to stimulate quicker and more frequent purchases of a product.

sales team A group of salespeople and possibly other employees who act collectively to sell to a customer.

sales tracking A type of research that follows changes in sales during and after promotional programs to see how the marketing efforts affected the company's sales.

samples Trial portions of a product used to encourage product purchase.

Samples Trial portions of a product used to encourage product purchase.

sampling The process of selecting a subset of the population that is representative of the population as a whole.

search engine optimization (SEO) The process of driving traffic to a company's website from "free" or "organic" search results using search engines.

seasonal discounts Price reductions given to customers purchasing goods or services out of season.

secondary data Data that have been collected for purposes other than answering the firm's particular research question.

segmentation bases Broad categories of consumer characteristics that influence consumers' buying behavior and help to divide the market into segments.

selective distribution A distribution strategy that allows brands to sell through fewer channels but still capture economies of scale in production.

self-regulation A structure to settle disagreements in business without having to include government resources.

self-regulation A structure to settle disagreements in business without having to include government resources.

sender Starts the communication process by transmitting a message.

sentiment analysis A measurement that indicates whether people are reacting favorably or unfavorably to products or marketing efforts.

service failure When a customer's perceptions of a service fall below his or her expectations to a degree that leads to customer dissatisfaction.

Service Gaps Model This model provides companies with a framework for understanding the differences between customers' perceptions of a service outcome and the customers' initial expectations.

service quality A *measurement activity* of how good a service experience is or was.

service recovery Actions taken by a company in response to a service failure to improve the customer's level of satisfaction.

service retailers Stores that mostly sell services rather than merchandise.

share of customer Measures the quantity of purchase dollars each customer spends on the company's products rather than the number of customers.

shopping products Products that require considerable time to compare brands and features prior to making a purchase decision.

simple random sampling A type of sampling in which everyone in the target population has an equal chance of being selected.

situational analysis The systematic collection of data to identify the trends, conditions, and competitive forces that have the potential to influence the performance of the firm and the choice of appropriate strategies.

situational and personal influences Factors such as time, personality, lifestyle, values, and social surroundings that serve as an interface between the consumers and their decision-making process.

snowball sampling A type of sampling in which a set of participants is selected based on the referral of other participants who know they have some knowledge of the subject in question.

social media Electronic, two-way communication that allows users to share information, content, ideas, and messages to create a customizable experience.

social media influencers Consumers who have a large following and credibility within a certain market segment.

social media marketing The most popular form of digital marketing that utilizes online social networks and applications as a method to communicate mass and personalized messages about brands and products.

social media marketing campaign A coordinated marketing effort to advance marketing goals using one or more social media platforms.

social media platform A website-based media channel used to facilitate communication and connection.

sociocultural The combination of social and cultural factors that affect individual development.

specialty products Products that require a great deal of consumer search and for which substitutes will not be accepted.

specialty retailers Stores that concentrate on a specific product category.

sponsorship Firms that often sponsor YouTube or Instagram celebrities who in turn endorse the firms' products. These so-called online influencers are often compensated in multiple ways for their endorsements.

statistical analysis The mathematical classification, organization, presentation, and interpretation of numerical data.

stockout When a company does not have enough inventory available to fill an order.

store brands Products developed by a retailer and sold only by that specific retailer. Also known as *private label brands.*

straight rebuy When a business customer signals its satisfaction by agreeing to purchase the same product at the same price.

strategic greening A sustainable marketing strategy that requires a holistic approach that integrates and coordinates all the firm's activities on environmental issues across every functional area.

strategic planning The process of thoughtfully defining a firm's objectives and developing a method for achieving those objectives

strategic planning The process of thoughtfully defining a firm's objectives and developing an approach for achieving those objectives.

strategy The set of actions taken to accomplish organizational objectives.

strategy The set of actions taken to accomplish organizational objectives.

strengths Internal capabilities that help the company achieve its objectives.

supercenters Stores that offer traditional grocery items as well as apparel, beauty products, home goods, toys, hardware, electronics, and various other merchandise.

supermarkets Large, self-service retailers concentrating in supplying a wide variety of food, beverage, and kitchen products.

supply chain A set of three or more companies directly linked by one or more of the upstream and downstream flows of products, services, finances, and information from a source to a customer.

supply chain management The actions the firm takes to coordinate the various flows within a supply chain.

supply chain management The actions the firm takes to coordinate the various flows within a supply chain.

supply chain orientation A management philosophy that guides the actions of company members toward the goal of actively managing the upstream and downstream flows of goods, services, finances, and information across the supply chain.

surveys A data collection tool that poses a sequence of questions to respondents.

survival pricing A pricing strategy that involves lowering prices to the point at which revenue just covers costs, allowing the firm to endure during a difficult time.

sustainable competitive advantage When a product exhibits characteristics that are valuable, rare, and hard to imitate, and exhibits four elements of excellence: customer, operational, product, and locational.

sustainable marketing The process of creating, communicating, and delivering value to customers through the preservation and protection of the natural systems that provide the natural resources upon which our society and economy depend.

sustainable value innovation An environmental marketing strategy that entails reshaping the industry through the creation of differential value for consumers and through making contributions to society in the form of both reduced costs and reduced environmental impact.

sweepstakes Sales promotions based on chance such that entry is the only requirement to win.

Sweepstakes Sales promotions based on chance such that entry is the only requirement to win.

SWOT Analysis An evaluation of a firm's strengths, weaknesses, opportunities, and threats.

T

tactical greening A sustainable marketing strategy that involves implementing limited change within a single area of the organization, such as purchasing or advertising.

tangibles The appearance of physical facilities, equipment, personnel, and communication materials.

target market The group of customers toward which an organization has decided to direct its marketing efforts.

target marketing The second step of the segmentation, targeting, and positioning process. Occurs when marketers evaluate each market segment and determine which segment or segments present the most attractive opportunity to maximize sales. Also known as *targeting*.

targeting The second step of the segmentation, targeting, and positioning process. Occurs when marketers evaluate each market segment and determine which segment or segments present the most attractive opportunity to maximize sales. Also known as *target marketing*.

tariffs Taxes on imports and exports between countries.

tariffs Taxes on imports and exports between countries.

technical quality The quality of the service result or output.

television home shopping A non-store retailing method that allows customers to see a product demonstrated on television and purchase the product by calling a number or purchasing online.

test marketing Introducing a new product in its final form to a geographically limited market to see how well the product sells and get reactions from potential users.

threats Current and potential external factors that may challenge the firm's short- and long-term performance.

touchpoint Any point at which a customer and the company come into contact.

trade sales promotions Promotion tools directed at B2B firms, including wholesalers and retailers, rather than individual consumers.

trade sales promotions Are promotion tools directed to B2B firms, including wholesalers and retailers, rather than individual consumers.

transactional relationship Where both parties protect their own interests and where partners do things for each other purely as an exchange.

U

unbundling Separating out the individual goods, services, or ideas that make up a product and pricing each one individually.

underpricing Charging customers less than they are willing to pay.

undifferentiated targeting A targeting strategy that approaches the marketplace as one large segment. Because the firm doesn't segment the market further, it can approach all consumers with the same product offering and marketing mix.

United States-Mexico-Canada Agreement (USMCA) Regulates trade among the United States, Mexico, and Canada. The USMCA replaced the North American Free Trade Agreement (NAFTA) on July 1, 2020.

unsought products Products that are unknown to consumers and that they often don't look for, unless needed.

upstream flow The movement of payment (and other things, like information, returns, etc.) from the customer toward intermediaries and the original source in a supply chain.

users Individuals who will use the product once it is acquired.

users Individuals who will use the product once it is acquired.

V

validity How well the data measure what the researcher intended them to measure.

VALS Framework The most commonly used psychographic segmentation tool, which stands for Values and Lifestyles Framework. VALS classifies adults age 18 years and older into eight psychographic segments—Innovators, Thinkers, Believers, Achievers, Strivers, Experiencers, Makers, and Survivors—according to how they respond to a set of attitudinal and demographic questions. It measures two dimensions: primary motivation and resources.

Value the customer's perception that a certain product offers a better relationship between costs and benefits than competitors' products do.

value creation Results from a consumer's use of a product that satisfies a need and/or want.

values A consumer's belief that specific behaviors

are socially or personally preferable to other behaviors.

variable costs Costs that vary depending on the number of units produced or sold.

vending machines A non-store retail method that dispenses merchandise to consumers via an automated vending machine.

viral campaigns Promotional messages spread quickly by social media users forwarding promotional messages throughout their social networks.

volume maximization A pricing strategy that involves setting prices low to encourage a greater volume of purchases. Also known as *penetration pricing.*

W

wants Something that you would like to have; typically shaped by a customer's personality, family, job, background, and previous experiences.

warehouse retailers Stores that offer food and general merchandise products, usually in larger quantities and at discounted prices.

warehousing Storage, movement, and production of materials and finished goods.

weaknesses Internal limitations that may prevent or disrupt the firm's ability to meet its stated objectives.

weighted preference The sum (or score) of the weights applied to decision-making criteria that will be used to make a purchase decision.

weights Measures of importance or preference that consumers apply to various criteria used to make a product purchase decision.

wholesaler A firm that sells goods to anyone other than an end-user consumer.

wholesaler A firm that sells goods to anyone other than an end consumer.

wholesaler A firm that buys large quantities of goods from various producers or vendors, warehouses them, and resells to retailers.

wholesaling The sale of goods or merchandise to retailers; industrial, commercial, institutional, or other professional business users; or other wholesalers.

World Trade Organization (WTO) Regulates trade among participating countries and helps importers and exporters conduct their business.

Annotations

1. Daniel H. Pink, To Sell Is Human: The Surprising Truth about Moving Others (New York: Riverhead Books, 2013).

2. Anne Marie Mohan, "Store Brand Sales Grow Amidst Coronavirus Stock-Ups," Packaging World, May 1, 2020, https://www.packworld.com/issues/business-intelligence/article/21131043/store-brand-sales-grow-amidst-coronavirus-stockups.

3. Monte Burk, "Five Guys Burgers: America's Fastest Growing Restaurant Chain," Forbes, July 18, 2012, http://www.forbes.com/forbes/2012/0806/restaurant-chefs-12-five-guys-jerrymurrell-all-in-the-family.html.

9. Access Engineering, review of The Toyota Way: 14 Management Principles from the World's Greatest Manufacturer, by Jeffrey K. Liker (New York: McGraw-Hill Education, 2004). https://www.accessengineeringlibrary.com/browse/toyota-way-14-management-principles-from-the-worlds-greatest-manufacturer.

2. MBA Skool Team, "Airbnb Marketing Strategy & Marketing Mix (4Ps)," https://www.mbaskool.com/marketing-mix/services/17335-airbnb.html.

2. "How Is Southwest Different from Other Airlines?," Investopedia, 2015, http://www.investopedia.com/articles/investing/061015/how-southwest-different-other-airlines.asp.

15. Chloe Aiello, "Intel plunges on product delays and fears it's losing a near 'monopolistic position,'" July 27, 2018, CNBC, https://www.cnbc.com/2018/07/27/intel-plunges-on-competition-concerns-product-delays.html.

18. Sam Greenspan, "Coke Once Made an Intentionally Terrible Clear Cola to Take Down Crystal Pepsi." 11points.com, July 21, 2016, http://11points.com/coke-made-intentionally-terrible-clear-cola-take-crystal-pepsi/ (updated March 22, 2018).

15. "Parenting in America," Pew Research Center, December 17, 2015, http://www.pewsocialtrends.org/2015/12/17/1-the-american-family-today/.

8. AP News, "Miss. Producer Distributor Has Military Contract," Bloomberg Businessweek, March 29, 2013, http://www.businessweek.com/ap/2013-03-29/miss-dot-producedistributor-has-military-contract.

9. Business Queensland, "Benefits of Direct Marketing," https://www.business.qld.gov.au/running-business/marketing-sales/marketing-promotion/direct-marketing/using/benefits.

3. Ibid.

11. The Dieline Awards 2018, "About," http://www.thedielineawards.com/about/.

2. Stephanie Clifford, "Revamping, Home Depot Woos Women," The New York Times, January 28, 2011, http://www.nytimes.com/2011/01/29/business/29home.html.

9. Home Depot, "DIY Projects and Ideas," https://www.homedepot.com/c/diy_projects_and_ideas.

8. https://www.mbusa.com/en/vehicles/build/s-class/maybach/s650x.

2. Sam's Club, "Scan & Go Helps Create a Safe Shopping Experience for Sam's Club Members."-Retrieved from https://corporate.samsclub.com/newsroom/2020/04/30/scan-go-helps-create-a-safe-shopping-experience-for-sams-club-members

20. https://www.wired.com/2014/09/apples-devious-u2-album-giveaway-even-worse-spam/

3. Sebastian Weiss, "Husband and Wife Team Ties into Rising Tide of E-Business," San Antonio Business Journal, December 12, 1999, http://www.bizjournals.com/sanantonio/stories/1999/12/13/story5.html?page=all.

20. https://www.wired.com/2014/09/apples-devious-u2-album-giveaway-even-worse-spam/

5. 8451. (2021). "Break through the noise and be relevant with your customers." https://www.8451.com/

2. Blake Droesch, "Here Are the Top 10 US Ecommerce Companies for 2021—Plus 6 Key Takeaways from Our Latest Forecast," Insider Intelligence, March 24, 2021, https://www.emarketer.com/content/top-10-us-ecommerce-companies-2021-plus-6-key-takeaways-our-latest-forecast.

12. Oreo Press Release, "Oreo Announces Casting Call for National 'Milk's Favorite Jingle' Contest," June 7, 2005, http://www.prnewswire.com/news-releases/oreor-announces-castingcall-for-national-milks-favorite-jingle-contest-54538802.html.

10. Ryan Poe, "Ikea in Birmingham? Metro Falls Short of Retailer's Population Target," Birmingham Business Journal, October 29, 2012, http://www.bizjournals.com/birmingham/ blog/2012/10/ikea-in-birmingham-dont-hold-your.html.

19. Seeking Alpha, "Caterpillar: Weak Yen Adds to Company's Problems and Boosts Japanese Competitor," March 23, 2015, https://seekingalpha.com/article/3020046-caterpillar-weak-yen-adds-to-companys-problems-and-boosts-japanese-competitor.

3. https://www.businessinsider.com/then-and-now-mcdonalds-menu-items-2020-2#however-in-2021-mcdonalds-released-its-newest-entry-into-the-infamous-fast-food-chicken-sandwich-wars-the-crispy-chicken-sandwich-17

5. Jacquie McNish and Sean Silcoff, "The Inside Story of How the iPhone Crippled BlackBerry," The Wall Street Journal, May 22, 2015, https://www.wsj.com/articles/behind-the-rise-and-fall-of-blackberry-1432311912.

5. Danielle Sacks, "Scenes from the Culture Clash," Fast Company, January 1, 2006, http://www.fastcompany.com/54444/scenes-culture-clash.

11. https://www.marketwatch.com/story/the-9-most-misleading-product-claims-2014-06-11

16. https://researchguides.library.tufts.edu/c.php?g=248798&p=1659738.

6. OkCupid website, https://theblog.okcupid.com/we-experiment-on-human-beings-5dd9fe280cd5.

2. "Innovative Products," Kimberly-Clark, http://www.kimberly-clark.com/ourcompany/innovations.aspx.

3. Emily Bryson York, "Five Guys: An America's Hottest Brands Case Study, Advertising Age, November 16, 2009, http://adage.com/article/special-report-americas-hottest-brands-2009/guys-america-s-hottest-brands-case-study/140470/.

9. PR Newswire, "McDonalds Goes Blue This Summer with the Launch of The Smurfs," Bloomberg, July 29, 2011, http://www.bloomberg.com/apps/news?pid=conewsstory&tkr=MCD:US &sid=aaBly7Nclp98.

11. Molly Buccini, "How Much Are You Investing in Sales Training?," Brainshark, September 2016, https://www.brainshark.com/ideas-blog/2016/september/investing-in-sales-training.

6. Greg McFarlane, "Why Airlines Aren't Profitable," Investopedia, 2014, http://www.investopedia.com/stock-analysis/031714/why-airlines-arent-profitable-dal-ual-aal-luv-jblu.aspx.

7. Mike Esterl, "'Share a Coke' Credited with a Pop in Sales," The Wall Street Journal, September 25, 2014, https://www.wsj.com/articles/share-a-coke-credited-with-a-pop-in-sales-1411661519.

9. Jonathan Camhi and Stephanie Pandolph, "Amazon's Earnings Point to International Expansion," Business Insider, May 1, 2017, https://www.businessinsider.com/amazons-earnings-point-to-international-expansion-2017-5.

5. Jonathan Camhi and Stephanie Pandolph, "Amazon's Earnings Point to International Expansion," Business Insider, May 1, 2017, http://www.businessinsider.com/amazons-earnings-point-to-international-expansion-2017-5.

21. Stuart Leung, "5 Tips for a Successful Sales Promotion Strategy," Salesforce.com, August 20, 2015, https://www.salesforce.com/blog/2015/08/5-key-elements-successful-sales-promotion.html.

3. Natalie Jarvey, "Hulu Grows to 28 Million Subscribers," Hollywood Reporter, May 1 2019, https://www.hollywoodreporter.com/live-feed/hulu-grows-28-million-subscribers-1206322.

31. World Trade Organization, "The WTO," https://www.wto.org/english/thewto_e/thewto_e.htm.

2. "Coca-Cola Product Placement Photos," Product Placement Blog, https://productplacementblog.com/tag/coca-cola/

6. Evan Tarver, "What Makes the 'Share a Coke' Campaign So Successful?" Investopedia, http://www.investopedia.com/articles/markets/100715/what-makes-share-coke-campaign-so-successful.asp.

15. Sam Fiorella, "10 Examples of Innovative Marketing Campaigns," SENSEI Marketing, June 13, 2017, https://senseimarketing.com/10-examples-innovative-influencer-marketing-campaigns/.

* http://www.salesharvest.com.au/f.ashx/pdf/Sales-Harvest-How-To-Recruit-Sales-Stars.pdf

16. Raisa Bruner, "This Beauty Startup Has Become So Popular That It Has 10,000 People on a Waitlist for Lipstick," Business Insider, May 24, 2016, http://www.businessinsider.com/how-glossier-became-so-popular-2016-5.

5. Kerry Windle, "Share a Coke and a Word: How Coca-Cola Captured Millennials through Word of Mouth Marketing," Medium, May 19, 2016, https://medium.com/@kwindle/share-a-coke-and-a-word-how-coca-cola-captured-millennials-through-word-of-mouth-marketing-44896573d21c.

1. https://fortune.com/most-important-private-companies/2016/airbnb/.

8. Yoni Heisler, "A Look at Google's Many, Many Product Failures," BGR, August 30, 2016, http://bgr.com/2016/08/30/google-failures-buzz-plus-glass-wave/.

8. "The Great Recession," Investopedia, n.d., http://www.investopedia.com/terms/g/great-recession.asp.

3. "How Packaging Gives Apple's Buyers a Sensory Experience That Reinforces Brand," Personalics, February 3, 2016, https://www.personalics.com/2016/02/03/sensory-design-packaging/.

29. Tweriod website, http://www.tweriod.com/.

*Public Agenda, "Social Security," n.d., http://www.publicagenda.org/articles/social-security; Immersion Active, "50+ Facts and Fiction," n.d., http://www.immersionactive.com/resources/50-plus-facts-and-fiction/; Ellen Byron, "From Diapers to 'Depends': Marketers Discreetly Retool for Aging Boomers," The Wall Street Journal, February 5, 2011, http://online.wsj.com/article/SB10001424052748704013604576104394209062996.html.

2. "It All Began in 1905 in a Small Minnesota Town Called Red Wing," Red Wing Shoes, http://www.redwingheritage.com/us/USD/page/history.

4. FBA Help, "Locations of Amazon Fulfillment Centers." Retrieved from https://fba.help/list-of-amazon-fulfillment-centers

27. Liz Reid, "Steelworkers Rally against Alleged Illegal 'Dumping' of South Korean Steel Pipes," 90.5 WESA, May 19, 2014, http://wesa.fm/post/steelworkers-rally-against-alleged-illegal-dumping-south-korean-steel-pipes#stream/0.

4. "What Is 'Share a Coke'?" Coca-Cola Journey, https://www.coca-cola.ie/faq/share-a-coke

3. Paresh Dave, "Dollar Shave Club Succeed with Razors, but the Rest of the Bathroom Is a Challenge," Los Angeles Times, September 1, 2017, http://www.latimes.com/business/technology/la-fi-tn-dollar-shave-club-unilever-anniversary-20170828-htmlstory.html.

9. https://www.campaignlive.co.uk/article/mattel-turned-too-perfect-unrelatable-barbie-symbol-female-empowerment/1663884.

7. The World Bank, "GDP per capita (2018)," https://data.worldbank.org/indicator/ny.gdp.pcap.cd.

13. "The Polaroid Pop Instant Digital Camera Offers a Modern Take on the Classic Polaroid Instant Print," Polaroid, January 5, 2017, https://www.polaroid.com/news/polaroid-pop-instant-digital-camera.

3. Amazon, "Amazon Services," https://services.amazon.com/global-selling/overview.html/ref=asus_ags_snav_over.

1. BlendJet, "Blendjet," https://blendjet.com/products/blendjet-one?variant=7922727452738&currency=USD&gclid=EAIaIQobChMIgaS-ovj44wlVxODICh2B8g2hEAYYASABEgL2wfD_BwE.

27. Michelle Dipardo, "Expedia.ca Responds to Angry Feedback with New Ads," Marketing, January 24, 2014, http://marketingmag.ca/news/marketer-news/expedia-ca-responds-to-angry-social-media-feedback-with-new-ads-99039.

David Mildenburg, "Census finds Hurricane Katrina Left New Orleans Richer, Whiter, Emptier," Bloomberg, February 3, 2011, http://www.bloomberg.com/news/2011-02-04/censusfinds-post-katrina-new-orleans-richer-whiter-emptier.html.

8. Smarty, Ann, (2013), "The Birth and Evolution of Digital Content Marketing," Internet Marketing Ninjas, https://www.internetmarketingninjas.com/blog/content/the-birth-and-evolution-of-digital-content-marketing/

*Gene Marchial, "Discount Retailer Dollar General Taking Away Market Share from No. 1 Wal-Mart," Forbes, December 29, 2011, http://www.forbes.com/sites/genemarcial/2011/12/29/discount-retailer-dollar-general-taking-away-market-sharefrom-no-1-wal-mart/; "The Substance of Style," The Economist, September 17, 2009, http://www.economist.com/node/14447276.

3. "Teens Prove Their #Streettread," Ketchum, https://www.ketchum.com/work/teens-prove-their-streettread/.

3. https://www.businessinsider.com/then-and-now-mcdonalds-menu-items-2020-2#however-in-2021-mcdonalds-released-its-newest-entry-into-the-infamous-fast-food-chicken-sandwich-wars-the-crispy-chicken-sandwich-17.

10. Joshua Porter, "The Button Color A/B Test: Red Beats Green," https://blog.hubspot.com/blog/tabid/6307/bid/20566/the-button-color-a-b-test-red-beats-green.aspx.

12. Clare O'Connor, "Forbes Top Influencers: How YouTuber Lilly Singh Is Going Mainstream—and Making Millions," Forbes, June 20, 2017, https://www.forbes.com/forbes/welcome/?toURL; https://www.forbes.com/sites/clareoconnor/2017/06/20/forbes-top-influencers-lilly-singh-superwoman-youtube/&refURL=https://en.wikipedia.org/&referrer=https://en.wikipedia.org/#1cea120e33e7.

6. Eliana Dockterman, "FDA Warns the Parmesan You Eat May Be Wood Pulp," Time, February 16, 2016, http://time.com/4226321/parmesan-wood-pulp/.

14. Walmart, "Walmart International," https://corporate.walmart.com/our-story/our-business

5. Christian Grönroos, Service Management and Marketing, 3rd ed. (Chichester, NY: Wiley, 2007).

21. https://appleinsider.com/articles/18/09/09/the-free-u2-album-songs-of-innocence-was-a-debacle-for-apple-fans-on-september-9-2018

11. Tracey Miller, "'Got Milk' Ad Campaign Retired after 20 Years of Milk Moustaches," New York Daily News, February 25, 2014, http://www.nydailynews.com/life-style/health/milk-ads-retired-20-years-milk-moustaches-article-1.1701064.

8. Deloitte US, The CMO Survey: Rising to Meet the Challenge:

Observations, Insights, and Benchmarks for the Modern Marketer,

https://www2.deloitte.com/us/en/pages/chief-marketing-officer/articles/cmo-survey.html.

*"Are You Winning with Hispanics?," Adweek, April 27, 2011, http://www.adweek.com/sa-article/are-you-winning-hispanics-131093.

22. Licensing International, "Global Sales of Licensed Products and Services Reach US $280.3 Billion, Fifth Straight Year of Growth for the Licensing Industry," June 4, 2019, https://licensinginternational.org/news/global-sales-of-licensed-products-and-services-reach-us-280-3-billion-fifth-straight-year-of-growth-for-the-licensing-industry/.

10. Lucy Handley, "We Need 'Fewer' Ads, Says Consumer Goods Company That Spent $7.2 billion on Advertising in 2016," CNBC, April 5, 2017, https://www.cnbc.com/2017/04/05/we-need-fewer-ads-says-pg-which-spent-7-2bn-on-adverts-in-2016.html.

5. https://info.hanesbrandsb2b.com/about-hanesbrandsb2b/index.html.

7. http://www.thedielineawards.com/about/

6. Kashmir Hill, "How Target Figured Out a Teen Girl Was Pregnant before Her Father Did," Forbes, February 16, 2012, https://www.forbes.com/sites/kashmirhill/2012/02/16/how-target-figured-out-a-teen-girl-was-pregnant-before-her-father-did/2/#8df38d447e85.

7. Joseph R. Hair Jr. et al., Essentials of Marketing Research, 2nd ed. (New York: McGraw-Hill/Irwin, 2010).

8. American Marketing Association, "Dictionary," https://www.ama.org/resources/pages/dictionary.aspx?dLetter=D.

26. Hannah Givertz, "Chile: The Largest Consumer of Mayo in the Americas," North by Northwestern, October 18, 2016, http://www.northbynorthwestern.com/story/chile-the-largest-consumer-of-mayo-in-the-americas/.

7. Gavin Clarke, "Oracle Tools Up for Cloud Wars with Sales Re-Org," The Register, July 22, 2016, https://www.theregister.co.uk/2016/07/22/oracle_sales_reorganisation_for_cloud/.

2. E. Mazareanu, "Monthly Users of Uber's Ride-Sharing App Worldwide 2016–2019," Statista, August 9 2019, https://www.statista.com/statistics/833743/us-users-ride-sharing-services/.

3. https://www.theatlantic.com/entertainment/archive/2016/05/legos/484115/.

8. "Share a Coke," Revolvy, https://www.revolvy.com/page/Share-a-Coke.

2. OkCupid, "How Match Search Works," https://okcupid.desk.com/customer/en/portal/articles/2152314-how-match-search-works.

9. Abbas Ali, "Amazon Officially Launches in the UAE," Techradar, May 1, 2019, https://www.techradar.com/news/amazon-officially-launches-in-the-uae.

4. Oracle, "Oracle Class of—North America," http://www.oracle.com/webfolder/joinclassof/index.html.

2. "Teens Prove Their #Streettread," Ketchum, https://www.ketchum.com/work/teens-prove-their-streettread/.

5. Abraham Maslow, Motivation and Personality (New York: Harper, 1954), p. 236.

4. https://www.restaurantbusinessonline.com/food/brief-history-chicken-sandwich-wars.

30. Caroline Cakebread, "Snapchat Is American Teens' Favorite Social Media App—and Facebook Can't Be Happy about That," Business Insider, October 16, 2017, http://www.businessinsider.com/for-us-teens-snapchat-is-more-popular-than-instagram-twitter-charts-2017-10.

7. Panos Mourdoukoutas, "By the Time Wal-Mart Catches Up with Amazon, There Will Be No Neighborhood Stores," Forbes, August 13, 2016, https://www.forbes.com/sites/panosmourdoukoutas/2016/08/13/by-the-time-wal-mart-catches-up-with-amazon-there-will-be-no-neighborhood-stores/#70758b967ac3.

1. Avery Hartmans, "15 Fascinating Facts You Probably Didn't Know about Amazon," Business Insider, April 9, 2017, http://www.businessinsider.com/jeff-bezos-amazon-history-facts-2017-4.

https://www.securityweek.com/beauty-and-breach-est%C3%A9e-lauder-exposes-440-million-records-unprotected-database

8. Joseph R. Hair Jr. et al., Essentials of Marketing Research, 2nd ed. (New York: McGraw-Hill/Irwin, 2010).

3. Steve Dudash, "WeWork Faces Challenge Before IPO," USNews.com, July 22, 2019, https://money.usnews.com/money/blogs/the-smarter-mutual-fund-investor/articles/wework-faces-challenge-before-ipo.

4. Marc de Swaan Arons, Frank van den Driest, and Keith Weed, "The Ultimate Marketing Machine," Harvard Business Review, July–August 2014, https://hbr.org/2014/07/the-ultimate-marketing-machine.

3. Marc de Swaan Arons, Frank van den Driest, and Keith Weed, "The Ultimate Marketing Machine," Harvard Business Review, July–August 2014, https://hbr.org/2014/07/the-ultimate-marketing-machine.

1. "The Five Guys Story," Five Guys, http://www.fiveguys.com/Fans/The-Five-Guys-Story.

24. Zach Brooke, "Five Tourism Campaigns That Backfired," American Marketing Association, April 7, 2016, https://www.ama.org/publications/eNewsletters/Marketing-News-Weekly/Pages/tourism-ad-marketing-fails-backfire.aspx.

11. https://channelsignal.com/blog/as-barbie-turns-60-mattel-aims-to-stay-on-trend-with-consumers/.

6. Shelley Frost, "The Disadvantages of Advertising to Kids," Chron, http://smallbusiness.chron.com/disadvantages-advertising-kids-25801.html.

2. Kian Bakhtiari, "How Will the Pandemic Change Consumer Behavior?" May 18, 2020, https://www.forbes.com/sites/kianbakhtiari/2020/05/18/how-will-the-pandemic-change-consumer-behavior/?sh=2c20dd1866f6.

2. Investopedia. (2015). How Is Southwest Different From Other Airlines? Investopedia. http://www.investopedia.com/articles/investing/061015/how-southwest-different-other-airlines.asp

3. Jackie Quintana, "5 Companies Who Are Doing Integrated Marketing Right in 2017," lonely brand, March 7, 2017, https://lonelybrand.com/blog/3-companies-integrated-marketing-right/.

14. https://www.inc.com/geoffrey-james/the-20-worst-brand-translations-of-all-time.html.

8. Yoni Heisler, "Inside Apple's Secret Packaging Room," Network World, January 24, 2012, https://www.networkworld.com/article/2221536/data-center/inside-apple-s-secret-packaging-room.html.

28. Kyodo News, "Time Is Ripe for Banana Vending Machine: Dole," The Japan Times, July 31, 2010, https://www.japantimes.co.jp/news/2010/07/31/national/time-is-ripe-for-banana-vending-machine-dole/#.WhjVUEpKtPY.

6. Nicola Fumo, "When Did Shipping Boxes Get Pretty?", Racked, Jul 7 2016. https://www.racked.com/2016/7/7/12002156/packaging-commerce-shipping-boxes-glossier-warby-parker.

21. BEHR, "ColorSmart by BEHR® Mobile App," BEHR.com, http://www.behr.com/consumer/colors/mobile-apps.

1. Ryan Nakashima, "Disney Joins NBC and News Corp. with Hulu Stake," San Diego Union-Tribune, April 30 2019, https://www.sandiegouniontribune.com/sdut-us-disney-hulu-043009-2009apr30-story.html.

12. Joseph R. Hair Jr. et al., Essentials of Marketing Research, 2nd ed. (New York: McGraw-Hill/Irwin, 2010).

1. Michele Kayal, "America's Patchwork of Potato Chip Varieties," Huff Post, August 8, 2012, http://www.huffingtonpost.com/2012/08/14/regional-potatochip-varieties_n_1775098.html.

4. Inc., "2012 Inc.com 5000 List," n.d., http://www.inc.com/inc5000/list/2012/400/employees/ascend.

11. U.S. Census Bureau, "U.S. and World Population Clock," https://www.census.gov/popclock/.

3. https://www.sec.gov/Archives/edgar/data/1359841/000095014406009152/g03506e10vk.htm.

1. Kayleigh Moore, "Red Bull's Approach to Marketing: Then and Now," https://hashtagpaid.com/banknotes/red-bulls-approach-to-marketing-then-and-now.

1. Oracle, "About Oracle," https://www.oracle.com/corporate/index.html.

2. Coca-Cola, "About Us—Coca-Cola History," https://www.worldofcoca-cola.com/about-us/coca-cola-history/.

1. Instagram, "Everlane," https://www.instagram.com/everlane/?hl=en, accessed April 5, 2018.

1. Coca-Cola, "About Us—Coca-Cola History," https://www.worldofcoca-cola.com/about-us/coca-cola-history/.

35. Mitali Shaw, "How UNIQLO Is Set to Become the World's Top Fashion Retailer," April 3, 2021, https://thestrategystory.com/2021/04/03/uniqlo-business-strategy/.

9. Duo, "Secure Access Starts with (Zero) Trust," https://duo.com/.

5. "How Is Southwest Different from Other Airlines?," Investopedia, 2015, http://www.investopedia.com/articles/investing/061015/how-southwest-different-other-airlines.asp.

3. https://www.campaignlive.co.uk/article/mattel-turned-too-perfect-unrelatable-barbie-symbol-female-empowerment/1663884.

6. "Walmart: 3 Keys to Successful Supply Chain Management Any Business Can Follow," FlashGlobal, April 12, 2018, http://flashglobal.com/blog/supply-chain-management-walmart.

4. Paresh Dave, "Dollar Shave Club Succeed with Razors, but the Rest of the Bathroom Is a Challenge," Los Angeles Times, September 1, 2017, http://www.latimes.com/business/technology/la-fi-tn-dollar-shave-club-unilever-anniversary-20170828-htmlstory.html.

13. Federal Trade Commission, "Advertising and Marketing," Federal Trade Commission, https://www.ftc.gov/tips-advice/business-center/advertising-and-marketing.

7. Issie Lapowsky, "Why GoPro's Success Isn't Really about the Cameras," Wired, June 26, 2014, https://www.wired.com/2014/06/gopro/.

7. Citeman. "Decoding Problems." October 11, 2010. Accessed at https://www.citeman.com/11065-decoding-problems.html

1. Jia Wertz, "Changes in Consumer Behavior Brought On by the Pandemic," January 31, 2021, https://www.forbes.com/sites/jiawertz/2021/01/31/changes-in-consumer-behavior-brought-on-by-the-pandemic/?sh=1f3a11cc559e.

3. Amazon Seller Central, "Where and What to Sell with Amazon Global Selling," https://sellercentral.amazon.com/gp/help/external/G201468340.

Joshua Caucutt, "Cracker Barrel Is Cracklin'," InvestorGuide, May 25, 2010, http://www.investorguide.com/article/6467/ cracker-barrel-is-cracklin-cbrl/.

20. https://www.wired.com/2014/09/apples-devious-u2-album-giveaway-even-worse-spam/

2. Dan Blystone, "The Story of Uber," Investopedia, 2017, http://www.investopedia.com/articles/personal-finance/111015/story-uber.asp.

8. Catherine Morin, "Patagonia's Customer Base and the Rise of an Environmental Ethos," June 24, 2020, https://crm.org/articles/patagonias-customer-base-and-the-rise-of-an-environmental-ethos.

3. https://news.airbnb.com/about-us/

5. Jeremiah D. McWilliams, "Coca-Cola No. 3 on Most Valuable Brands Ranking," Coca-Cola, October 5, 2016, http://www.coca-colacompany.com/coca-cola-unbottled/coca-cola-no-3-on-most-valuable-brands-ranking.

19. Barbara De Lollis, "IHG's Loyalty Club to Raise Point Rates for 25% of Hotels," USA Today, January 6, 2012, http://travel.usatoday.com/hotels/post/2012/01/ihg-priority-club-to-raisepoint-rates-but-phase-in-change/591898/1.

6. Statista, "Annual net revenue of Amazon from 2006 to 2020, by segment." Retrieved from https://www.statista.com/statistics/266289/net-revenue-of-amazon-by-region/

21. Jake Rheude, "Why Your Business Should Get into the Export Market," Forbes, July 6, 2017, https://www.forbes.com/sites/forbescommunicationscouncil/2017/07/06/why-your-business-should-get-into-the-export-market/.

5. https://www.nytimes.com/2021/02/18/magazine/amazon-workers-employees-covid-19.html.

10. McGraw Hill Marketing Insights Podcast, "Walk It, Like They Talk It Brands (Diversity, Equity & Inclusion in Branding), https://podcasts.apple.com/us/podcast/walk-it-like-they-talk-it-brands-diversity-equity-inclusion/id1378417469?i=1000492069911.

10. https://www.uber.com/us/en/about/

3. "Company History," Toyota, http://corporatenews.pressroom.toyota.com/corporate/company+history/.

4. Ellen Byron, "Bounty Puts a New Spin on Spills," The Wall Street Journal, February 17, 2009, http://online.wsj.com/article/SB123482144767494581.html.

2. Thomas A. Stewart and Anand P. Raman, "Lessons from Toyota's Long Drive," Harvard Business Review 85, no. 7 (2007), p. 8.

9. Carson, Biz, (2016), "This Is the True Story of How Mark Zuckerberg Founded Facebook, and It Wasn't to Find Girls," Business Insider, http://www.businessinsider.com/the-true-story-of-how-mark-zuckerberg-founded-facebook-2016-2.

1. Nike, "Get Help," https://help-en-us.nike.com/app/answer/a_id/3393.

11. https://www.marketwatch.com/story/the-9-most-misleading-product-claims-2014-06-11

10. Brian Grow, "The Great Rebate Runaround," Bloomberg Businessweek, November 22, 2005, http://www.businessweek.com/stories/2005-11-22/the-great-rebate-runaround.

6. "Best Global Brands 2017 Rankings," Interbrand, http://interbrand.com/best-brands/best-global-brands/2017/ranking/.

3. Peter Drucker, Management: Tasks, Responsibilities, Practices (New York: Truman Talley Books, 1986), pp. 58–69.

2. Keith Naughton, "The Great Wal-Mart of China," Newsweek, October 29, 2006, http://www.thedailybeast.com/newsweek/ 2006/10/29/the-great-wal-mart-of-china.html.

10. J.D. Power, "Most Owners Still in Love with Their Three-Year-Old Vehicles," J.D. Power Finds, February 13, 2019, https://www.jdpower.com/business/press-releases/2019-us-vehicle-dependability-studyvds.

1. WeWork, "Our Mission," www.wework.com/mission.

7. Lauren Thomas, "Sears Is Closing 63 More Stores," CNBC, November 3, 2017, https://www.cnbc.com/2017/11/03/sears-is-closing-63-more-stores.html.

2. Russ W, "Red Bull's $30 Million Marketing Stunt Almost Didn't Happen," June 24, 2020, https://bettermarketing.pub/red-bulls-30-million-marketing-stunt-almost-didn-t-happen-88d24fefdeff.

6. "Company History," Toyota, http://corporatenews.pressroom.toyota.com/corporate/company+history/.

7. https://www.bizjournals.com/triad/news/2020/03/13/hanesbrands-partners-with-amazon-for-exclusive.html.

6. Christian Grönroos, Service Management and Marketing, 3rd ed. (Chichester, NY: Wiley, 2007).

6. Statista, "Annual Net Revenue of Amazon from 2006 to 2020, by Segment," https://www.statista.com/statistics/266289/net-revenue-of-amazon-by-region/.

7. Mourdoukoutas, Panos. (2016). By The Time Wal-Mart Catches Up With Amazon, There Will Be No Neighborhood Stores. Forbes. https://www.forbes.com/sites/panosmourdoukoutas/2016/08/13/

by-the-time-wal-mart-catches-up-with-amazon-there-will-be-no-neighborhood-stores/#70758b967ac3

2. Dale Buss, "Ford Thinks It Has a Better Idea about How to Handle Facebook," Forbes, May 15, 2012, http://www.forbes.com/sites/dalebuss/2012/05/15/ford-thinks-it-has-betteridea-about-how-to-handle-facebook/.

24. Rapaport News, "Tiffany Loses Appeal on $450M Swatch Ruling," April 27, 2017, http://www.diamonds.net/News/NewsItem.aspx?ArticleID=58823.

7. "Why You Need to Measure Brand Equity—and How to Do It," Neilpatel, https://blog.kissmetrics.com/measure-brand-equity/.

5. "Best Global Brands 2017 Rankings,"Interbrand, http://interbrand.com/best-brands/best-global-brands/2017/ranking/.

26. Richard Koch, "How Uber Used a Simplified Business Model to Disrupt the Taxi Industry," Entrepreneur, January 3, 2017, https://www.entrepreneur.com/article/286683.

3. Adbrands, "The Us Top Advertisers in 2015 by Expenditure," http://www.adbrands.net/us/top_us_advertisers.htm.

https://www.digitalcommerce360.com/article/us-ecommerce-sales/

9. Nicola Fumo, "When Did Shipping Boxes Get Pretty?," Racked, July 7, 2016, https://www.racked.com/2016/7/7/12002156/packaging-commerce-shipping-boxes-glossier-warby-parker.

17. "XE Currency Charts, EUR to USD," XE, https://www.xe.com/currencycharts/?from=EUR&to=USD&view=2Y.

Lindsay M. Howden and Julie A. Meyer, "Age and Sex Composition: 2010," U.S. Census Bureau, May 2011, http:// www.census.gov/prod/cen2010/briefs/c2010br-03.pdf .

4. Jay LaBella, "Chase Bank Commercial Walking Baby Pig," YouTube, May 5 2016, https://www.youtube.com/watch?v=NaFlfaXJW4Y.

2. https://www.theatlantic.com/entertainment/archive/2016/05/legos/484115/.

24. Anders Vinderslev, "The Top 10 Examples of BuzzFeed Doing Native Advertising," Native Advertising Institute, https://nativeadvertisinginstitute.com/blog/10-examples-buzzfeed-native-advertising/.

1. Whole Foods, "We Satisfy and Delight Our Customers," n.d., http://www.wholefoodsmarket.com/mission-values/core-values/we-satisfy-delight-and-nourish-our-customers.

1. Monte Burk, "Five Guys Burgers: America's Fastest Growing Restaurant Chain," Forbes, July 18, 2012, http://www.forbes. com/forbes/2012/0806/restaurant-chefs-12-five-guys-jerrymurrell- all-in-the-family.html .

2. Issie Lapowsky, "4 Takeaways from the Iconic 'Got Milk?' Ad Campaign," Inc.(2014), https://www.inc.com/issie-lapowsky/marketing-tips-got-milk.html.

9. "The World's Most Valuable Brands, 2019 Ranking," Forbes, https://www.forbes.com/powerful-brands/list/.

7. Akshay Heble, "Case Study on Coca-Cola 'Share A Coke' Campaign,"Digital Vidya, November 11, 2019, https://www.digitalvidya.com/blog/case-study-on-coca-colas-share-a-coke-campaign/.

8. "The Us Top Advertisers in 2015 by Expenditure," Adbrands, http://www.adbrands.net/us/top_us_advertisers.htm.

2. https://www.kicksonfire.com/history-of-nike/.

7. Nick Bunkley, "Sales Decline 20%, but GM Sees a Bright Spot," The New York Times, September 3, 2008, http://www.nytimes.com/2008/09/04/business/04auto.html.

9. https://chrisrmark217.wordpress.com/2018/05/02/lego-building-a-successful-future-through-market-research-one-block-at-a-time/.

32. A. J. Agrawal, "How to Utilize Real-Time Data in Your Marketing Efforts," Forbes, January 13, 2016, https://www.forbes.com/sites/ajagrawal/2016/01/13/how-to-utilize-real-time-data-in-your-marketing-efforts/#3c1c44a05741.

5. Walmart, "Supporting the Dignity of Workers Everywhere," 2017, https://corporate.walmart.com/2017grr/sustainability/supporting-the-dignity-of-workers-everywhere; Walmart, "Promoting Responsibility," 2017, https://corporate.walmart.com/sourcing/promotingresponsibility.

1. Coca-Cola, "About Us," World of Coca-Cola, https://www.worldofcoca-cola.com/about-us/coca-cola-history/.

3. Oracle, "About Oracle," https://www.oracle.com/corporate/index.html.

4. "Wholesaler," Business Dictionary, http://www.businessdictionary.com/definition/wholesaler.html.

1. "Southwest Corporate Fact Sheet," Southwest Airlines, https://www.swamedia.com/pages/corporate-fact-sheet.

6. https://info.hanesbrandsb2b.com/why-become-retailer/index.html.

1. Planet Fitness, "About Planet Fitness," PlanetFitness.com, 2022. https://www.planetfitness.com/about-planet-fitness

1. Jason Del Rey, "The Key to the Business of Love? Data, of Course," Advertising Age, February 12, 2012, http://adage.com/article/digital/okcupid-s-key-business-love-data/232686/.

1. Uber, "Our Trip History," 2017, https://www.uber.com/our-story/.

7. Stuart Elliott, "Super Bowl Commercial Time Is a Sellout," The New York Times, January 8, 2013, http://www.nytimes .com/2013/01/09/business/media/a-sellout-for-super-bowlcommercial-time.html?_r 5 0.

Monte Burk, "Five Guys Burgers: America's Fastest Growing Restaurant Chain," Forbes, July 18, 2012, http://www.forbes. com/forbes/2012/0806/restaurant-chefs-12-five-guys-jerrymurrell- all-in-the-family.html.

11. Interactive Advertising Bureau, "The Native Advertising Playbook," Interactive Advertising Bureau, December 4, 2013, https://www.iab.com/wp-content/uploads/2015/06/IAB-Native-Advertising-Playbook2.pdf.

4. "Bags Fly Free," Southwest Airlines, 2017, https://www.southwest.com/html/air/bags-fly-free.html.

2. Akshay Heble, "Case Study on Coca-Cola 'Share A Coke' Campaign," Digital Vidya, November 11, 2019, https://www.digitalvidya.com/blog/case-study-on-coca-colas-share-a-coke-campaign/.

10. Nicola Fumo, "When Did Shipping Boxes Get Pretty?," Racked, July 7, 2016, https://www.racked.com/2016/7/7/12002156/packaging-commerce-shipping-boxes-glossier-warby-parker.

6. https://www.greenamerica.org/amazon-labor-exploitation-home-and-abroad.

6. https://www.forbes.com/sites/cherylrobinson/2020/03/17/how-global-head-of-barbie-focuses-on-the-brands-girl-empowerment-messaging/?sh=1b489bd5190f.

4. Hulu, "What Are the Costs and Commitments for Hulu?," Hulu, August 2 2019, https://help.hulu.com/s/article/how-much-does-hulu-cost?language=en_US.

11. H. O. Maycotte, "Beacon Technology: The Where, What, Who, How and Why," Forbes, September 1, 2015, https://www.forbes.com/sites/homaycotte/2015/09/01/beacon-technology-the-what-who-how-why-and-where/#51ac16d41aaf.

3. Walmart, "Truck Fleet," 2017, http://corporate.walmart.com/global-responsibility/environment-sustainability/truck-fleet.

6. Evan Tarver, "Why the 'Share a Coke' Campaign Is so Successful," Investopedia, September 24, 2019, https://www.investopedia.com/articles/markets/100715/what-makes-share-coke-campaign-so-successful.asp.

29. David Vinjamuri, "Ethics and the Five Deadly Sins of Social Media," Forbes, November 3, 2011, https://www.forbes.com/sites/davidvinjamuri/2011/11/03/ethics-and-the-5-deadly-sins-of-social-media/#4b1d3ea3e1fb.

6. Tom Huddleston, "Here's How Many People Watched the Super Bowl," Fortune, February 6, 2017, http://fortune.com/2017/02/06/super-bowl-111-million-viewers/.

6. Sawhney Ravi and Deepa Prahalad, "The Role of Design in Business," Bloomberg Businessweek, February 1, 2010, http://www.businessweek.com/stories/2010-02-01/the-roleof-design-in-businessbusinessweek-business-news-stockmarket-and-financial-advice.

9. Penelope Green, "Under One Roof, Building for Extended Families," The New York Times, November 29, 2012, http://www.nytimes.com/2012/11/30/us/building-homes-for-modern-multigenerational-families.html?pagewanted=1.

*D'Arcy Maine, "Serena Williams Gives Ode to Her Daughter about the Power of Sports in New Gatorade Ad," The Buzz, November 21, 2017, http://www.espn.com/espnw/culture/the-buzz/article/21488041/serena-williams-delivers-powerful-message-impact-sports-gatorade-ad.

6. Jonathan Camhi and Stephanie Pandolph, "Amazon's Earnings Point to International Expansion," Business Insider, May 1, 2017, http://www.businessinsider.com/amazons-earnings-point-to-international-expansion-2017-5.

5. Handley, Lucy "We need 'fewer' ads, says consumer goods company that spent $7.2 billion on advertising in 2016," CNBC, https://www.cnbc.com/2017/04/05/we-need-fewer-ads-says-pg-which-spent-7-2bn-on-adverts-in-2016.html

2. Piper MacDougall, "The Brand That Invented the Hoodie Is Finally Cool Again," FASHION magazine, April 26, 2019, https://fashionmagazine.com/style/champion-is-cool-again/.

8. https://www.businessinsider.com/inside-barbies-brand-comeback-with-global-gross-billings-up-19-percent.

1. Dan Blystone, "The Story of Uber," Investopedia, June 25 2019, https://www.investopedia.com/articles/personal-finance/111015/story-uber.asp.

7. Jonathan Camhi and Stephanie Pandolph, "Amazon's Earnings Point to International Expansion," Business Insider, May 1, 2017, https://www.businessinsider.com/amazons-earnings-point-to-international-expansion-2017-5.

8. OkCupid website, https://theblog.okcupid.com/we-experiment-on-human-beings-5dd9fe280cd5.

5. "2018 Annual Report," RE/MAX, https://s2.q4cdn.com/001766218/files/doc_financials/annual_reports/2019/RMAX-2018-Annual-Report-Bookmarked.pdf.

10. "How Artificial Intelligence (AI) Is Helping Retailers Predict Prices," Intelligence Node, June 7, 2017, http://www.intelligencenode.com/blog/retailers-predict-prices-using-ai/.

2. Tessa Berenson, "Millennials Don't Know What Fabric Softener Is For," Fortune, December 16, 2016, http://fortune.com/2016/12/16/millennials-fabric-softener-downy/.

9. OkCupid website, https://theblog.okcupid.com/we-experiment-on-human-beings-5dd9fe280cd5.

8. Avery Hartmans and Nathan McAlone, "The Story of How Travis Kalanick Built Uber into the Most Feared and Valuable Startup in the World,"Business Insider, August 1, 2016, http://www.businessinsider.com/ubers-history.

8. Nat Levy, "Amazon Completes $580M Acquisition of Middle East E-Commerce Gian Souq.com," GeekWire, July 3, 2017, https://www.geekwire.com/2017/amazon-completes-580m-acquisition-middle-east-e-commerce-giant-souq-com/.

U.S. Census Bureau. "Table 14, State Population—Rank, Percent Change, and Population Density: 1980–2010," hun61094_Statistical Abstract of the United States: 2012, n.d., http:// www.census.gov/compendia/statab/2012/tables/12s0014.pdf.

7. "Facial Recognition: Getting to Know Your Customers Just Got More Interesting," Figment, March 14, 2017, https://www.figmentagency.com/blog/facial-recognition-getting-to-know-your-customers-just-got-more-interesting/.

https://www.digitalcommerce360.com/article/us-ecommerce-sales/

1. Nikki Ekstein, "You Can Now Rent Hotel Rooms by the Minute," Bloomberg, April 24, 2017, https://www.bloomberg.com/news/articles/2017-04-24/you-can-now-rent-hotel-rooms-by-the-minute-with-the-recharge-app.

7. Jonathan Camhi and Stephanie Pandolph, "Amazon's Earnings Point to International Expansion," Business Insider, May 1, 2017, https://www.businessinsider.com/amazons-earnings-point-to-international-expansion-2017-5.

20. Polina Marinova, "Why China Is Crucial for Airbnb's Global Ambitions," Fortune, December 5, 2017, http://fortune.com/2017/12/05/airbnb-china-growth/.

4. Tameem Rahman, "How Champion Became Cool Again," Medium, May 28, 2020, https://medium.com/better-marketing/how-champion-became-cool-again-c4e58fa46671.

*Stephanie Startz, "Hyundai Formula: Inconspicuous Luxury Plus Empathy," BrandChannel, September 22, 2009, http://www.brandchannel.com/home/post/2009/09/22/Hyundai-Formula-Inconspicuous-Luxury-Plus-Empathy.aspx#.

17. Oreo Press Release, "Oreo Announces Casting Call for National 'Milk's Favorite Jingle' Contest," PR Newswire, June 7, 2005, http://www.prnewswire.com/news-releases/oreor-announces-castingcall-for-national-milks-favorite-jingle-contest-54538802.html.

1. Hayley Peterson, "This Is What the Average Walmart Shopper Looks Like," Business Insider, October 7, 2016, http://www.businessinsider.com/walmart-shopper-demographics-2016-10; https://corporate.walmart.com/.

13. Kashmir Hill, "How Target Figured Out a Teen Girl Was Pregnant before Her Father Did," Forbes, February 16, 2012, https://www.forbes.com/sites/kashmirhill/2012/02/16/how-target-figured-out-a-teen-girl-was-pregnant-before-her-father-did/#1983dc256668.

4. Yoni Heisler, "Inside Apple's Secret Packaging Room," Network World, Jan 24, 2012. https://www.networkworld.com/article/2221536/data-center/inside-apple-s-secret-packaging-room.html.

6. Zsolt Katona, "How to Identify Influence Leaders in Social Media," Bloomberg, February 26, 2012, http://www.bloomberg.com/news/2012-02-27/how-to-identify-influence-leaders-insocial-media-zsolt-katona.html.

6. Rob Dickens, "Old Spice Guy, Viral Media Coup: Social Media Game-Changers," BizCommunity, July 19, 2010, https://www.bizcommunity.com/Article/196/16/50138.html.

3. Evan Tarver, "Why the 'Share a Coke' Campaign Is So Successful," Investopedia, May 16, 2020, https://www.investopedia.com/articles/markets/100715/what-makes-share-coke-campaign-so-successful.asp.

4. FBA Help, "Locations of Amazon Fulfillment Centers," https://fba.help/list-of-amazon-fulfillment-centers.

3. Sheree Johnson, "New Research Sheds Light on Daily Ad Exposures," SJ Insights, September 29, 2014, https://sjinsights.net/2014/09/29/new-research-sheds-light-on-daily-ad-exposures/.

10. https://www.today.com/health/lawsuit-claims-healthy-juice-labels-are-misleading-t103617

11. Patrick M. Dunne, Robert F. Lusch, and James R. Carver, Retailing, 7th ed. (Mason, OH: South-Western Cengage, 2011), p. 295.

5. https://www.statista.com/statistics/457786/number-of-children-in-the-us-by-age/#:~:text=In%202020%2C%20there%20were%20about,aged%20zero%20to%20five%20years.

3. Madeline Farber, "Google Tops Apple as the World's Most Valuable Brand," Fortune, February 2, 2017, http://fortune.com/2017/02/02/google-tops-apple-brand-value/.

12. Kenzai, "About Us," https://www.kenzai.com/about-us/.

7. Valarie A. Zeithaml, A. Parasuraman, and Leonard L. Berry, Delivering Quality Service (New York: Free Press, 1990).

6. Peter Pham, "The Impacts of Big Data That You May Not Have Heard Of," Forbes, August 28, 2015, https://www.forbes.com/sites/peterpham/2015/08/28/the-impacts-of-big-data-that-you-may-not-have-heard-of/#747a15d76429.

1. Ramona Sukhraj, "How Dollar Shave Club Grew from Viral Video to $1 Billion Acquisition," Impact, July 21, 2016, https://www.impactbnd.com/blog/how-dollar-shave-club-grew-from-just-a-viral-video-to-a-615m-valuation-brand.

17. Witt Wells and Hannah Nemer, "Step inside the Hub: Coke's Social Media HQ," Coca-Cola, January 12, 2017, http://www.coca-colacompany.com/stories/step-inside-the-hub-coke-s-social-media-lab.

1. Jason Del Ray, "The Key to the Business of Love? Data of Course," February 13, 2012, https://adage.com/article/digital/okcupid-s-key-business-love-data/232686.

4. https://www.prnewswire.com/news-releases/kimberly-clark-looking-for-innovation-partners-to-create-solutions-for-digitally-savvy-consumers-300385645.html

3. Coca-Cola, "About Us—Coca-Cola History," https://www.worldofcoca-cola.com/about-us/coca-cola-history/.

5. 8451. (2021). "Break through the noise and be relevant with your customers." https://www.8451.com/

9. Peter Adams, "Starbucks' Stumble on Black Lives Matter Shows Rising Stakes for Brands in Addressing Race," June 22, 2020, https://www.marketingdive.com/news/starbucks-stumble-black-lives-matter-rising-stakes-race/580131/.

2. "Walmart: 3 Keys to Successful Supply Chain Management Any Business Can Follow," FlashGlobal, April 12, 2018, http://flashglobal.com/blog/supply-chain-management-walmart/.

6. Pew Research, "The Rise of Asian Americans," Social & Demographic Trends, June 19, 2012, http://www.pewsocialtrends.org/2012/06/19/the-rise-of-asian-americans/.

7. Akshay Heble, "Case Study on Coca-Cola 'Share A Coke' Campaign," Digital Vidya, November 11, 2019, https://www.digitalvidya.com/blog/case-study-on-coca-colas-share-a-coke-campaign/.

13. Nathan Furr and Daniel Snow, "The Prius Approach," Harvard Business Review 93, no. 11 (2015).

12. Frito-Lay, "Welcome to Frito-Lay Careers," https://fritolayemployment.com/.

20. Federal Trade Commission, https://www.ftc.gov/news-events/press-releases/2012/05/skechers-will-pay-40-million-settle-ftc-charges-it-deceived.

25. Macworld Staff, "Apple Pay FAQ: Apple, Goldman Sachs to Issue an Apple Pay Credit Card," Macworld, May 10, 2018, https://www.macworld.com/article/2834669/ios/apple-pay-faq-the-ultimate-guide-on-how-and-where-to-use-apples-payment-platform.html.

1. Robert Moritz, "Guts, Glory, and Megapixels: The Story of GoPro," Popular Mechanics, June 11, 2012, http://www.popularmechanics.com/adventure/sports/a7703/guts-glory-and-megapixels-the-story-of-gopro-8347639/.

8. V. A. Zeithaml, M. J. Bitner, and D. D. Gremler, Services Marketing: Integrating Customer Focus across the Firm, 6th ed. (New York: McGraw-Hill, 2012), p. 88.

16. Jefferson Graham, "Samsung Galaxy Note 7 Recall by the Numbers," USA Today, October 10, 2016, https://www.usatoday.com/story/tech/2016/10/10/samsung-galaxy-note-7-recall-cost/91876162/.

4. Ben Levitt, "Downy: Long Live Laundry—Print," n.d., http://cargocollective.com/benlevitt/Downy-Long-Live-Laundry-Print.

4. Bureau of Labor Statistics, "Industries at a Glance," www.bls.gov/iag/tgs/iag454.htm (accessed April 5, 2018).

3. Subway, "Franchises," https://www.entrepreneur.com/franchises/subway/282839.

3. Monte Burk, "Five Guys Burgers: America's Fastest Growing Restaurant Chain," Forbes, July 18, 2012, http://www.forbes.com/forbes/2012/0806/restaurant-chefs-12-five-guys-jerry-murrell-all-in-the-family.html.

9. https://channelsignal.com/blog/as-barbie-turns-60-mattel-aims-to-stay-on-trend-with-consumers/

2. Sam's Club, "Scan & Go Helps Create a Safe Shopping Experience for Sam's Club Members," https://corporate.samsclub.com/newsroom/2020/04/30/scan-go-helps-create-a-safe-shopping-experience-for-sams-club-members.

5. https://www.qsrmagazine.com/fast-food/whos-really-winning-chicken-sandwich-wars.

* Rackham, N. 1988. SPIN Selling. New York, NY: McGraw-Hill.

15. Julie Horowitz, "Some Companies Are Part of the Coronavirus Economy. The Rest Are in Trouble," CNN Business, May 12, 2020, https://www.cnn.com/2020/05/12/business/coronavirus-company-earnings/index.html.

8. Toyota, "Customer First and Quality First Measures," http://www.toyota-global.com/sustainability/society/quality/.

4. Planeterra Foundation, "What We Do," https://planeterra.org/what-we-do/.

5. Lindsay M. Howden and Julie A. Meyer, "Age and Sex Composition: 2010," U.S. Census Bureau, May 2011, http://www.census.gov/prod/cen2010/briefs/c2010br-03.pdf.

2. "Camera Campaign Uses Actors Posing as Tourists," Los Angeles Times, August 5, 2002, http://articles.latimes.com/2002/aug/05/business/fi-techbrfs5.1.1.

6. Kaylene C. Williams and Robert A. Page, "Marketing to the Generations," Journal of Behavioral Studies in Business, no.3 (April 2011), pp. 1–17.

32. International Monetary Fund, "About the IMF," http://www.imf.org/en/About.

1. Russell Redman, "Kroger Tests 'Smart' Shopping Cart from Caper," January 19, 2021, https://www.supermarketnews.com/technology/kroger-tests-smart-shopping-cart-caper.

2. Mark Borden, "The Team Who Made Old Spice Smell Good Again Reveals What's behind Mustafa's Towel," Fast Company, July 14, 2010, https://www.fastcompany.com/1670314/team-who-made-old-spice-smell-good-again-reveals-whats-behind-mustafas-towel.

4. https://www.forbes.com/sites/cherylrobinson/2020/03/17/how-global-head-of-barbie-focuses-on-the-brands-girl-empowerment-messaging/?sh=1b489bd5190f.

4. Derek Thompson, "A Case for College: The Unemployment Rate for Bachelor's-Degree Holders Is 3.7%," The Atlantic, February 1, 2013, http://www.theatlantic.com/business/archive/2013/02/a-case-for-college-the-unemployment-ratefor-bachelors-degree-holders-is-37-percent/272779/.

1. "The Roll That Changed History: Disposable Toilet Tissue Story," Kimberly-Clark, http://www.cms.kimberly-clark.com/umbracoimages/UmbracoFileMedia/ProductEvol_ToiletTissue_umbracoFile.pdf.

5. Citeman, "Decoding Problems," October 11, 2010, https://www.citeman.com/11065-decoding-problems.html.

3. Dale Buss, "Ford Thinks It Has a Better Idea about How to Handle Facebook," Forbes, May 15, 2012, http://www.forbes.com/sites/dalebuss/2012/05/15/ford-thinks-it-has-betteridea-about-how-to-handle-facebook/.

1. Kroger, "Feed the Human Spirit," TheKrogerCo.com, 2017 https://www.thekrogerco.com/about-kroger/

10. https://www.securityweek.com/beauty-and-breach-est%C3%A9e-lauder-exposes-440-million-records-unprotected-database.

3. Phil Mooney, "What It's Worth," Coca-Cola Journey, January 1, 2012, https://www.coca-colacompany.com/stories/collectors-columns-what-its-worth.

4. "Kimberly'Clark Looking for Innovation Partners to Create Solutions for Digitally Savvy Consumers," PR Newswire, January 4, 2017, https://www.prnewswire.com/news-releases/kimberly-clark-looking-for-innovation-partners-to-create-solutions-for-digitally-savvy-consumers-300385645.html.

5. "Distributor," Business Dictionary, http://www.businessdictionary.com/definition/distributor.html.

12. Bryan Kirschner, "The Not-So-Secret Strategy behind Walgreens' Ecosystem Advantage," CIO, February 6, 2017, https://www.cio.com/article/3161786/leadership-management/the-not-so-secret-strategy-behind-walgreens-ecosystem-advantage.html.

19. https://www.buzzfeednews.com/article/mbvd/urban-outfitters-features-vintage-red-stained-kent-state-swe

6. "What Was the 'Share a Coke' Campaign?," https://www.coca-colacompany.com/au/faqs/what-was-the-share-a-coke-campaign; Akshay Heble, "Case Study on Coca-Cola 'Share A Coke' Campaign,"Digital Vidya, November 11, 2019, https://www.digitalvidya.com/blog/case-study-on-coca-colas-share-a-coke-campaign/.

4. "Company History," Toyota, http://corporatenews.pressroom.toyota.com/corporate/company+history/.

*Louis Lataif, "Universities on the Brink," Forbes, February 1, 2011, http://www.forbes.com/2011/02/01/college-educationbubble- opinions-contributors-louis-lataif.html.

2. "Apple's Marketing Strategy," Marketing Minds, 2017, http://www.marketingminds.com.au/apple_branding_strategy.html.

9. https://www.usatoday.com/story/money/2020/12/16/39-most-outrageous-product-claims-of-all-time/115127274/

19. https://www.buzzfeednews.com/article/mbvd/urban-outfitters-features-vintage-red-stained-kent-state-swe

16. Patrick M. Dunne, Robert F. Lusch, and James R. Carver, Retailing, 7th ed. (Mason, OH: South-Western Cengage, 2011), p. 295.

9. "Air Cargo Market in the U.S.: Statistics & Facts," Statista, https://www.statista.com/topics/2815/air-cargo-market-in-the-united-states/.

6. Business Insider Intelligence, "Sears Tanked Because the Company Failed to Shift to Digital," Business Insider, August 26, 2016, http://www.businessinsider.com/sears-tanked-because-the-company-failed-to-shift-to-digital-2016-8.

4. www.levistrauss.com/sustainabilit/planet/water

1. "Apple's Marketing Strategy," Marketing Minds, 2017,
http://www.marketingminds.com.au/apple_branding_strategy.html.

10. https://www.securityweek.com/beauty-and-breach-est%C3%A9e-lauder-exposes-440-million-records-unprotected-database

13. https://www.vice.com/en/article/434gqw/i-made-my-shed-the-top-rated-restaurant-on-tripadvisor

11. Kimberly-Clark, "Changing Consumers' Minds and Habits," https://www.vml.com/united-states/our-work/kimberly-clark.

14. Barbara De Lollis, "IHG's Loyalty Club to Raise Point Rates for 25% of Hotels," USA Today, January 6, 2012, http://travel.usatoday.com/hotels/post/2012/01/ihg-priority-club-to-raisepoint-rates-but-phase-in-change/591898/1.

4. https://bettermarketing.pub/lego-marketing-lessons-from-one-of-the-strongest-brands-of-all-time-f5ab6a2a2396.

5. Kicks on Fire, "History of Nike," https://www.kicksonfire.com/history-of-nike/.

1. https://nrf.com/sites/default/files/2018-10/Uniquely-Gen-Z_Jan2017.pdf.

4. Phil Mooney, "What It's Worth," Coca-Cola, January 1, 2012, http://www.coca-colacompany.com/stories/collectors-columns-what-its-worth.

2. Joseph Pigato, "How 9 Successful Companies Keep Their Customers," Entrepreneur, April 3, 2015, https://www.entrepreneur.com/article/243764.

31. Tanya Dua, "Instagram Stories Has Emerged as a Clear Favorite for Marketers over Snapchat," Business Insider, August 2, 2017, http://www.businessinsider.com/instagram-stories-has-emerged-as-a-clear-favorite-for-marketers-over-snapchat-2017-8.

3. Zach Gottlieb, "Old Spice Man Is Here . . . for You," Wired, July 15, 2010, https://www.wired.com/2010/07/old-spice-man-is-here-for-you/.

14. Maggie Majstrova, "What Do Effective Integrated Marketing Campaigns Have in Common?" Smart Insights, October 8, 2014, https://www.smartinsights.com/traffic-building-strategy/integrated-marketing-case-studies/.

*N. Rackham, SPIN Selling (New York: McGraw-Hill, 1988).

18. Katherine Glover, "Burger King Apologizes for Hindu Goddess Ad," Moneywatch, July 9, 2009, https://www.cbsnews.com/news/burger-king-apologizes-for-hindu-goddess-ad/.

5. Natalie Jarvey, "Hulu Grows to 28 Million Subscribers," Hollywood Reporter, May 1 2019, https://www.hollywoodreporter.com/live-feed/hulu-grows-28-million-subscribers-1206322.

13. Mae Anderson, "Dr Pepper's New Brand Is a Manly Man's Soda," Associated Press, October 10, 2011, http://www.msnbc.msn.com/id/44849414/ns/business-us_business/t/dr-peppers-new-brand-manly-mans-soda/.

12. Suneera Tandon, "Uniqlo Is Finally Coming to India," Quartz, May 9, 2018, https://qz.com/1273211/uniqlo-is-finally-coming-to-india/.

23. https://www.vox.com/culture/2019/12/3/20993432/peloton-new-commercial-horror-movie

5. Joseph R. Hair Jr., Robert P. Bush, and David J. Ortinau, Marketing Research in a Digital Information Environment, 4th ed. (New York: McGraw-Hill/Irwin, 2009).

30. Statista, "European Union: Share in Global Gross Domestic Product Based on Purchasing-Power-Parity from 2014 to 2024," https://www.statista.com/statistics/253512/share-of-the-eu-in-the-inflation-adjusted-global-gross-domestic-product/.

34. A. J. Agrawal, "How to Utilize Real-Time Data in Your Marketing Efforts," Forbes, January 13, 2016, https://www.forbes.com/sites/ajagrawal/2016/01/13/how-to-utilize-real-time-data-in-your-marketing-efforts/#3c1c44a05741.

34. Zoe-Lee Skelton, "An SEO Beginner Guide: What Is Call Tracking and Why Should I Care?," Search Engine Journal, January 3, 2014, https://www.searchenginejournal.com/seo-beginner-guide-call-tracking-care/83173/.

5. Pew Research, "The Rise of Asian Americans," Social & Demographic Trends, June 19, 2012, http://www.pewsocialtrends.org/2012/06/19/the-rise-of-asian-americans/.

1. Randall Craig, "Learning Strategy from McDonald's . . . and Five Guys," D!gitalist, May 6, 2014, http://www.digitalistmag.com/lob/human-resources/2014/05/06/learning-strategy-from-mcdonalds-and-five-guys-01250174.

12. Mark Hutcheon, "Consumers Expect Brands to Address Climate Change," The Wall Street Journal, April 20, 2021.

1. https://simconblog.wordpress.com/2016/06/19/nike-brand-analysis.

3. "Southwest Airlines Co SuccessStory," SuccessStory.com, 2017, https://successstory.com/companies/southwest-airlines-co.

6. "The Five Guys Story," Five Guys, http://www.fiveguys.com/Fans/The-Five-Guys-Story.

Berenson, Tessa. (2016). Millennials Don't Know What Fabric Softener Is For. Fortune. http://fortune.com/2016/12/16/millennials-fabric-softener-downy/

4. Kirk McElhearn, "Wrapping It Right: In Praise of Apple's Packaging," Macworld, August 20, 2015, http://www.macworld.com/article/2973339/tech-events-dupe/wrapping-it-right-in-praise-of-apple-s-packaging.html.

14. PR Newswire, "McDonalds Goes Blue This Summer with the Launch of The Smurfs," Bloomberg, July 29, 2011, http://www. bloomberg.com/apps/news?pid=conewsstory&tkr=MCD:US &sid=aaBly7Nclp98.

8. Colin Barr, "'Mass Affluent' Are Strapped Too, BofA Finds," CNNMoney, January 24, 2011, http://finance.fortune.cnn.com/2011/01/24/mass-affluent-are-strapped-too-bofa-finds/.

1. John Battelle, "The Birth of Google," Wired, 2005, https://www.wired.com/2005/08/battelle/.

* Toyota website, http://www.kokomo-toyota.com/blog/toyota-sienna-called-swagger-wagon/.

9. Stephen Bruce, "Hiring the Wrong Salesperson Is a $2-Million Mistake," HR Daily Advisor, December 3, 2014,

http://hrdailyadvisor.blr.com/2014/12/03/hiring-the-wrong-salesperson-is-a-2-million-mistake/.

22. Gonzalo E. Mon and Collier Shannon Scott, "Marketing Sweepstakes via the Mail," DMNews.com, September 25, 2002, http://www.dmnews.com/marketing-strategy/marketing-sweepstakes-via-the-mail/article/78726/.

5. Jaclyn Trop, "How Dollar Shave Club's Founder Built a $1 Billion Company That Changed the Industry," Entrepreneur, March 28, 2017, https://www.entrepreneur.com/article/290539.

9. E. J. Schultz, "'Got Milk' Dropped as National Milk Industry Changes Tactics," Ad Age, February 24, 2014, http://adage.com/article/news/milk-dropped-national-milk-industry-tactics/291819/.

1. Larry Olmstead, "Great American Bites: Why Burger Lovers Flock to Five Guys," USA Today, October 10, 2013, https://www.usatoday.com/story/travel/columnist/greatamericanbites/2013/10/10/five-guys-burger-hot-dog-fries/2952349/.

2. Rani Molla, "Amazon Could Be Responsible for Nearly Half of U.S. E-Commerce Sales in 2017," recode, October 24, 2017, https://www.recode.net/2017/10/24/16534100/amazon-market-share-ebay-walmart-apple-ecommerce-sales-2017.

34. AJ Agrawal. (2016, Jan 13). "How to Utilize Real-time Data In Your Marketing Efforts." Forbes. Retrieved from https://www.forbes.com/sites/ajagrawal/2016/01/13/how-to-utilize-real-time-data-in-your-marketing-efforts/#3c1c44a05741.

12. Ann Smarty, "The Birth and Evolution of Digital Content Marketing," Internet Marketing Ninjas, July 29, 2013, https://www.internetmarketingninjas.com/blog/content/the-birth-and-evolution-of-digital-content-marketing/.

1. "Toyota Motor Sales Reports, Toyota, December 2018 and Year-End Sales," https://www.prnewswire.com/news-releases/toyota-motor-north-america-reports-december-2018-year-end-sales-300772433.html

20. Avi Singhal and Clara De Martel, "Hooked: Examining the Rise of an App," The Weekender, November 6, 2016, http://www.dailycal.org/2016/11/06/hooked-examining-rise-app/.

https://www.greenamerica.org/amazon-labor-exploitation-home-and-abroad

5. "What Was the 'Share a Coke' Campaign?," https://www.coca-colacompany.com/au/faqs/what-was-the-share-a-coke-campaign.

2. Nicholas Woodman, "A Letter from Our Founder and CEO," https://gopro.com/about-us.

10. U.S. Census Bureau, "The Majority of Children Live with Two Parents, Census Bureau Reports," November 17, 2016, https://www.census.gov/newsroom/press-releases/2016/cb16-192.html.

2. SmartBrief, "Need a Pit Stop? Recharge Offers Rooms by the Minute," SmartBrief, June 26, 2017, http://www.smartbrief.com/s/2017/06/need-pit-stop-recharge-offers-rooms-minute.

2. Sarah Ashley O'Brien, "WeWork Isn't Just Selling Desk Space. It's Selling a New Way of Life," CNN.com, May 3, 2019, https://www.cnn.com/2019/05/03/tech/wework-culture/index.html.

2. Peter Romeo, "Five Guys Founder Jerry Murrell's Approach to Leadership," Restaurant Business, July 8, 2015, http://www.restaurantbusinessonline.com/operations/leaders/five-guys-founder-jerry-murrell-s-approach-leadership.

3. Derek Thompson, "A Case for College: The Unemployment Rate for Bachelor's-Degree Holders Is 3.7%," The Atlantic, February 1, 2013, http://www.theatlantic.com/business/archive/2013/02/a-case-for-college-the-unemployment-ratefor-bachelors-degree-holders-is-37-percent/272779/.

18. HGTV, "Congrats to the HGTV Dream Home 2013 Winner!," HGTV, http://www.hgtv.com/hgtv-dream-home-2013-giveaway/package/index.html.

3. Personal interview, Steve Spinelli, 2022.

1. https://www.forbes.com/sites/cherylrobinson/2020/03/17/how-global-head-of-barbie-focuses-on-the-brands-girl-empowerment-messaging/?sh=1b489bd5190f.

11. "We're Guessing You Love Ice Cream," Halo Top, https://www.halotop.com/.

7. MayeCreate, "3 Marketing Lessons from the 'Share a Coke' Campaign," http://mayecreate.com/blog/3-marketing-lessons-from-the-share-a-coke-campaign/.

6. Miller, Tracey, "'Got Milk' ad campaign retired after 20 years of milk moustaches," NY Daily News, http://www.nydailynews.com/life-style/health/milk-ads-retired-20-years-milk-moustaches-article-1.1701064

5. Nicola Fumo, "When Did Shipping Boxes Get Pretty?", Racked, Jul 7 2016.

4. Schultz, E.J., "'Got Milk' Dropped as National Milk Industry Changes Tactics," Advertising Age, http://adage.com/article/news/milk-dropped-national-milk-industry-tactics/291819/

7. U.S. Department of Transportation, Federal Railroad Administration, "Freight Rail Overview," https://www.fra.dot.gov/Page/P0362.

9. "BA's Price-Fix Fine Reaches £270m," BBC News, August 1, 2007, http://news.bbc.co.uk/2/hi/business/6925397.stm.

3. Bert Markgraf, "Examples of Questionable Marketing Ethics," Chron, smallbusiness.chron.com/examples-questionable-marketing.ethics-60520; Amanda Vanallen, Vanessa Weber, and Sarah Kunin, "Are Tweens Too Young for Makeup?," ABC News, https://abcnews.go.com/US/tweens-young-makeup/story?id=12777008.

1. L'Oréal, "L'Oréal's Supply Chain Ranked 9th Worldwide and 4th in Consumer Products Industry," www.loreal.com, (n.d.). https://www.loreal.com/en/news/commitments/loreal-ranked-4th-worldwide-for-its-supply-chain/

4. "Intro to PlantBottle Packaging," Coca-Cola Journey, http://www.coca-colacompany.com/plantbottle-technology.

7. Jonathan Camhi and Stephanie Pandolph, "Amazon's Earnings Point to International Expansion," Business Insider, May 1, 2017, http://www.businessinsider.com/amazons-earnings-point-to-international-expansion-2017-5.

10. Abbas Ali, "Amazon Officially Launches in the UAE," Techradar, May 1, 2019, https://www.techradar.com/news/amazon-officially-launches-in-the-uae.

2. Ibid.

3. Anne Marie Mohan, "Store Brand Sales Grow amidst Coronavirus Stock-Ups," Packaging World, May 1, 2020, https://www.packworld.com/issues/business-intelligence/article/21131043/store-brand-sales-grow-amidst-coronavirus-stockups.

24. Kirsten Korosec, "Why Rooftop Solar Might Get a Lot More Expensive in the U.S.," Fortune, September 22, 2017, http://fortune.com/2017/09/22/solar-costs-tariffs/.

3. "Working Age Population," OECD, https://data.oecd.org/pop/working-age-population.htm.

10. https://www.wsj.com/articles/mattel-closes-factories-as-toy-slump-weighs-on-supply-chain-11581256800.

4. Monte Burk, "Five Guys Burgers: America's Fastest Growing Restaurant Chain," Forbes, July 18, 2012, http://www.forbes.com/forbes/2012/0806/restaurant-chefs-12-five-guys-jerry-murrell-all-in-the-family.html.

5. Amazon, "Global Selling," https://sell.amazon.in/grow-your-business/amazon-global-selling.

* http://www.salesharvest.com.au/f.ashx/pdf/Sales-Harvest-How-To-Recruit-Sales-Stars.pdf

1. https://www.yahoo.com/now/u-weight-loss-market-2022-115900788.html

12. Nielsen Company, https://www.nielsen.com/us/en.

2. SIMCON Blog, https://simconblog.wordpress.com/2016/06/19/nike-brand-analysis/.

29. Geoffrey James, "20 Epic Fails in Global Branding," Inc., October 29, 2014, https://www.inc.com/geoffrey-james/the-20-worst-brand-translations-of-all-time.html; Chad Brooks, "Lost in Translation: 8 International Marketing Fails," Business News Daily, October 7, 2013, https://www.businessnewsdaily.com/5241-international-marketing-fails.html.

4. Michel Anteby and Rakesh Khurana, "The Human Relations Movement: Harvard Business School and the Hawthorne Experiments (1924–1933)," n.d., https://www.library.hbs.edu/hc/hawthorne/intro.html.

4. Kroger, "Feed the Human Spirit," TheKrogerCo.com, 2017 https://www.thekrogerco.com/about-kroger/

2. "Number of Planet Fitness locations in the United States in 2022," ScrapeHero.com, 2022. https://www.scrapehero.com/location-reports/Planet%20Fitness-USA/

26. "The Power of Pinterest: 4 of the Best Pinterest Marketing Campaigns," EM Marketing, August 11, 2015, https://e-m-marketing.com/blog/2015/08/the-power-of-pinterest-4-of-the-best-pinterest-marketing-campaigns/.

2. L'Oréal, "How Supply Chain Evolves to Meet Our Consumers' Demands," www.loreal.com, (n.d.). https://www.loreal.com/en/articles/operations/supply-chain-transformation/

8. Jonathan Camhi and Stephanie Pandolph, "Amazon's Earnings Point to International Expansion," Business Insider, May 1, 2017, https://www.businessinsider.com/amazons-earnings-point-to-international-expansion-2017-5.

37. UNIQLO, "Planet," https://www.uniqlo.com/jp/en/contents/sustainability/planet/.

19. Taylor Holland, "5 New Ways Marketers Are Leveraging Location Data in 2016," Skyword.com, April 5, 2016, https://www.skyword.com/contentstandard/creativity/5-new-ways-marketers-are-leveraging-location-data-in-2016/.

3. Levi Strauss & Co., "Sustainability—Planet: Water," www.levistrauss.com/sustainability/planet/water.

10. https://www.today.com/health/lawsuit-claims-healthy-juice-labels-are-misleading-t103617

2. "Quick Real Estate Statistics," National Association of Realtors, May 11, 2018, https://www.nar.realtor/research-and-statistics/quick-real-estate-statistics.

4. https://www.restaurantbusinessonline.com/food/brief-history-chicken-sandwich-wars

1. https://fuelcycle.com/blog/market-research-done-right-how-market-research-gave-lego-a-facelift/.

7. https://www.forbes.com/sites/cherylrobinson/2020/03/17/how-global-head-of-barbie-focuses-on-the-brands-girl-empowerment-messaging/?sh=1b489bd5190f.

11. Joshua Caucutt, "Cracker Barrel Is Cracklin'," InvestorGuide, May 25, 2010, http://www.investorguide.com/article/6467/cracker-barrel-is-cracklin-cbrl/.

3. Katherine Cullen, "Consumers embrace physical retail as U.S. emerges from the pandemic," July 15, 2021. Retrieved from https://nrf.com/blog/consumers-embrace-physical-retail-us-emerges-pandemic

5. Forbes, "10 Companies That Pay Salespeople Really Well," https://www.forbes.com/pictures/efkk45jgd/oracle-4/#4f6a05204a75.

7. Kristina Monllos, "Brands Are Throwing Out Gender Norms to Reflect a More Fluid World," Adweek, October 17, 2016, http://www.adweek.com/brand-marketing/brands-are-throwing-out-gender-norms-reflect-more-fluid-world-174070/.

9. Steve W. Martin, "Seven Personality Traits of Top Salespeople," SalesHarvest, June 27, 2011, http://www.salesharvest.com.au/f.ashx/pdf/Sales-Harvest-How-To-Recruit-Sales-Stars.pdf.

3. Kroger, "Feed the Human Spirit," TheKrogerCo.com, 2017 https://www.thekrogerco.com/about-kroger/

1. Sysco, "About," http://www.sysco.com/about-sysco.html.

7. OkCupid website, https://theblog.okcupid.com/we-experiment-on-human-beings-5dd9fe280cd5.

1. Press Release, "lululemon Unveils 'Power of Three' Strategic Plan to Accelerate Growth," lululemon.com, April 24, 2019. https://corporate.lululemon.com/media/press-releases/2019/04-24-2019-090026432

2. PRWeb, "Jen Selter, Social Media Star Helps BlendJet Sell over 200,000 Units in 195 Countries," PRWeb.com, April 30 2019, https://www.prweb.com/releases/jen_selter_social_media_star_helps_blendjet_sell_over_200_000_units_in_195_countries/prweb16276534.htm.

25. Mallory Schlossberg, "26 Crazy McDonald's Items You Can't Get in America," Business Insider, July 2015, http://www.businessinsider.com/mcdonalds-international-menu-items-2015-7/#donalds-canadas-poutine-13.

36. Christine at Talkwalker, "UNIQLO Marketing Srtategy," July 27, 2020. https://www.talkwalker.com/blog/uniqlo-marketing-strategy.

1. Southwest. (2017b). Southwest Coporate Fact Sheet.

   *Joseph B. White, "Hitching the New Small Wagon to Better Fuel Efficiency," The Wall Street Journal, December 24, 2012, http://online.wsj.com/article/SB10001424127887324907204578187560383059762.html.

3. Ellen Byron, "Bounty Puts a New Spin on Spills," The Wall Street Journal, February 17, 2009, http://online.wsj.com/article/SB123482144767494581.html.

5. "Toyota Introduces 'Let's Go Places,'" Advertising Age, September 11, 2012, http://adage.com/article/news/toyota-introduces-places/237138/.

28. Witt Wells and Hannah Nemer, "Step inside the Hub: Coke's Social Media HQ," Coca-Cola, January 12, 2017, http://www.coca-colacompany.com/stories/step-inside-the-hub--coke-s-social-media-lab.

13. HGTV, "Congrats to the HGTV Dream Home 2013 Winner!" n.d., http://www.hgtv.com/hgtv-dream-home-2013-giveaway/package/index.html.

22. Samsung website, http://www.samsung.com/us/news/socialcenter/.

12. Jaclyn Jaeger, "AT&T Might Be the Next Wells Fargo (and Doesn't Seem to Be Doing Anything About It)," Compliance Week, October 15, 2019. https://www.complianceweek.com/boards-and-shareholders/atandt-might-be-the-next-wells-fargo-and-doesnt-seem-to-be-doing-anything-about-it/27872.article

9. Sharon McLoone, "Getting Government Contracts," The New York Times, October 7, 2009, http://www.nytimes.com/2009/10/08/business/smallbusiness/08contracts.html?_r50.

4. YouTube, "GoPro: Fireman Save Kitten," https://www.youtube.com/watch?v=CjB_oVeq8Lo.

10. Sangram Vajre, "The B2B Sales Funnel Is Dead—and Here's the Proof," Salesforce.com blog, April 10, 2015, https://www.salesforce.com/blog/2015/04/the-b2b-sales-funnel-is-dead-heres-the-proof.html.

6. Stephanie Clifford, "Revamping, Home Depot Woos Women," The New York Times, January 28, 2011, http://www.nytimes.com/2011/01/29/business/29home.html.

7. Citeman. "Decoding Problems." October 11, 2010. Accessed at https://www.citeman.com/11065-decoding-problems.html

5. David Griner, "Hey Old Spice Haters, Sales Are up 107%," Adweek, July 27, 2010, https://www.adweek.com/creativity/hey-old-spice-haters-sales-are-107-12422/.

1. Ketchum, "Ketchum and Michelin Win Best Public Relations Campaign of the Year," PR Newswire, June 7, 2019, https://www.prnewswire.com/news-releases/ketchum-and-michelin-win-best-public-relations-campaign-of-the-year-300864141.html.

1. Jeff Goodby, "20 Years of 'Got Milk?,'" Adweek, October 25, 2013, http://www.adweek.com/creativity/20-years-got-milk-153399/.

2. Issie Lapowsky, "4 Takeaways from the Iconic 'Got Milk?' Ad Campaign," Inc., February 27, 2014, https://www.inc.com/issie-lapowsky/marketing-tips-got-milk.html.

8. Selena Larson, "The Hacks That Left Us Exposed in 2017," CNN Tech, December 20, 2017, http://money.cnn.com/2017/12/18/technology/biggest-cyberattacks-of-the-year/index.html.

17. Adam Sarhan, "Planned Obsolescence: Apple Is Not the Only Culprit," Forbes, December 22, 2017, https://www.forbes.com/sites/adamsarhan/2017/12/22/planned-obsolescence-apple-is-not-the-only-culprit/#6f8db3e23cf2.

1. "ANA 2010: Marc Pritchard," YouTube, October 15, 2010, https://www.youtube.com/watch?v=tazRr0CNKLU.

10. Amy Schade, "Designing for 5 Types of E-Commerce Shoppers," Nielsen Norman Group, March 2, 2014, https://www.nngroup.com/articles/ecommerce-shoppers/.

3. Kicks on Fire, "History of Nike," https://www.kicksonfire.com/history-of-nike/.

7. Christine Alemany, "Marketing in the Age of Resistance," September 3, 2020, https://hbr.org/2020/09/marketing-in-the-age-of-resistance.

4. Marc de Swaan Arons, Frank van den Driest, and Keith Weed, "The Ultimate Marketing Machine," Harvard Business Review, July–August 2014, https://hbr.org/2014/07/the-ultimate-marketing-machine.

2. "The Five Guys Story," Five Guys, http://www.fiveguys.com/Fans/The-Five-Guys-Story.

6. Elka Torpey, "Same Occupation, Different Pay: How Wages Vary," Bureau of Labor Statistics, 2015, https://www.bls.gov/careeroutlook/2015/article/wage-differences.htm.

4. Akshay Heble, "Case Study on Coca-Cola 'Share A Coke' Campaign,"Digital Vidya, November 11, 2019, https://www.digitalvidya.com/blog/case-study-on-coca-colas-share-a-coke-campaign/.

7. https://fortune.com/2021/07/30/jeff-bezos-net-worth-amazon-stock-amzn-earnings-update/#:~:text=Amazon%20shares%20are%20down%207,person%2C%20ahead%20of%20Elon%20Musk.

8. Jonathan Camhi and Stephanie Pandolph, "Amazon's Earnings Point to International Expansion," Business Insider, May 1, 2017, https://www.businessinsider.com/amazons-earnings-point-to-international-expansion-2017-5.

4. Old Spice, "The Man Your Man Could Smell Like," YouTube, February 4, 2010, http://www.youtube.com/watch?v=owGykVbfgUE.

23. "Skechers Launches Joint Venture in South Korea," Business Wire, November 10, 2016, http://www.businesswire.com/news/home/20161110005045/en/SKECHERS-Launches-Joint-Venture-South-Korea.

13. Michelle Sanusi et al., "The Swoosh of Creativity," Business Today, July 6, 2014, https://www.businesstoday.in/magazine/lbs-case-study/nike-marketing-strategies-global-brand/story/207237.html.

1. Ed Caesar, "Nike's Quest to Beat the Two-Hour Marathon Comes Up Oh So Short," Wired, May 6, 2017, https://www.wired.com/2017/05/two-hour-marathon-failed-nike/; ClickZ, "18 Lessons We Can Learn from Airbnb, Nike, and IKEA in 2017," ClickZ, January 2, 2018, https://www.clickz.com/18-lessons-airbnb-nike-ikea-2017/205144/.

6. "We Are RE/MAX," RE/MAX, https://www.joinremax.com/.

6. https://www.ftc.gov/news-events/news/press-releases/1997/05/joe-camel-advertising-campaign-violates-federal-law-ftc-says

1. Grete Grivina, "Champion Brand Clothing: Origins and Revival," Printful, May 18, 2020, https://www.printful.com/blog/champion-brand/.

8. American Trucking Associations, "Reports, Trends, & Statistics," http://www.trucking.org/News_and_Information_Reports_Industry_Data.aspx.

1. https://www.yahoo.com/now/u-weight-loss-market-2022-115900788.html

7. Zsolt Katona, "How to Identify Influence Leaders in Social Media,"Bloomberg, February 26, 2012, http://www.bloomberg.com/news/2012-02-27/how-to-identify-influence-leaders-insocial-media-zsolt-katona.html.

4. Phil Wahba, "The Silver Lining in Walmart's Slowing E-Commerce Growth," Fortune, May 20, 2016, http://fortune.com/2016/05/20/the-silver-lining-in-walmarts-slowing-e-commerce-growth/.

1. https://www.tasteatlas.com/chicken-sandwich#:~:text=It%20is%20believed%20that%20the,it%20Chick%2Dfil%2DA.

1. https://www.statista.com/statistics/257076/number-of-the-gap-inc-employees-worldwide/.

14. Bill Mears, "California Ban on Sale of 'Violent' Video Games to Children Rejected," CNNU.S., June 27, 2011, http://www.cnn.com/2011/US/06/27/scotus.video.games/index.html.

4. Amazon, "Amazon Services," https://services.amazon.com/global-selling/overview.html/ref=asus_ags_snav_over.

7. Wade Thiel, "U.S. Auto Sales Brand Rankings—December 2018 YTD," goodcarbadcar, January 3, 2019, http://www.goodcarbadcar.net/2019/01/u-s-auto-sales-brand-rankings-december-2018-ytd/.

9. Alexandre Tanzi, "Top 3% of U.S. Taxpayers Paid Majority of Income Tax in 2016," October 14, 2018, https://www.bloomberg.com/news/articles/2018-10-14/top-3-of-u-s-taxpayers-paid-majority-of-income-taxes-in-2016.

2. https://www.chick-fil-a.com/about/history.

9. Nicole Silberstein, "5 Ways COVID-19 Has Reshaped Consumer Behavior," February 12, 2021, https://retailtouchpoints.com/features/trend-watch/5-ways-covid-19-has-reshaped-consumer-behavior.

3. Katherine Cullen, "Consumers Embrace Physical Retail as U.S. Emerges from the Pandemic," July 15, 2021, https://nrf.com/blog/consumers-embrace-physical-retail-us-emerges-pandemic.

15. http://www.businessinsider.com/apples-delayed-the-homepod-only-the-latest-of-its-broken-promises-2017-11

1. Ekaterina Walter, "The Top 30 Stats You Need to Know When Marketing to Women," The Next Web, January 24, 2012, http://thenextweb.com/socialmedia/2012/01/24/the-top-30-stats-you-need-to-know-when-marketing-towomen/.

18. Pew Research Center, "Mobile Fact Sheet," Pew Research Center, February 5, 2018, http://www.pewinternet.org/fact-sheet/mobile/.

3. Jade Scipioni, "P&G's Downy Facing Stiff Reality: Millennials Don't Use Fabric Softener," Fox Business, December 19, 2016, http://www.foxbusiness.com/features/2016/12/19/p-g-s-downy-facing-stiff-reality-millennials-don-t-use-fabric-softener.html.

1. Avery Hartmans, "15 Fascinating Facts You Probably Didn't Know about Amazon," Business Insider, June 17, 2019, https://www.businessinsider.com/jeff-bezos-amazon-history-facts-2017-4.

1. "What Was the 'Share a Coke' Campaign?," https://www.coca-colacompany.com/au/faqs/what-was-the-share-a-coke-campaign.

7. Budweiser 2017 Super Bowl Commercial, "Born the Hard Way," YouTube, January 31, 2017, https://www.youtube.com/watch?v=HtBZvl7dlu4.

5. Shelley Frost, "The Disadvantages of Advertising to Kids," Chron, http://smallbusiness.chron.com/disadvantages-advertising-kids-25801.html.

7. https://fortune.com/2021/07/30/jeff-bezos-net-worth-amazon-stock-amzn-earnings-update/#:~:text=Amazon%20shares%20are%20down%207,person%2C%20ahead%20of%20Elon%20Musk

2. https://www.forbes.com/sites/cherylrobinson/2020/03/17/how-global-head-of-barbie-focuses-on-the-brands-girl-empowerment-messaging/?sh=1b489bd5190f.

5.

10. Statista, "Desktop E-Commerce Spending in the United States on Cyber Monday from 2005 to 2017 (in Million U.S. Dollars)," Statista, November 2017, https://www.statista.com/statistics/194643/us-e-commerce-spending-on-cyber-monday-since-2005/.

12. Heidi Daitch, (posted 2017, Dec 14), "2017 Data Breaches—the Worst So Far." Identity Force. Retrieved from https://www.identityforce.com/blog/2017-data-breaches.

6. Joseph R. Hair Jr., Robert P. Bush, and David J. Ortinau, Marketing Research in a Digital Information Environment, 4th ed. (New York: McGraw-Hill/Irwin, 2009).

15. Brian Grow, "The Great Rebate Runaround," Bloomberg Businessweek, November 22, 2005, http://www.businessweek. com/stories/2005-11-22/the-great-rebate-runaround.

8. "The 2020 World's Most Valuable Brands," Forbes, 2020, https://www.forbes.com/the-worlds-most-valuable-brands/.

8. Jonathan Camhi and Stephanie Pandolph, "Amazon's Earnings Point to International Expansion," Business Insider, May 1, 2017, https://www.businessinsider.com/amazons-earnings-point-to-international-expansion-2017-5.

5. "Kimberly'Clark Looking for Innovation Partners to Create Solutions for Digitally Savvy Consumers," PR Newswire, January 4, 2017, https://www.prnewswire.com/news-releases/kimberly-clark-looking-for-innovation-partners-to-create-solutions-for-digitally-savvy-consumers-300385645.html.

5. https://www.barbie.com/en-gb/dream-gap.

8. "Our Training Calendar," RE/MAX, https://www.gateway2realestate.com/our-training-calendar.

Joshua Caucutt, "Cracker Barrel Is Cracklin'," InvestorGuide, May 25, 2010, http://www.investorguide.com/article/6467/ cracker-barrel-is-cracklin-cbrl/ .

13. Biz Carson, "This Is the True Story of How Mark Zuckerberg Founded Facebook, and It Wasn't to Find Girls," Business Insider, February 28, 2016, http://www.businessinsider.com/the-true-story-of-how-mark-zuckerberg-founded-facebook-2016-2.

Lindsay M. Howden and Julie A. Meyer, "Age and Sex Composition: 2010," U.S. Census Bureau, May 2011, http:// www.census.gov/prod/cen2010/briefs/c2010br-03.pdf .

14. Kif Leswing, "Apple Is Struggling to Make and Meet Promises about When It Will Ship New Products—and the HomePod Is Only the Latest Example," Business Insider, November 18, 2017, http://www.businessinsider.com/apples-delayed-the-homepod-only-the-latest-of-its-broken-promises-2017-11.

7. "How to Go Viral: Old Spice's 'Responses'," L&T, October 3, 2017, https://landt.co/2017/10/go-viral-old-spices-responses/.

11. Sean Rossman, "What Is French Yogurt and Is It the New Greek?," USA Today, July 6, 2017, https://www.usatoday.com/story/money/nation-now/2017/07/06/what-french-yogurt-and-new-greek/439935001/.

9. https://chrisrmark217.wordpress.com/2018/05/02/lego-building-a-successful-future-through-market-research-one-block-at-a-time/

5. https://www.ftc.gov/news-events/news/press-releases/1997/05/joe-camel-advertising-campaign-violates-federal-law-ftc-says

9. Jonathan Camhi and Stephanie Pandolph, "Amazon's Earnings Point to International Expansion," Business Insider, May 1, 2017, https://www.businessinsider.com/amazons-earnings-point-to-international-expansion-2017-5.

2. Kroger, "Feed the Human Spirit," TheKrogerCo.com, 2017 https://www.thekrogerco.com/about-kroger/

15. "Lobbying, Legal and PR Efforts Continue into 2017," American Suntanning Association, December 1, 2016, https://americansuntanning.org/lobbying-legal-and-pr-efforts-continue-into-2017/.

4. https://www.forbes.com/sites/cherylrobinson/2020/03/17/how-global-head-of-barbie-focuses-on-the-brands-girl-empowerment-messaging/?sh=1b489bd5190f

4. Google, "Company," http://www.google.com/about/company.

4. Walmart, "Sustainability," 2017, http://corporate.walmart.com/global-responsibility/sustainability/.

10. Stephen Bruce, "Hiring the Wrong Salesperson Is a $2-Million Mistake," HR Daily Advisor, December 3, 2014,

http://hrdailyadvisor.blr.com/2014/12/03/hiring-the-wrong-salesperson-is-a-2-million-mistake/.

10. Federal Trade Commission, Code of Federal Regulations, Title 16, Volume 1, Part 500, Rev. January 1, 2000.

9. Nancy Trejos, "Marriott Debuts Hotel Brand for Millennials," USA Today, December 8, 2014, https://www.usatoday.com/story/travel/hotels/2014/12/08/marriott-ac-hotels-millennials-new-orleans/20066811/.

3. https://www.campaignlive.co.uk/article/mattel-turned-tfoo-perfect-unrelatable-barbie-symbol-female-empowerment/1663884

1. Jade Scipioni, "P&G's Downy Facing Stiff Reality: Millennials Don't Use Fabric Softener," Fox Business, December 19, 2016, http://www.foxbusiness.com/features/2016/12/19/p-g-s-downy-facing-stiff-reality-millennials-don-t-use-fabric-softener.html.

13. Lydia Dishman, "How Outsiders Get Their Products to the Innovation Big League at Procter & Gamble," Fast Company, July 13, 2012, http://www.fastcompany.com/1842577/how-outsiders-get-their-products-innovation-big-leagueprocter-gamble.

7. "RE/MAX Training Program Recognized for Custom Content," RE/MAX, September 30, 2014, https://www.remax.com/newsroom/press-releases/remax-training-program-recognized-for-custom-content.htm.

7. Sandi Glass, "What Were They Thinking? Orbitz, The Lava Lamp of Soft Drinks," *Fast Company*, October 18, 2011, https://www.fastcompany.com/1786387/what-were-they-thinking-orbitz-lava-lamp-soft-drinks.

2. David Ferry, "The New War on (Overpriced) Drugs," *Wired*, June 10, 2017, https://www.wired.com/story/fighting-high-drug-prices/.

2. https://www.internetlivestats.com/google-search-statistics/.

10. Abbas Ali, "Amazon Officially Launches in the UAE," *Techradar*, May 1, 2019, https://www.techradar.com/news/amazon-officially-launches-in-the-uae.

2. Newt Barrett, "Five Guys Burgers and Fries Really Understand in Person Content Marketing," *Content Marketing Today*, July 9, 2009, http://contentmarketingtoday.com/2009/07/09/five-guys-burgers-and-fries-really-understand-in-person-content-marketing/ [Access date: April 5, 2017.]

10. https://chrisrmark217.wordpress.com/2018/05/02/lego-building-a-successful-future-through-market-research-one-block-at-a-time/.

4. World Bank, "National Accounts Data, OECD National Accounts Data Files," 2017, https://data.worldbank.org/indicator/.

2. Lisa Waugh, "22 Surprising Facts about Hulu You Probably Didn't Know," *Ranker*, https://www.ranker.com/list/interesting-hulu-facts-and-statistics/lisa-waugh.

   MBA Skool Team. "Airbnb Marketing Strategy & Marketing Mix (4Ps)." https://www.mbaskool.com/marketing-mix/services/17335-airbnb.html

4. Abraham Maslow, *Motivation and Personality* (New York: Harper, 1954), p. 236.

2. Reference for Business, "Sysco Corporation—Company Profile, Information, Business Description, History, Background Information on Sysco Corporation," http://www.referenceforbusiness.com/history2/10/SYSCO-Corporation.html.

3. "The Roll That Changed History: Disposable Toilet Tissue Story," Kimberly-Clark, http://www.cms.kimberly-clark.com/umbracoimages/UmbracoFileMedia/ProductEvol_ToiletTissue_umbracoFile.pdf.

1. Jeff Goodby, "20 Years of 'Got Milk?'," *Adweek*(2013), http://www.adweek.com/creativity/20-years-got-milk-153399/.

4. "RE/MAX Sets New Record, Surpasses 123,000 Agents," RE/MAX, August 6, 2018, https://www.remax.com/newsroom/press-releases/remax-sets-new-record-surpasses-123000-agents.htm.

12. Jennifer Calfas, "Here's How Much It Costs to Buy a Commercial during Super Bowl 2019," *Money*, February 3, 2019, http://money.com/money/5633822/super-bowl-2019-commercial-ad-costs/.

2. Oracle, "About Oracle," https://www.oracle.com/corporate/index.html.

5. Planeterra Foundation, "Planeterra's Purpose," www.planeterra.org.

9. https://www.digitalcommerce360.com/article/us-ecommerce-sales/.

12. Trefis Team, "Echo to Remain a Key Growth Driver for Amazon Despite Competition from Apple," *Forbes*, June 8, 2017, https://www.forbes.com/sites/greatspeculations/2017/06/08/echo-to-remain-a-key-growth-driver-for-amazon-despite-competition-from-apple/#41d0a7b264f3.

3. http://www.sysco.com/about-sysco.html

8. https://www.ftc.gov/tips-advice/business-center/advertising-and-marketing